Fitness Professionals' Guide to Musculoskeletal Anatomy and Human Movement

Lawrence A. Golding, Ph.D., FACSM
University of Nevada, Las Vegas

Scott M. Golding, M.S.
CEO E2 Systems Inc.

Healthy Learning

Illustrations: Clint Smith

Layout design: Jennifer Bokelmann

Photo credits

Page 10: Matthew Stockman/Getty Images
Page 185: Brian Bahr/Allsport
Page 186: Ezra Shaw/Getty Images
Page 190: Andy Lyons/Allsport
Page 193: Sean Garnsworthy/Getty Images

Cover design: Jennifer Bokelmann

Library of Congress Number: 2002110886

ISBN: 1-58518-706-2

www.healthylearning.com

P.O. Box 1828
Monterey, CA USA 93942

Dedication

I would like to dedicate this book to my wife Carmen, who has had to live with the competition of my job, my profession, and the American College of Sports Medicine. Without her support through the many years of my professional life, I could not have accomplished many of the things I've undertaken and done. I would like to take this opportunity to give her my heartfelt thanks.

— Lawrence A. Golding

Contents

CONTENTS

Foreword

Knowledge of anatomy and human movement is a must for anyone wishing to be a fitness professional. What could be better than to have a unique text on musculoskeletal anatomy and human movement written by one of the best teachers of fitness professionals? Dr. Lawrence Golding has taught human anatomy for more than 40 years at major universities, and has been directly involved in the certification of fitness professionals through the Young Men's Christian Association (YMCA) and the American College of Sports Medicine (ACSM). He knows, from experience, what must be known and the depth to which it must be known in order for fitness professionals to function at a high level in the delivery of fitness programs. When this knowledge is coupled with his extensive teaching experience, a unique textbook results.

The text is loaded with "blank drawings" on which to practice what you think you have learned about the origin and insertion of muscles. In addition, a series of multiple choice quizzes that will test your knowledge of every aspect of anatomy and human movement presented in the text is provided. Finally, a CD-ROM containing special video sequences of movements and exercises, as well as most of the information in the text, is also included. Clearly, this combination of textbook and CD-ROM will make the path to knowledge easier for those learning the material for the first time, or for those needing a review. Teachers will also appreciate being able to project onto a screen the various elements in the text to help move a class along.

Dr. Golding is the author of numerous publications, including the classic YMCA Fitness and Assessment Manual. He is also the Editor-in-Chief of *ACSM's Health & Fitness Journal*. He has been recognized by his peers in the American College of Sports Medicine with that organization's prestigious Citation Award. This current text, representing a unique collaboration between father and son, is a "must have" in any fitness professional's library.

— Edward T. Howley, Ph.D., FACSM
Professor and Chair, Department of Exercise Science
University of Tennessee, Knoxville

PREFACE

This book is written primarily for anyone who is learning anatomy for the following reasons:

- To analyze movements, activities and exercises.
- To understand the muscles that are involved in athletic injuries.
- To prescribe exercises for particular muscle groups.
- To develop an exercise program designed to develop and strength the skeletal muscles.
- To determine the muscles that are involved in pathological conditions or orthopedic problems.

Most of the individuals who have the above reasons for learning musculo-skeletal anatomy are likely to be in the fields of exercise physiology, personal training, athletic training, physical education, physical therapy, coaching, or nursing. Students in these disciplines, during their academic curricula, take anatomy and physiology from biology departments who usually service these fields. The typical anatomy class at most universities includes the anatomy of all of the body's systems: the skeletal system (osteology); the system of joints and articulations (arthrology); the muscular system (myology); the vascular system (angiology), which includes the circulatory system and the lymphatic system; the nervous system (neurology), which includes the central nervous system (CNS) and the peripheral nervous system (PNS); the integumentary system; the alimentary system (often called the digestive system); the urogenital system; and the endocrine system. In exercise studies, the cardiorespiratory system is often referred to as indicating the two systems most involved in aerobic exercise: the angiology and respiratory systems. Because the typical anatomy class only lasts for 12-14 weeks, less than one week is normally spent on each system. Even a two-semester class would only double this amount of time. As a result, individuals in these fields who are interested in bones and the muscles that attach to them have limited time spent on the most important system for them: namely the musculoskeletal systems.

As a consequence, most disciplines usually teach an additional musculoskeletal anatomy course for their students, which teaches students about muscles, how muscles create movement, how they are used in movement, how they are developed, and how they are rehabilitated after injury. It is for individuals interested in this type of information that this book is written.

Relatively few textbooks are devoted exclusively to musculoskeletal anatomy. Students and instructors alike buy attractive muscle charts, which display all the surface muscles, and which are usually labeled, but are limited in scope. For example, if you are interested in muscle anatomy for the aforementioned practical reasons, then you must know where the muscles originate and where they insert so that their action can be determined. Examine the muscle charts so often purchased by those who want to know where the muscles are and what they look like. These charts do not show the muscle's origins or insertions because they cannot be seen or determined; only the belly of the muscle is shown.

This point can be illustrated by a simple example, looking at a common superficial muscle like the biceps. Most muscle charts clearly show the biceps – it's an upper, anterior, superficial arm muscle. Now ask yourself the question: where is its origin? The origin is on two places on the scapula (the coracoid process and the supraglenoid tubercle of the glenoid fossa). These can't be seen because the biceps is folded under the deltoid and the pectoralis major. Where is its insertion? The insertion is the radial tuberosity. That too cannot be seen because it is under the muscles and tendons of the forearm. In reality, the anatomy chart only shows that the bicep is an anterior, superficial, upper arm muscle, but it does not show where the muscle comes from, or where it goes.

One of the primary purposes of this book is to address that issue by presenting drawings of all the skeletal, locomotor muscles, so that at a glance, you will be able to clearly see the origin of the muscle, where it inserts, and hence be able to determine its action. If the origin and the insertion of a muscle are known, no need exists to learn its action because its action can be determined by knowing where it attaches, and what happens when it shortens.

The other phenomenon is that practitioners who use muscle names in their job enjoy impressing clients with their knowledge of the technical names of muscles. The human body has 680 muscles, and that's a lot of names to memorize. If you know the name of joint movements (which can typically be learned in about 20 minutes), then in less than a half-hour, you will also have a reasonable opportunity to learn the names of all the muscles, and in a more meaningful manner. Again, an illustration can help clarify this point.

A young exercise leader at a local health facility was explaining to me the arm curl on the Nautilus machine. He said that this was an excellent exercise for the biceps. So it is! But, his comment was also very discriminatory. What about the brachialis, the pronator teres, the supinator, the brachioradialis, and the flexor digitorum sublimus? These muscles are also used in the biceps curl. Why name the biceps? A number of possible reasons exist. First, he probably didn't know those other muscle's names, so he was showing off his limited knowledge. Second, who cares? Third, he wanted to sound educated.

Interestingly, the brachialis, which lies under the biceps and can't be seen on a muscle chart, is a truer flexor of the elbow than the biceps, since the biceps has two other actions involving the shoulder and radial-ulna joint, and the brachialis only has one action – elbow flexion. If this young man knew the joint movements, then naming all the muscles involved in the curl is simple. Bending the elbow involves a joint action called elbow flexion. Accordingly, when the young man was explaining the muscles involved in the curl, he simply could have said that performing the curl exercises the elbow flexors (of which there are six muscles). This way he would have mentioned all the muscles involved and didn't have to worry about remembering all the muscle's names.

Likewise, when explaining the muscles involved in the leg press (of which, there are many more than the arm curl) it could be explained that the muscles used are the knee extensors and the hip extensors. This book is designed to teach, illustrate, and explain all of the body's joint movements. Master these movements, and you've mastered all the skeletal muscle actions in the body. In addition, the muscles in each of the major groups of the body will be individually illustrated and shown.

Another desirable feature of this book is that it could be used as a coloring book of skeletal muscles. One of the best ways of learning the origin and insertion of muscles is to draw them on the skeleton on the individual bones. For this reason, the book includes several blank drawings so that each muscle can be drawn and can be seen in its entirety, not confused with or overlapping with any other muscles. Color-coding the origin and insertion further helps the learning process.

Understanding the joints actions also facilitates the learning of the muscle actions. In order for a muscle to work a joint, it must cross the joint. The question can be asked if the brachialis crosses the elbow joint, what is its action on the elbow joint? The elbow joint has only two actions: it either flexes or extends. Since it crosses the anterior of the joint, it can only flex the elbow. Since there are no other actions, there can be no other action to consider (unless it crosses another joint).

Another example will help clarify this factor. (Keep in mind that at this stage in the book, don't concern yourself with remembering these examples – they will be dealt with in detail as the book progresses. Right now, they are only being used to illustrate a particular point.) Take the middle deltoid, for example, which crosses the shoulder joint. The shoulder joint has six possible actions: flexion, extension, abduction, adduction, inward rotation, and outward rotation. Therefore, the middle deltoid must have one or some of these actions, but no others. Since the middle deltoid comes right over the top of the shoulder (from the acromion process of the scapula to the lateral middle humerus), it has no angle on the arm, therefore cannot flex or extend the arm. To flex the arm, it would have to be in front of the arm in order to pull it up into flexion. To extend the arm, it would have to be behind the arm, so it could pull the arm back into extension. By the same reasoning, with no angle on the humerus, it neither internally nor externally rotates the arm. Therefore, the middle deltoid only abducts the arm (comes over the top of the shoulder and pulls the lateral surface of the humerus into abduction). This approach offers a very easy way to learn muscle actions.

Bones have bony markings, which are named and illustrated. These bony markings are the origins and insertions locations of muscles. As a result, by first learning all the bony markings, the body's muscle origins and insertions are learned before the study of muscles is even started.

Using the Companion CD

The compact disc that is included with this book contains most of the same information in the book. It is designed for use on almost any Windows computer. Installation instructions for the CD can be found in Appendix B. The CD will install a program on your computer that will allow you to view all of the muscles shown in this text, and review the relationships between exercises, muscles, and activities. Over 300 exercises and 100 activities have been analyzed and are presented in this CD. The CD also includes video footage of all the exercises and body actions. Muscle drawings and blank skeletal drawings can be printed on any printer attached to the computer. Sample quizzes can also be printed along with other aids to help you learn the musculo-skeletal system faster and better. Used together, the book and CD offer an exceptional instructional tool for learning and fully understanding the musculoskeletal system of the body. If they help you in this regard, then the time, energy, and resources required to write this text and to develop the companion CD will have been well worth the effort.

— L.G., S.G.

Chapter 1 • Terminology

Most anatomy books have an extensive glossary of terms covering all the systems of the body. Since this book is primarily directed to the musculoskeletal system, this terminology section deals only with terms related to bones and muscles.

Anatomical Position

This is the starting position for all muscle actions. Refer to Figure 1-1. The body is standing upright with the arms at the sides and the palms of the hand facing forward (supinated). All muscle actions are given from this position, and although muscle actions can be done from other positions, the textbook action is from the starting position. For example: standing in the anatomical position, flex the elbows. The forearm comes forward in the sagittal plane. However, if you stand in the normal standing position with your arms by your side and your palms facing your thigh, and you flex your elbow, your forearm moves in the frontal plane, and you are standing with arms "akimbo". Likewise, if you abduct your whole arm so that it is parallel to the floor and flex your elbow, again the movement is in the frontal plane. These three examples all involve elbow flexion, but the resultant action is very different. The true action of elbow flexion, as with almost all actions, is from the anatomical position.

The true action of elbow flexion, as with almost all actions, is from the anatomical position.

All movements occur in a plane and around an axis.

Planes and Axes

All movements take place parallel to a plane. For example, elbow flexion, spine flexion, knee flexion, and hip flexion are all in the sagittal plane. Arm and leg abduction are in the frontal plane, while rotations occur in the horizontal plane. All movements also take place around an axis (a point of rotation—like an axle of a wheel), which is at right angles to the plane. For example, all movements in the sagittal plane occur around the frontal axis (refer to Figure 1-1).

The aforementioned point can be illustrated by standing and flexing and extending the arm like a pendulum. If you used a pencil to indicate the "axle" around which this movement occurs, the pencil would be coming in from the side into the middle of the shoulder (perpendicular to the sagittal plane), which would be the frontal axis. Now, stand and abduct and adduct the leg. The pencil "axle" would come in from the front to the top of the leg (perpendicular to the frontal plane), which would be the sagittal axis. Finally, rotate your head from right to left. Where would the pencil go to represent the "axle" around which this movement is occurring? The pencil would be coming from above into the middle of the top of the head, which would be the vertical axis.

Thus, all movements occur in a plane and around an axis. To summarize, movements in the sagittal plane occur around a frontal axis, movements in the frontal plane are around a sagittal axis, and movements in the horizontal plane occur around a vertical axis.

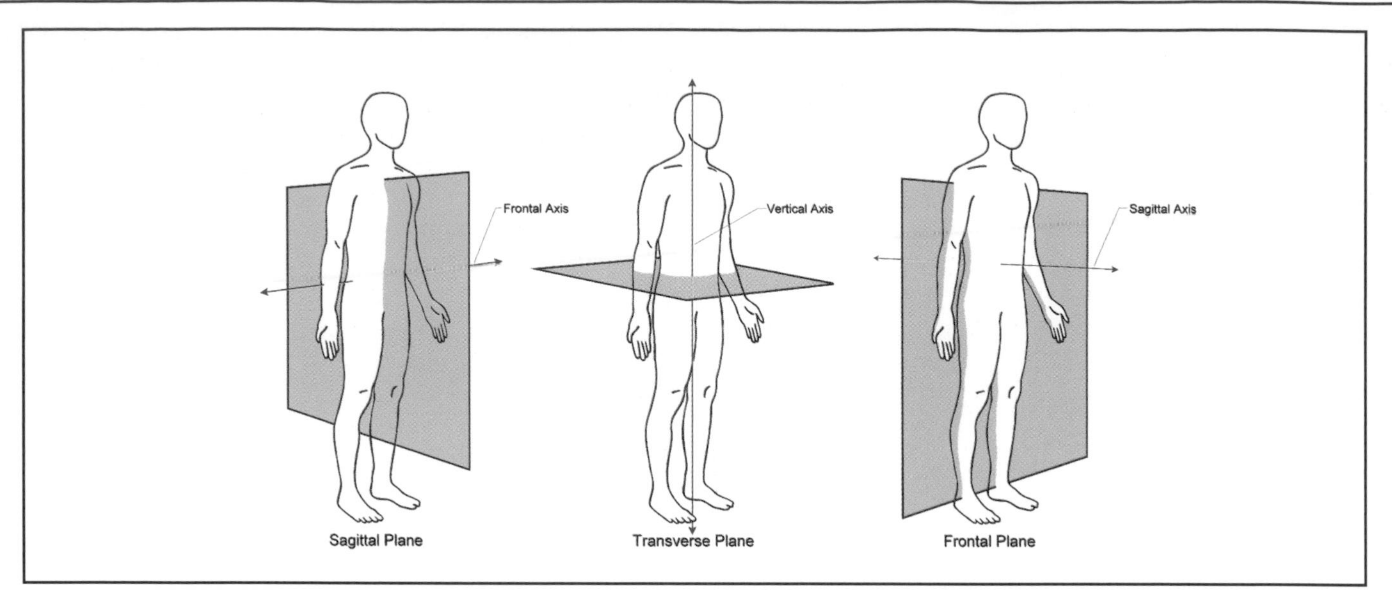

Figure 1-1. The body planes

Anatomical Direction

The following commonly used anatomy terms refer to direction. Refer to Figure 1-1.

- Median plane (or mid-sagittal plane) – a sagittal plane passing through the body from anterior to posterior, dividing the body into two equal halves. The Median plane is shown in the sagittal plane drawing above.
- Sagittal plane – any plane parallel to the median plane.
- Frontal plane – a plane passing through the body from side to side, dividing the body into an anterior and posterior part.
- Transverse plane (or horizontal plane) – any plane passing horizontally through the body.
- Anterior (or ventral) – toward the front of the body.
- Posterior (or dorsal) – toward the back of the body.
- Medial – closer to the median plane than some other point.
- Lateral – away from the median plane than some other point.
- Superior (or cranial) – toward the head, or closer to the head than some other point.
- Inferior (or caudal) – toward the tail, or farther from the head than some other point.
- Central – In the center of the body.
- Peripheral – toward the surface of the body.
- Distal – distant from a point of reference.
- Proximal – near to a point of reference.

Movements in the sagittal plane occur around a frontal axis.

Anatomical Descriptive Terms

The following descriptive terms are primarily prefixes that go before other roots. By knowing these terms, the meanings of words can be determined. For example, "cardio" refers to the heart; therefore a cardiologist is a physician specializing in the heart. Cardiovascular refers to the heart and circulation. Myocardium refers to the muscle of the heart. Epicardium is the inner most layer of the covering of the heart. Epicardial fat is the fat around the heart, and so on. This deductive reasoning can be applied with each of these terms.

- Cardio ....refers to the heart
- Cephalo ....refers to the head
- Chondro......................................refers to cartilage
- Encephalon ....refers to the brain
- Gastro ....refers to the stomach
- Hepato ....refers to the liver
- Myo ....refers to muscle
- Neuro ....refers to nerves
- Osseo ....refers to bone
- Pneumo.......................................refers to the lungs
- Reno ....refers to the kidneys
- Uro ....refers to the urinary system
- Vaso ....refers to vessels

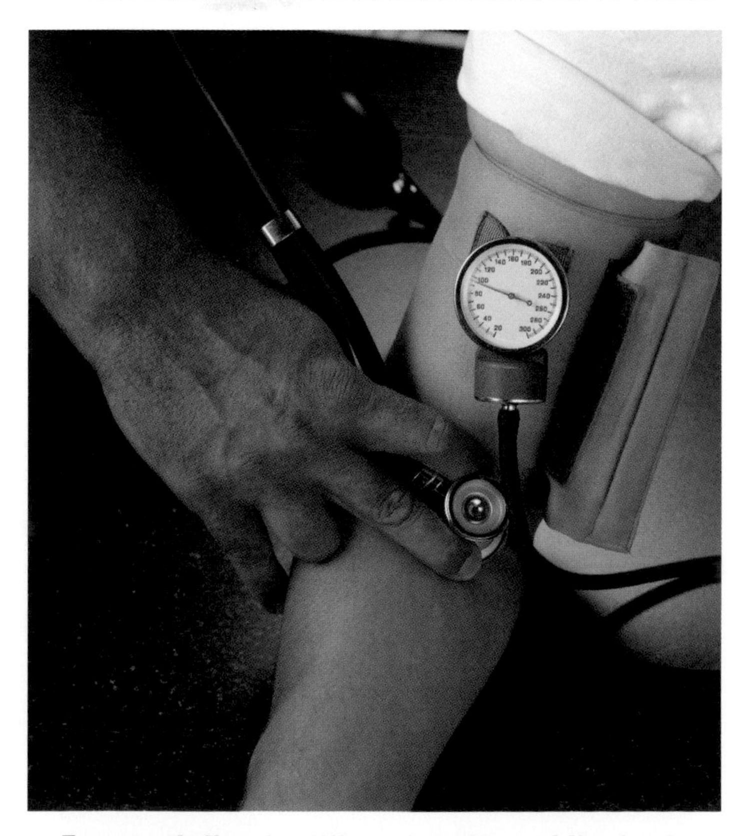

Terms relating to different regions of the body are often used in different contexts.

Anatomical Regions

The following terminology relating to the regions of the body is very important and is helpful when working with musculoskeletal systems. These terms refer to a part of the body and are used in many different contexts. For example, antibrachial refers to the area of the forearm (below or in front of the elbow). When someone speaks about drawing blood from the antibrachial region, the stethoscope diaphragm is placed in the antibrachial space when taking blood pressure. Brachial is obviously above the antibrachial region (above the elbow). In that regard, the body has both a brachial artery and a brachial pulse wave.

- Antibrachial . . . . forearm
- Axilla . . . . .armpit
- Brachial .shoulder to elbow
- Calcaneal . . . . .heel
- Carpal . . . . .wrist
- Cervical . . . . .neck
- Cubital . . . . .elbow
- Femoral . . . . .thigh
- Hallux . . . . .great toe
- Lumbar . . . . .small of the back
- Metacarpal . . . . .hand
- Metatarsal . . . . .foot
- Nuchae . . . . .back of the neck
- Pectoral . . . . .anterior chest
- Peroneal .area of the fibula
- Plantar . . . . .sole of the foot
- Pollex . . . . .thumb
- Popliteal .back of the knee
- Tarsal . . . . .ankle
- Thoracic . . . . .chest
- Vaso . . . . .vessels
- Volar . . . . .palm of the hand

Bony Markings

As with the other terminology, bony markings are general terms referring to characteristic marks on bones. When you learn the individual bones, the term used is specific to that bone. For example, the enlargement on the proximal end of the humerus is termed a tuberosity; the enlargement on the proximal end of the femur is called a trochanter. This is the same description generally – an enlargement on a bone, but different specifically, when speaking of a particular bone. The following terms are used when naming bony markings:

- Condyle . . . . . . .a smooth, rounded prominence covered with articular cartilage
- Crest . . . . .a ridge
- Epicondyle . . . . .above a condyle
- Facet . . . .a small, smooth cartilage covered area
- Foramen . . . . .a hole through a bone
- Fossa . . . . .a depression or hollow
- Head . . . . . . .a smooth, rounded end of a bone
- Line . . . . .a long narrow ridge
- Neck . .a constriction below the head of a bone
- Sinus . . . . .a mucous membrane lined cavity in a bone, filled with air
- Spine . . . . .a sharp projection
- Stylus . . . . .a pencil-point projection
- Tubercle, tuberosity, trochanter . . . .an eminence or enlargement

Chapter 2 • Bones

This chapter discusses bones and bony markings. By virtue of their name, the skeletal muscles attach to the skeleton and operate on the bone to cause movement. The markings on the bones (the bony markings) are the specific locations where muscles attach. Four different types of bones exist have a definite function in the body.

Types and Functions of Bones

Types of Bones

The human skeleton has four types of bones. Each type is described by its general shape: long, short, flat, and irregular. These names are very descriptive of what each type of bone looks like.

- *Long bones.* Long bones are long and are characterized by the fact they are longer than they are wide. They generally have tubular shafts and articular surfaces at each end. The function of long bones is locomotion. Muscles operating on long bones allow you to walk, run, crawl, climb, jump, kick, and hit. The major bones of the arms and legs are long bones that include the femur, tibia, fibula, humerus, radius, and ulna.
- *Short bones.* Short bones are shorter than long bones and also have generally tubular shafts and articular surfaces at each end. Short bones allow flexibility. For example, the reason you can make a fist is that the short bones of the finger can "roll" up into a fist. The short bones include all the metacarpals, metatarsals, and phalanges.
- *Flat bones.* Flat bones are flat and relatively thin and have broad, flat surfaces. Their function is protection. If the vital parts of the body were named, they would be the brain, the heart and lungs, and the reproductive organs. All these vital parts are protected by flat bones. The brain is completely encased by the cranium (six flat bones); the heart and lungs are protected by the flat bones of the rib cage and sternum; the reproductive organs are protected by the flat-boned pelvic girdle (the innominate - three flat bones).
- *Irregular bones.* Almost any bone that is not a long bone, a short bone, or a flat bone is an irregular bone. These bones are variable in size and shape. They are generally compact in nature and are distributed throughout the skeleton. These include the entire vertebral column, the eight carpal bones, the seven tarsal bones, and the patella.

Functions of Bones

As mentioned previously, bones have three primary functions – locomotion, flexibility, and protection. In addition, bones have three other very important functions – these additional functions include the following: the manufacture of blood cells; the storage of deposited of calcium and phosphorus; and support.

- Blood cells are made in the marrow of bones (in their centers) and released into the blood stream.
- When a body doesn't ingest enough calcium (as during pregnancy or in post menopausal women), the body uses stored calcium by taking it from the bones. This process has the effect of weakening the bones, which is why it is important to get enough calcium when the body's needs are high.
- The skeleton provides a support for organs and gives shape to the body.

Bony Markings

Bony markings (which are literally markings on the bone) may be characterized by ridges, depressions, holes, rough areas or smooth areas. These markings are usually the origin or insertion of muscles. By learning the bony markings, all of the origins and insertions of all of the skeletal muscles can be learned. Drawings of the bones showing the bony markings listed below are at the end of this section.

The bones and the markings on each bone that are used with the locomotor (skeletal) muscles discussed in this book are listed in the following section, by body region. The specific bone markings are noted below the bone on which they occur. Bones that do not have bony markings that are used for locomotor muscles have markings listed as "none." This designation does not mean that these bones do not have markings, just that they are not important for locomotion. The number of bones present in each particular body region (part) is denoted in parenthesis after the name of the bone.

Bones and Their Bony Markings

Bones of the Head

Although many muscles are attached to the head (e.g., muscles of expression and muscles used to talk and eat), only a limited number of these are locomotor muscles.

- *Frontal bone (1)*
  - none
- *Parietal bone (2)*
  - none
- *Temporal bone (2)*
  - mastoid process
- *Occipital bone (1)*
  - occipital protuberance
  - superior nuchal lines
  - inferior nuchal lines
  - medial nuchal line
- *Facial bones*
  - none

Bones of the Chest, Thorax, and Upper Trunk

- *Sternum (1)*
  - manubrium
  - body of sternum
  - xiphoid process
  - clavicular notch
- *Clavicle (2)*
  - sternal end
  - accromial end
- *Ribs (12 pairs)*
  - head
  - neck
  - angle
  - shaft
  - costal cartilage
- *Vertebra, cervical (7)*
  - body
  - superior articular process
  - inferior articular process
  - transverse process
  - transverse foramen
  - vertebral foramen
  - anterior tubercle (1st & 2nd)
  - posterior tubercle (1st & 2nd)
  - spiney process
- *Vertebra, thoracic (12)*
  - body
  - superior articular process
  - transverse process
  - inferior articular process
  - vertebral foramen
  - spiney process
  - pedlcle
  - superior demifacet
  - inferior demifacet
- *Vertebra, lumbar (5)*
  - body
  - superior articular process
  - inferior articular process
  - transverse process
  - vertebral foramen
  - spiney process
  - pedicle
- *Sacrum (five in children, one in adults)*
  - none
- Coccyx (four in children, one in adults)
  - none
- Scapula (2)
  - inferior angle
  - vertebral border
  - superior angle
  - superior border
  - scapula notch
  - coracoid process
  - glenoid fossa (cavity)
  - supraglenoid tubercle
  - infraglenoid tubercle
  - axillary border
  - spine
  - acromion process
  - supraspinatus fossa
  - infraspinatus fossa
  - subscapular fossa

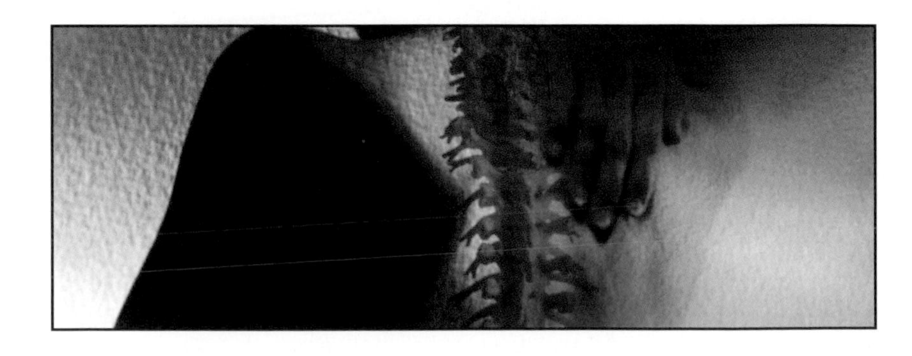

Bones of the Arm

- *Humerus (2)*
  - head
  - anatomical neck
  - greater tuberosity
  - lesser tuberosity
  - crest of the greater tuberosity
  - crest of the lesser tuberosity
  - surgical neck
  - deltoid tuberosity
  - lateral epicondyle
  - lateral supracondylar ridge
  - medial epicondyle
  - medial supracondylar ridge
  - trochlear
  - olecranon fossa
  - radial fossa
  - coronoid fossa
  - capitulum
- *Radius (2)*
  - head
  - radial tuberosity
  - styloid process
- *Ulna (2)*
  - olecranon process
  - radial notch
  - semilunal (trochlear) notch
  - coronoid process
  - ulna tuberosity
  - supinator crest
  - styloid process

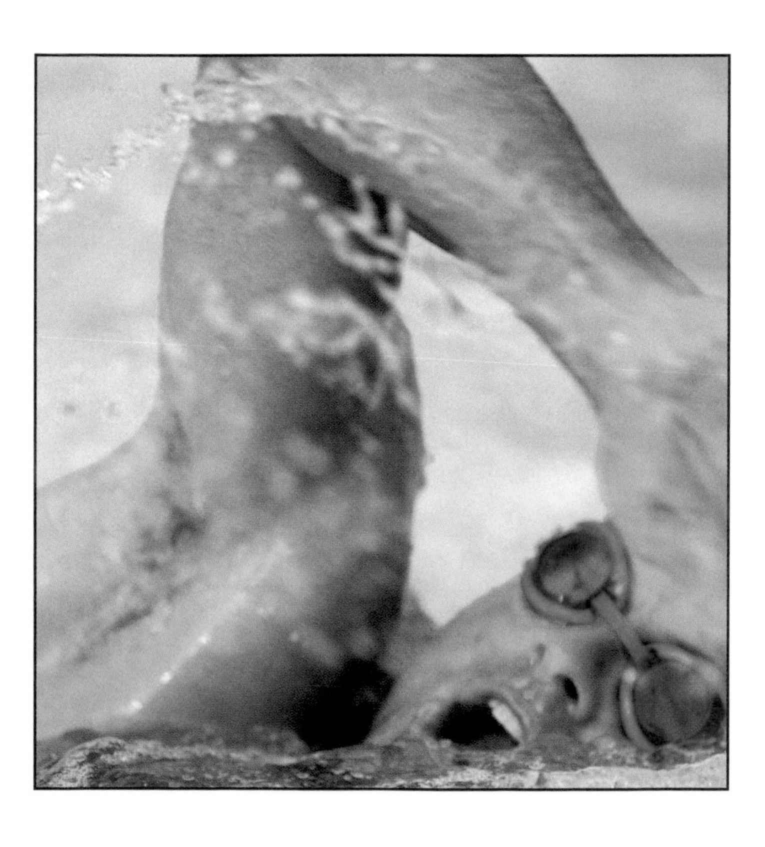

Bones of the Leg

- *Femur (2)*
  - head
  - neck
  - intertrochanteric line
  - greater trochanter
  - intertrochanteric crest
  - intercondylar fossa
  - lesser trochanter
  - gluteal line
  - pectineal line
  - linea aspera
  - medial epicondylar ridge
  - lateral epicondylar ridge
  - adductor tubercle
  - medial epicondyle
  - lateral epicondyle
  - medial condyle
  - lateral condlye
- *Tibia (2)*
  - lateral condyle
  - medial condyle
  - intercondyle emminence
  - anterior tuberosity
  - popliteal surface
  - popliteal line
  - anterior crest
  - medial malleous
- *Fibula (2)*
  - head
  - lateral malleous

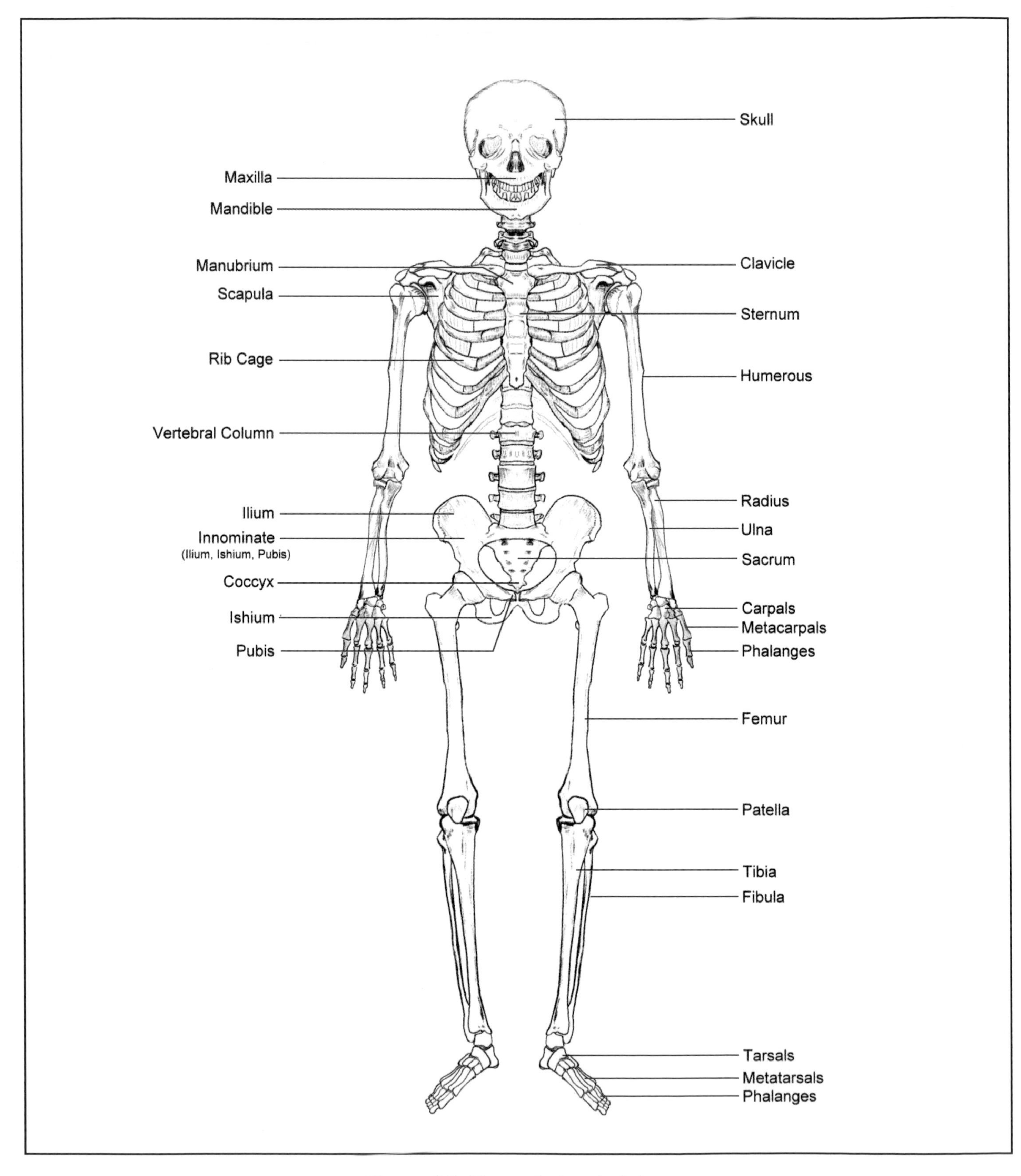

Figure 2-1. The anterior skeleton

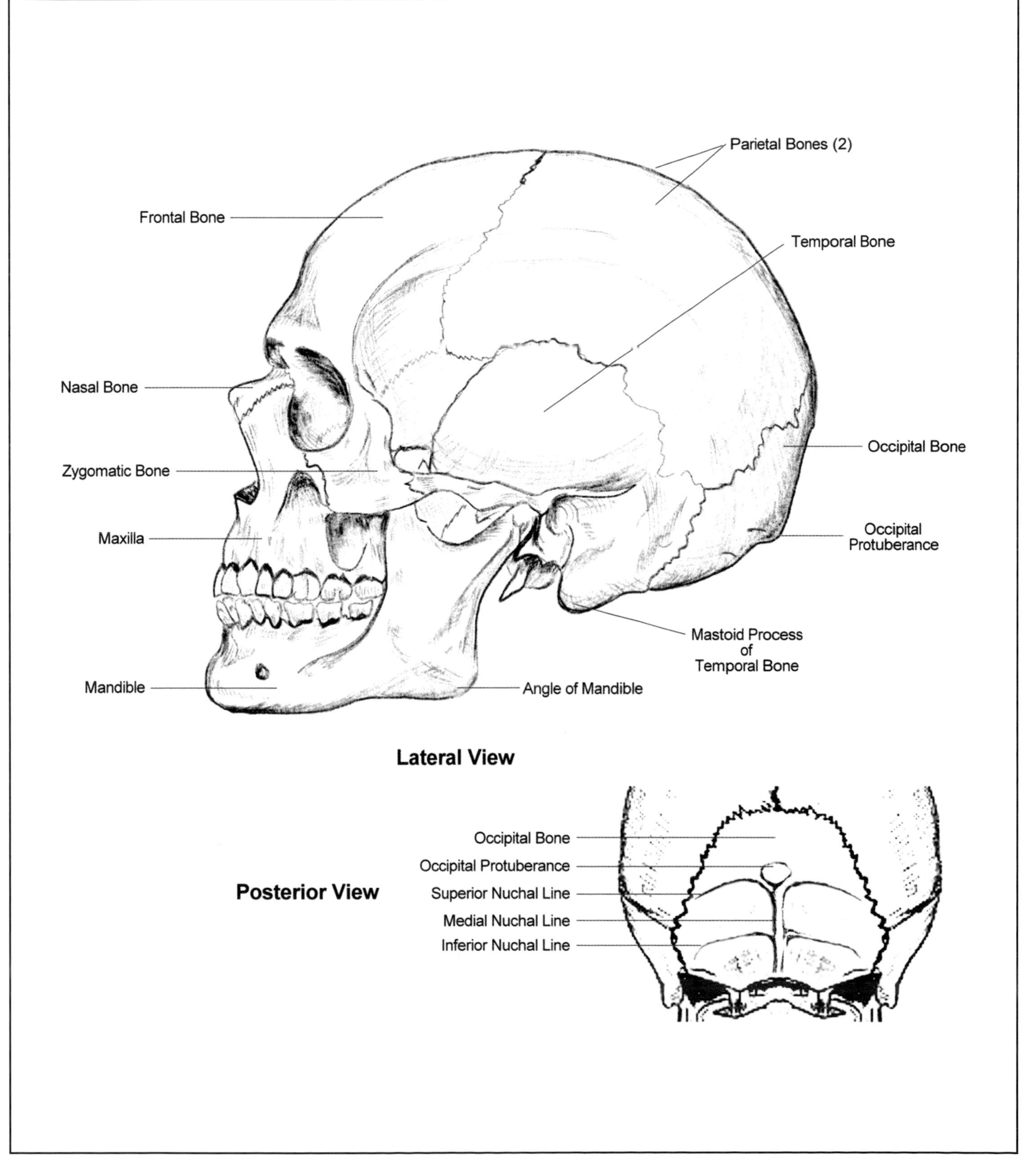

Figure 2-2. The skull

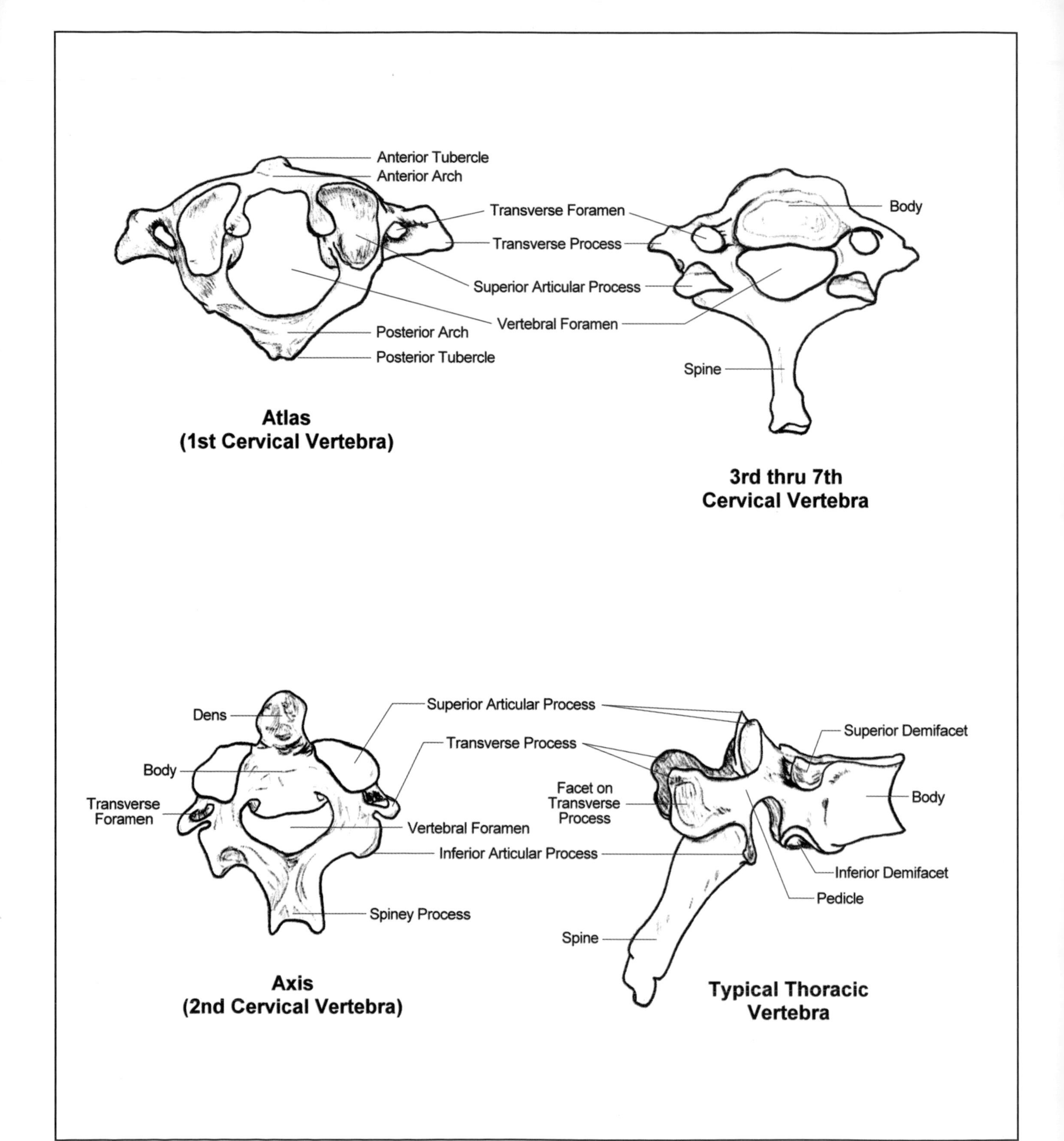

Figure 2-3. The vertebra

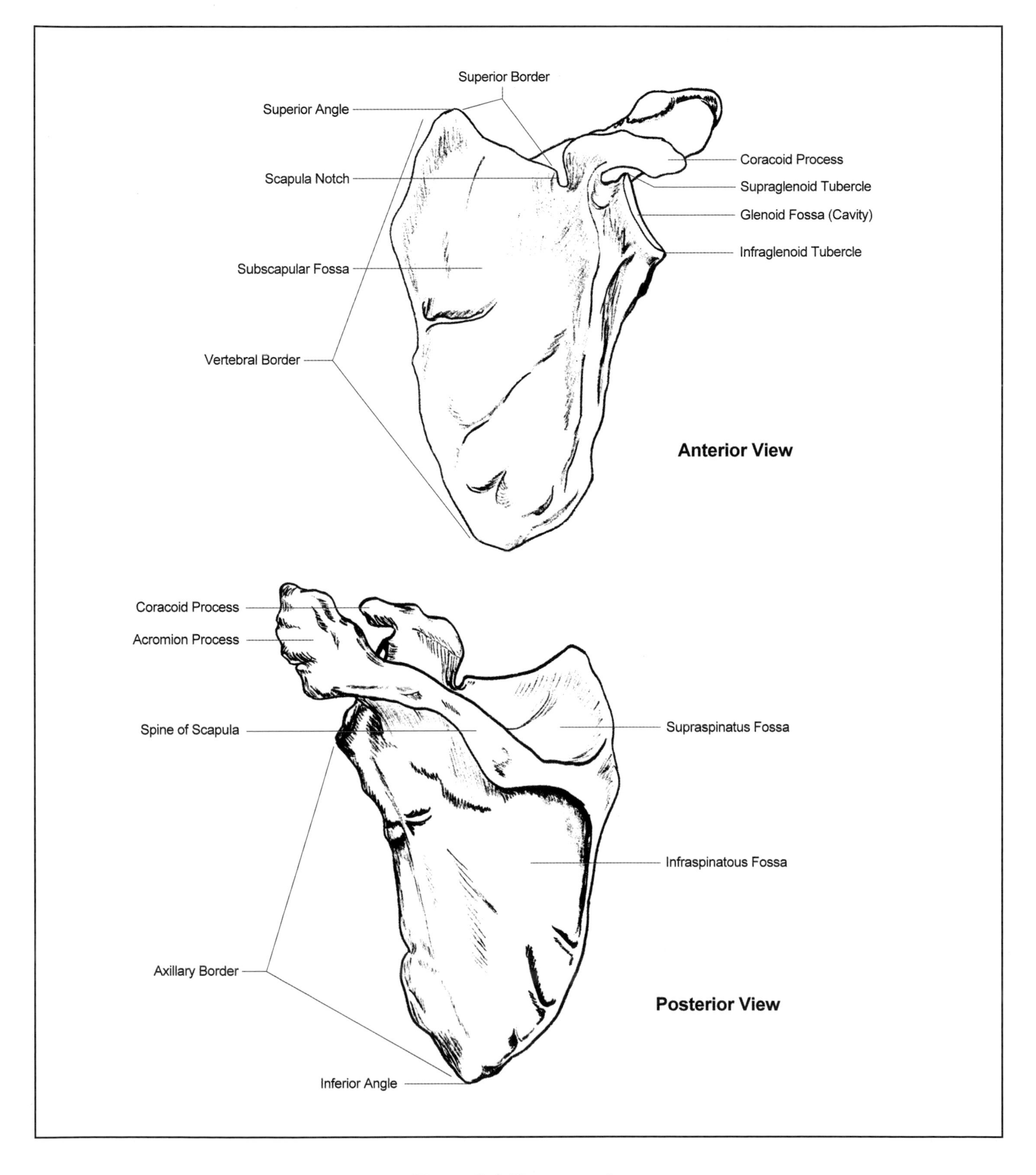

Figure 2-4. The scapula

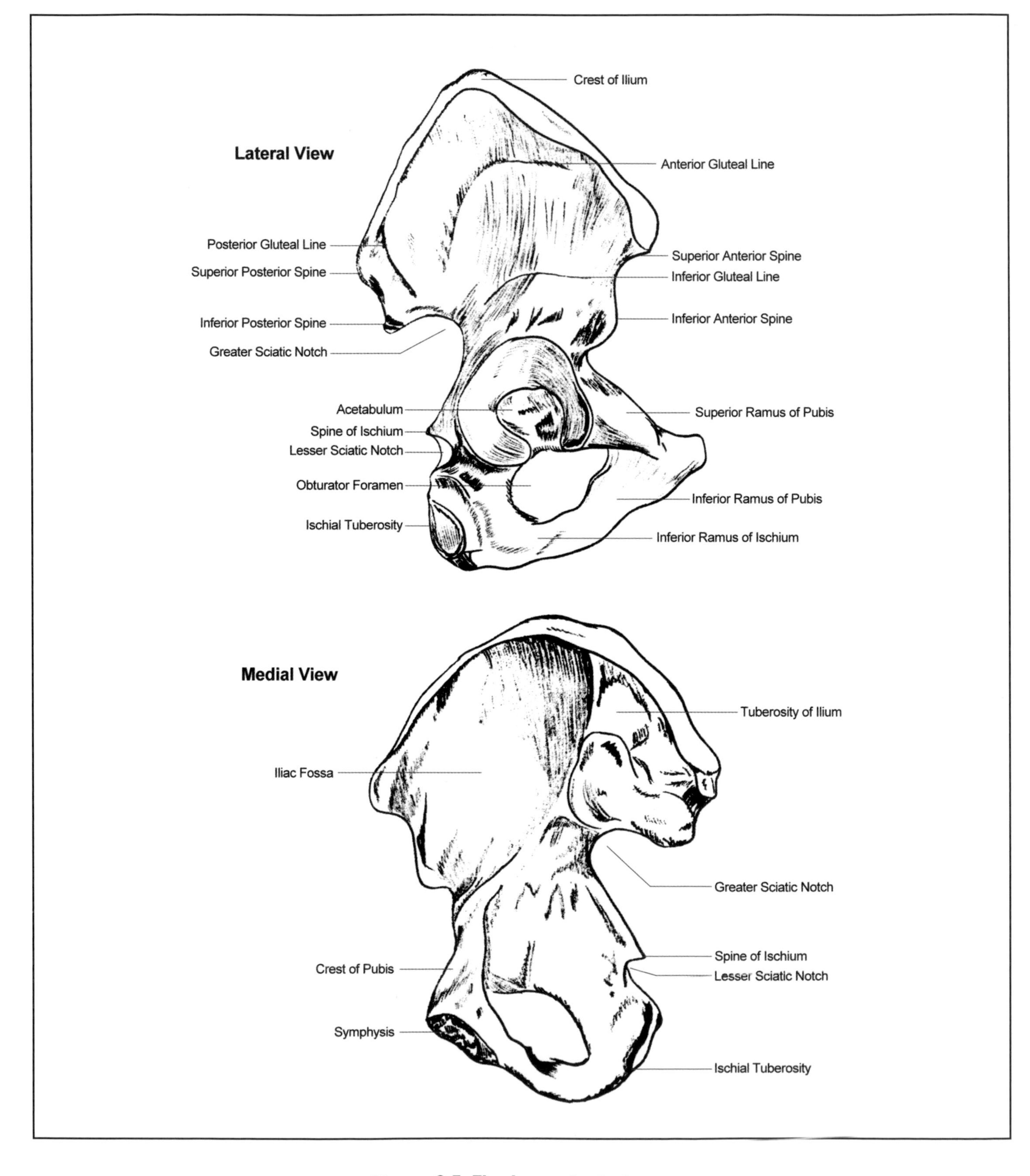

Figure 2-5. The innominate bone

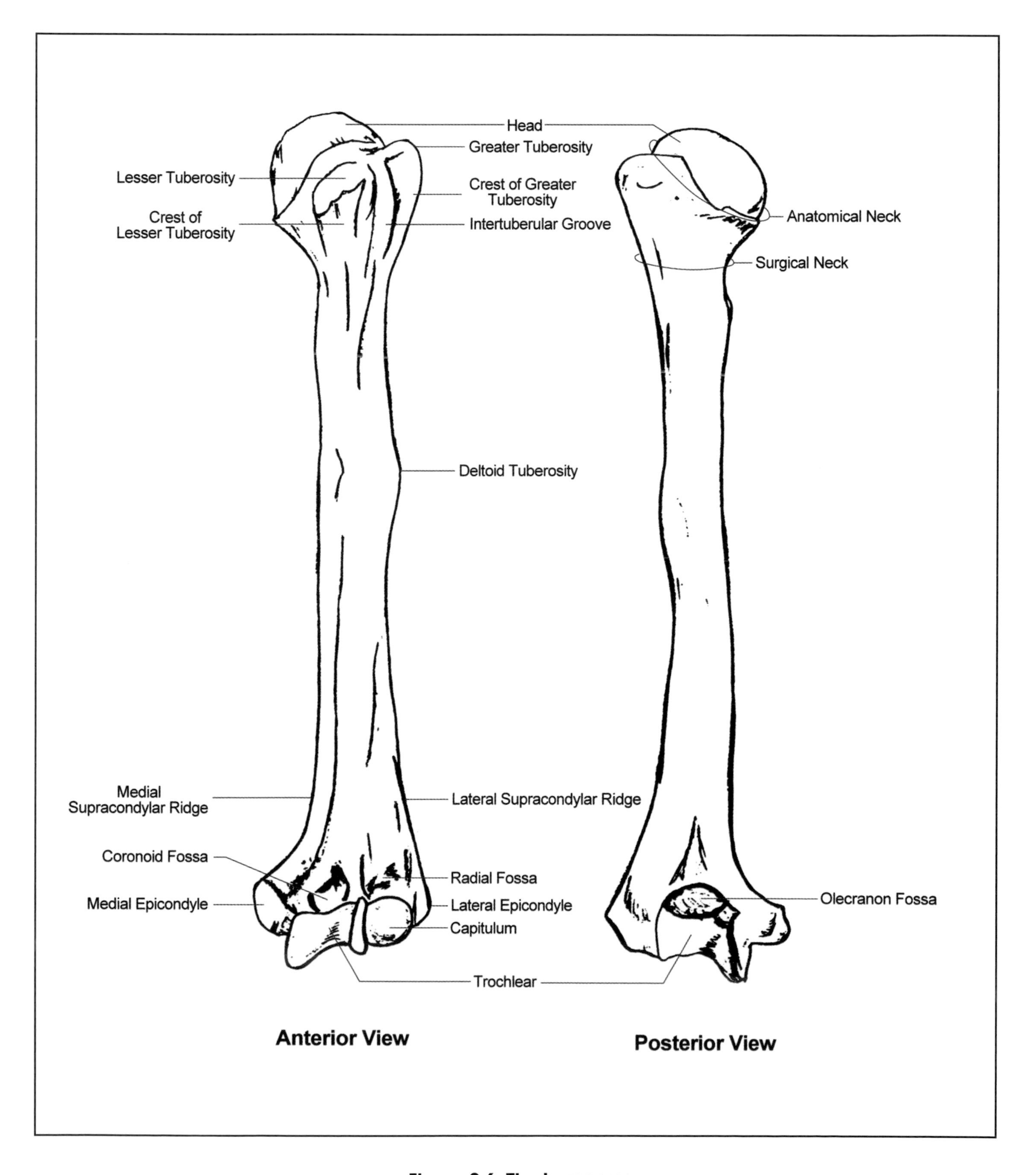

Figure 2-6. The humerus

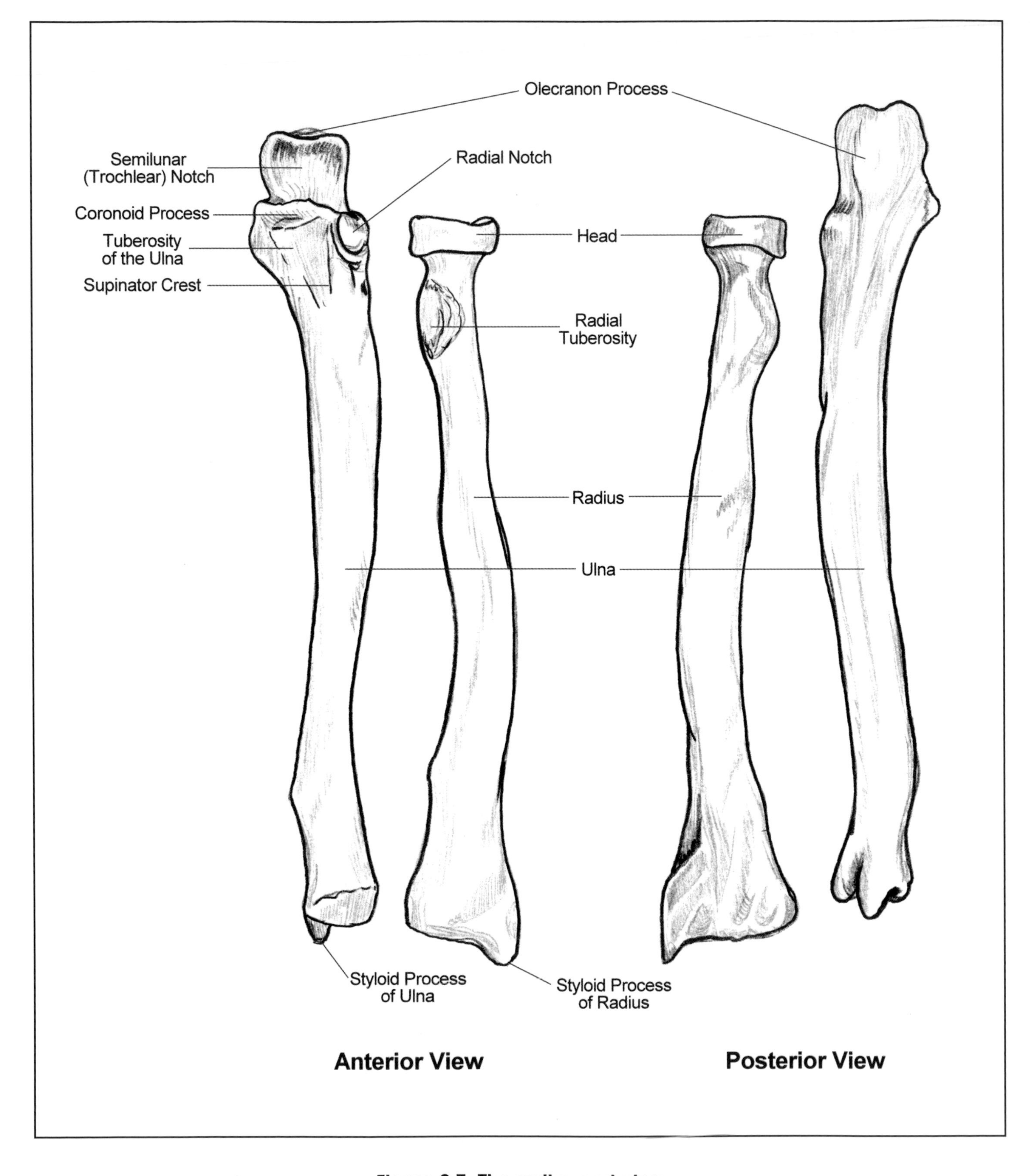

Figure 2-7. The radius and ulna

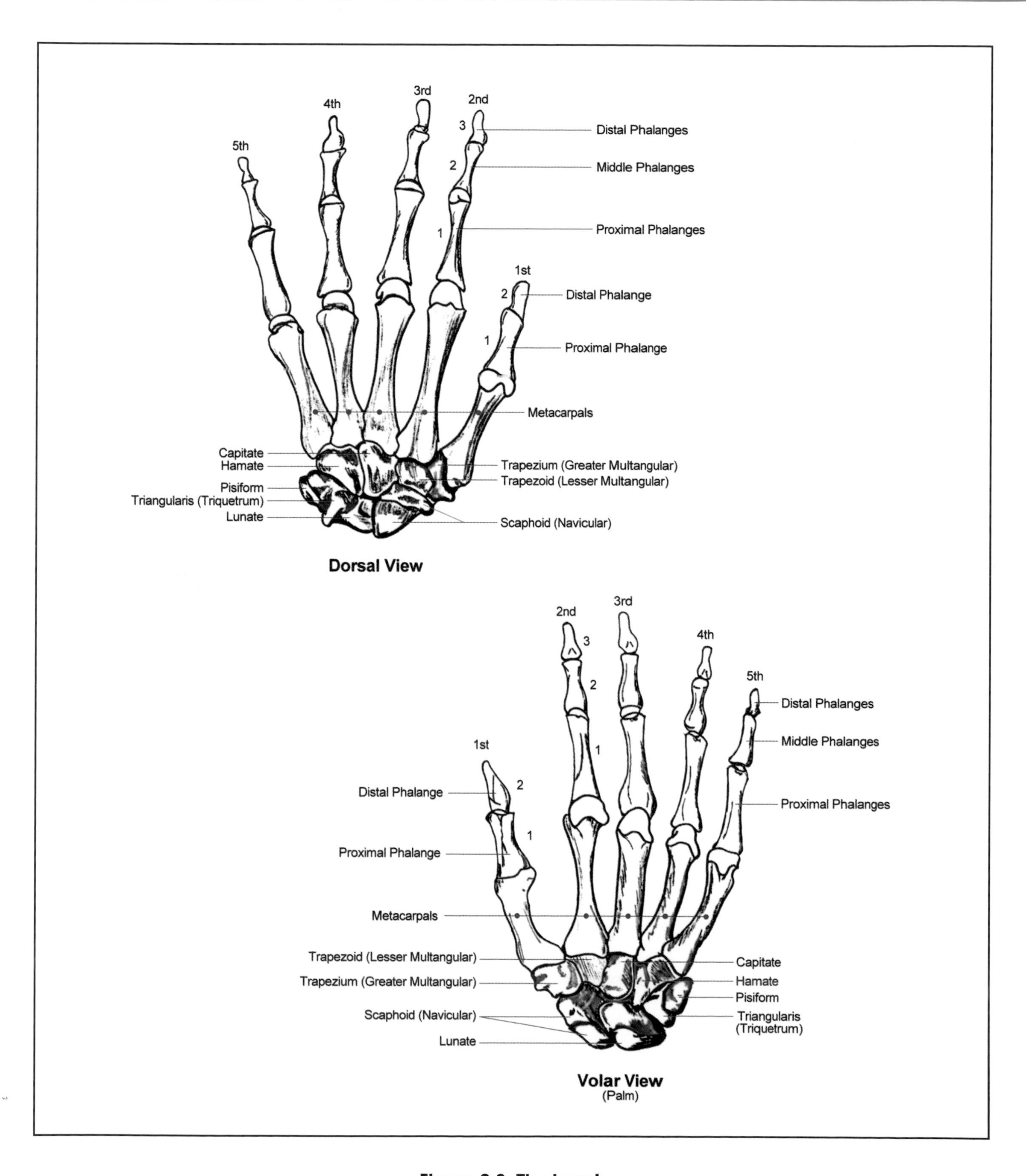

Figure 2-8. The hand

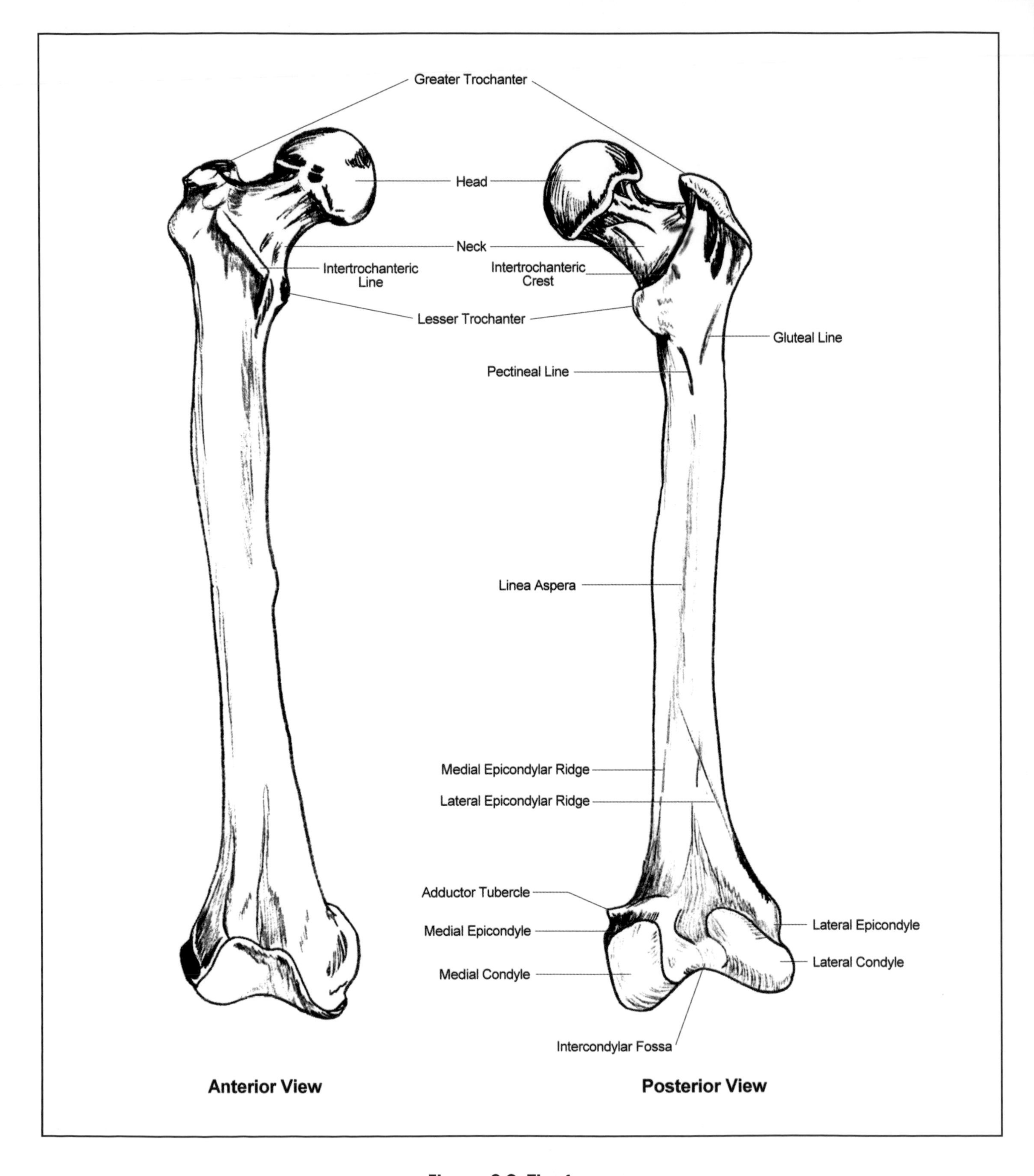

Figure 2-9. The femur

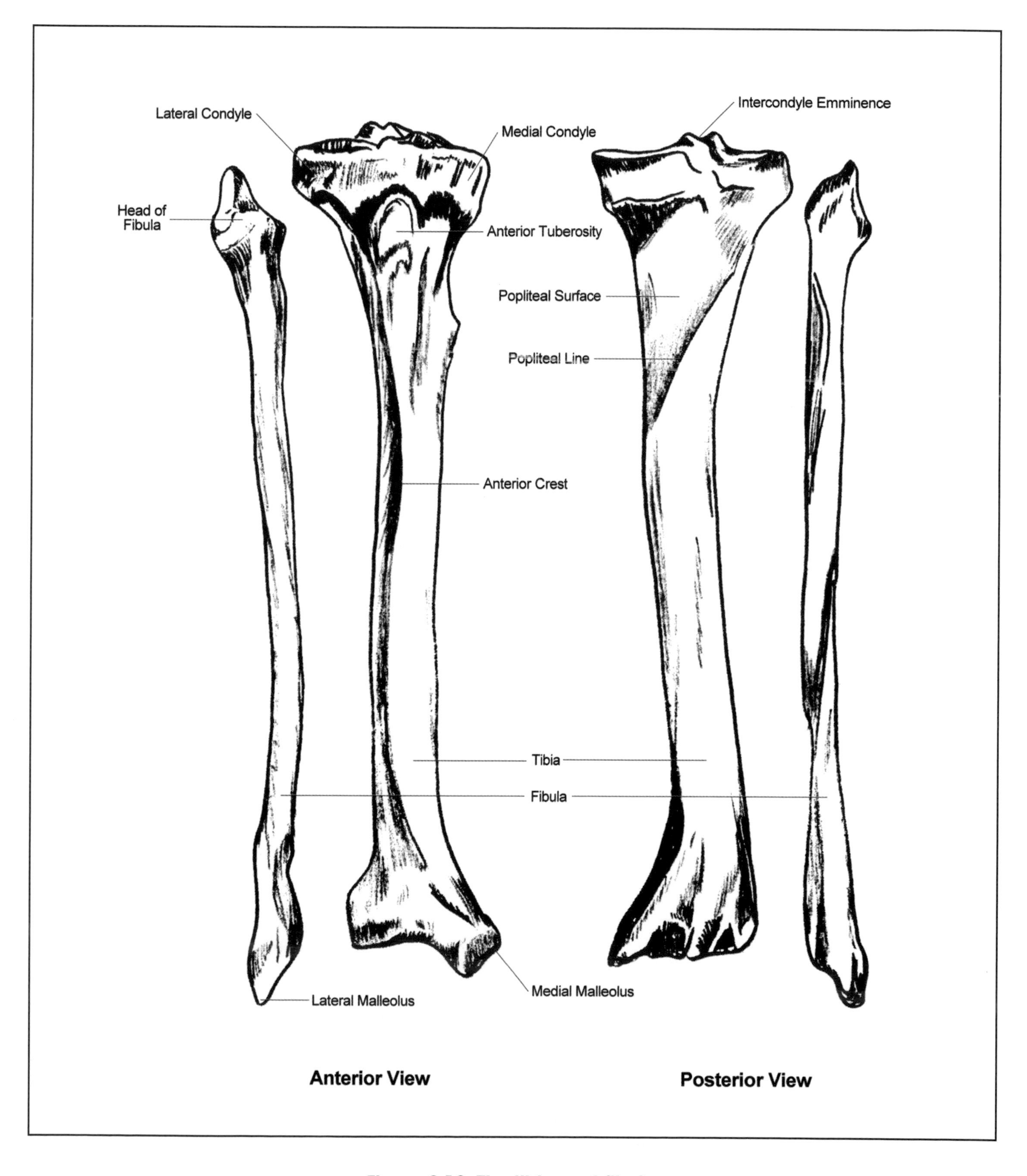

Figure 2-10. The tibia and fibula

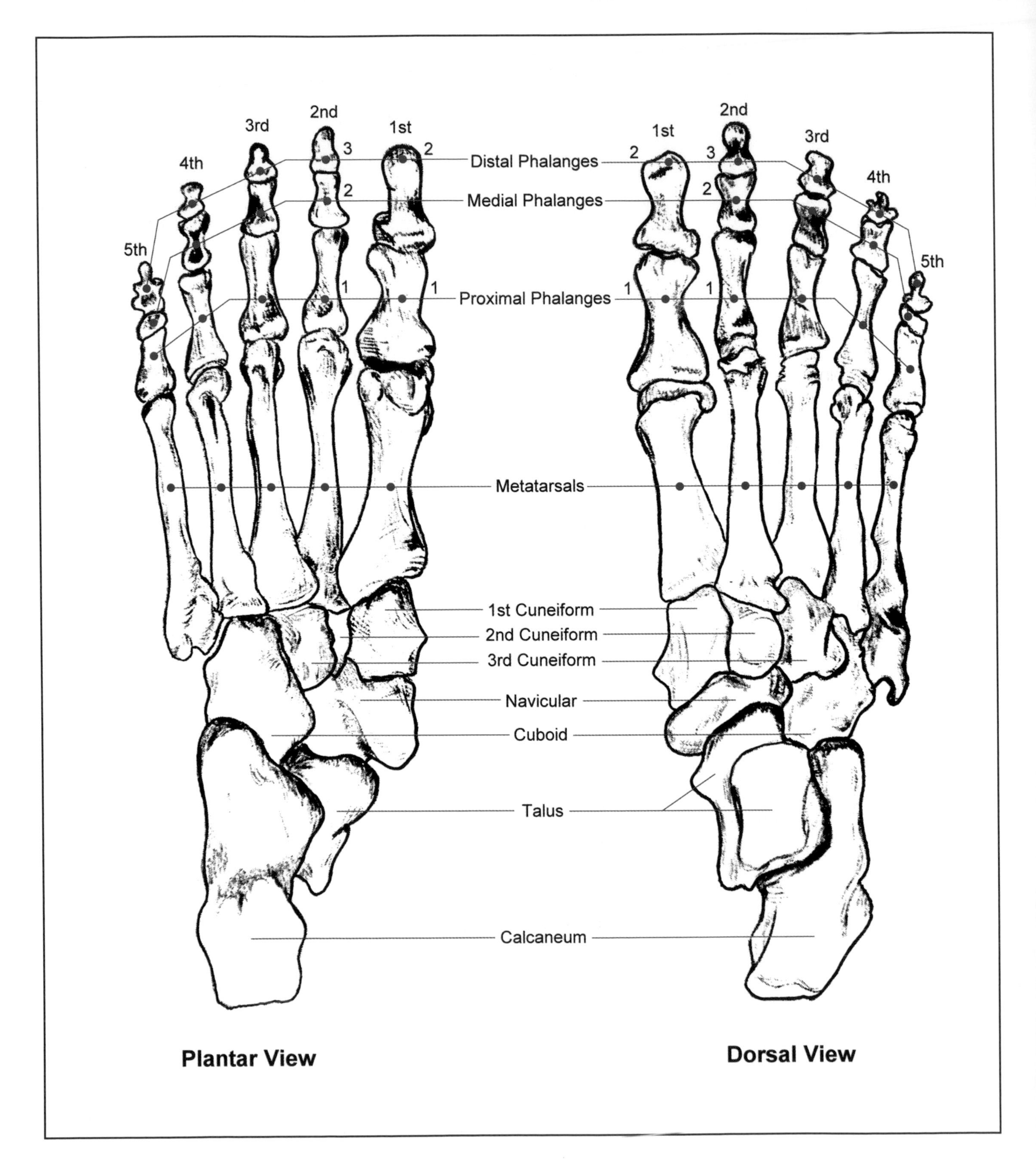

Figure 2-11. The foot

Chapter 3 • Joints and Body Actions

This chapter deals with the types and actions of the different joints in the body, and the body actions that each joint can make.

Arthrology

Arthrology is that branch of anatomy that deals with joints or articulations. A joint or articulation is the connection or junction between two or more bones or between cartilage and bone.

Classification of Joints

Three types of joints exist - each of which is differentiated by its structure and ability to move:

- ❑ Synarthrosis joints (juncturae fibrosae) - immoveable joints
- ❑ Amphiarthrosis joints (juncturae cartilagineae) - joints with slight movement
- ❑ Diarthrosis joints (juncturae synoviales) - free-movement joints

- *Synarthrosis joints* - two bones coming together but with no movement between the bones. In many cases, the bones have fused together. Three examples of synarthrosis joints are the innominate bone, the sacrum, and the cranium. The innominate bone is three bones (ischium, ilium and pubic bone) that have fused together to make one innominate bone. There is no movement between any of these bones. Where the bones join together is difficult to see. The sacrum is five sacral vertebra in the child, which fuse together in the adult into one bone - the sacrum. Likewise, with the coccyx, which lies below the sacrum. The cranium is made up of the frontal bone, two parietal bones, the occipital bone, and two temporal bones. The joints between these bones are called sutures.
- *Amphiarthrosis joints* - two bones coming together but with very slight movement between the bones. There are two types of amphiarthrosis joints - symphysis (cartilaginous) joints and syndesmosis (ligamentous) joints. Two examples of symphysis joints are the symphysis pubis (the junction between the two pubic bones that make up the pubic arch) and the junction between the upper part of the sternum, the manubrium, and body of the sternum. In forced inspiration this joint allows the sternum to "bend" a little. An example of a syndesmosis joint is the coraco-acromial joint (a ligament that joins the coracoid process with the acromion process).
- *Diarthrosis joints* - because of its freedom of movement it is often referred to as a true joint. Diarthrosis joints have several essential structures (characteristics), including the following:
  - Articular surfaces of bones. The ends of the bones are smooth and covered with cartilage.
  - Articular cartilage. The cartilage covering the end of the bone provides a smooth, cushioned surface for two bones to come together.
  - Articular disk. A disk (or meniscus) may be situated between dissimilar surfaces in order to facilitate free movement.
  - Articular capsule. A ligamentus capsule surrounds the joint and contains synovial fluid, which helps to lubricate the joint and serves as a nutrient source.
  - A synovial membrane lines the articular capsule containing the synovial fluid.
  - Ligaments which run from one bone to the other which bind and stabilize the joint.

There are six types of diarthrosis joints:

1. Gliding joint (arthrodia or articulatio plana). A gliding joint involves bones that lie next to one another and as movement takes place, they glide (or slide or rub) together. A good example is the carpal and tarsal bones, which lie next to one another. When the wrist or ankle moves, these bones move slightly against one another. This movement involves the intercarpal and intertarsal joints, respectively. Another example of a gliding joint is the movement between the head of the rib and the vertebra, or the junction of the clavicle with the sternum.
2. Hinge joint (ginglymus). The hinge joint is somewhat self descriptive, namely two bones acting like a hinge on a door. Hinge joints are primarily in one plane - flexion and extension. Good examples are the interphalangeal joints of the fingers and the humerus and ulna. The knee and ankle joints are also hinge joints, but are not as typical, since they allow slight rotation.
3. Pivot joint (trochoid). In the pivot joint, the movement is primarily rotation with one bone rotating in a ring of another. Two good examples of a pivot joint are the radioulnar joint (where the head of the radius rotates in the radial notch of the ulnar) and the dens of the axis rotating in the ring of the atlas vertebra.

4. Ball and socket joint (spheroidea or enarthrosis). Like the hinge joint, the name of this joint is self-descriptive and consists of a round head or projection of one bone fitting into a cup or socket of another bone. The hip and shoulder joints are good examples of a ball and socket joint. Movement is permitted in all planes. Because of its structure, the hip joint is more stable than the shoulder joint. The hip joint has a deep socket (the acetabulum of the innominate bone) and a very pronounced ball (the greater trochanter of the femur). In addition, the hip joint involves large strong muscles, which surround and support the joint. The shoulder has a very shallow socket (the glenoid fossa of the scapula), fitting against the greater tuberosity of the humerus.

5. Saddle joint (articulatio sellaris). In this joint, the opposing bones that come together are convex and concave. The surface of the joint (bone) is saddle-shaped (like a western saddle used in horse riding). The skeleton only has a few saddle joints. Saddle joints allow all movements except rotation. The best example of a saddle joint is the junction between the first metacarpal (of the thumb) and the greater multangular (trapezium) carpal (carpo-metacarpal joint of the thumb).

6. Condyloid joint (articulatio ellipsoidea). This joint is similar to the saddle joint in that a condyle of one bone fits into an elliptical cavity of another bone, allowing all movements except rotation. The wrist joint with the radius and the scaphoid (navicular) carpal is an example of a condyloid joint. Another example of a condyloid joint is the occipital bone and the superior articular process of the atlas (first cervical vertebra).

Body Actions

Grouped by the type of action of the joint, Table 3-1 presents a detailed list of body actions. Figures 3-3 to 3-7 provide drawings that clearly illustrate all of these body actions. The muscles used for each of these body actions are summarized in the second section of Chapter 5. The companion CD that accompanies this text includes video footage of these body actions.

As a matter of convention, the naming of body actions (except in the quizzes) is consistent throughout this book. However, it should be noted that other texts use other conventions that mean the same thing. For example, arm flexion is used in this text as an action, some texts call this Shoulder Flexion, and some may even call it arm-shoulder flexion. All are identical. In fact, while this book employs the term hip-leg for describing actions of that joint, using either hip or leg would also be acceptable. Other equivalent notations would be wrist or hand, foot or ankle, spine or trunk, head or neck, and so on. The important point to remember is the joint on which that the action is taking place.

❑ Body parts that flex, extend, and hyperextend:
- Arm
- Spine
- Hip-leg
- Head
- Finger
- Wrist
- Toe

❑ Body parts that only flex and extend:
- Elbow
- Knee
- Thumb

❑ Scapula actions:
- Adduction
- Abduction
- Elevation
- Depression
- Downward rotation
- Upward rotation

❑ Body parts that adduct and abduct:
- Arm
- Horizontal arm
- Hip-leg
- Finger
- Thumb
- Toe

❑ Body parts that rotate:
- Spine (left and right)
- Head (left and right)
- Arm (internal and external)
- Hip-leg (internal and external)

❑ Body parts that laterally flex:
- Head
- Spine

❑ Miscellaneous body actions:
- Forearm supination
- Forearm pronation
- Ankle dorsiflexion
- Ankle plantar flexion
- Ankle eversion
- Ankle inversion
- Wrist radial flexion
- Wrist ulnar flexion
- Thumb-finger opposition

Table 3-1. Body actions

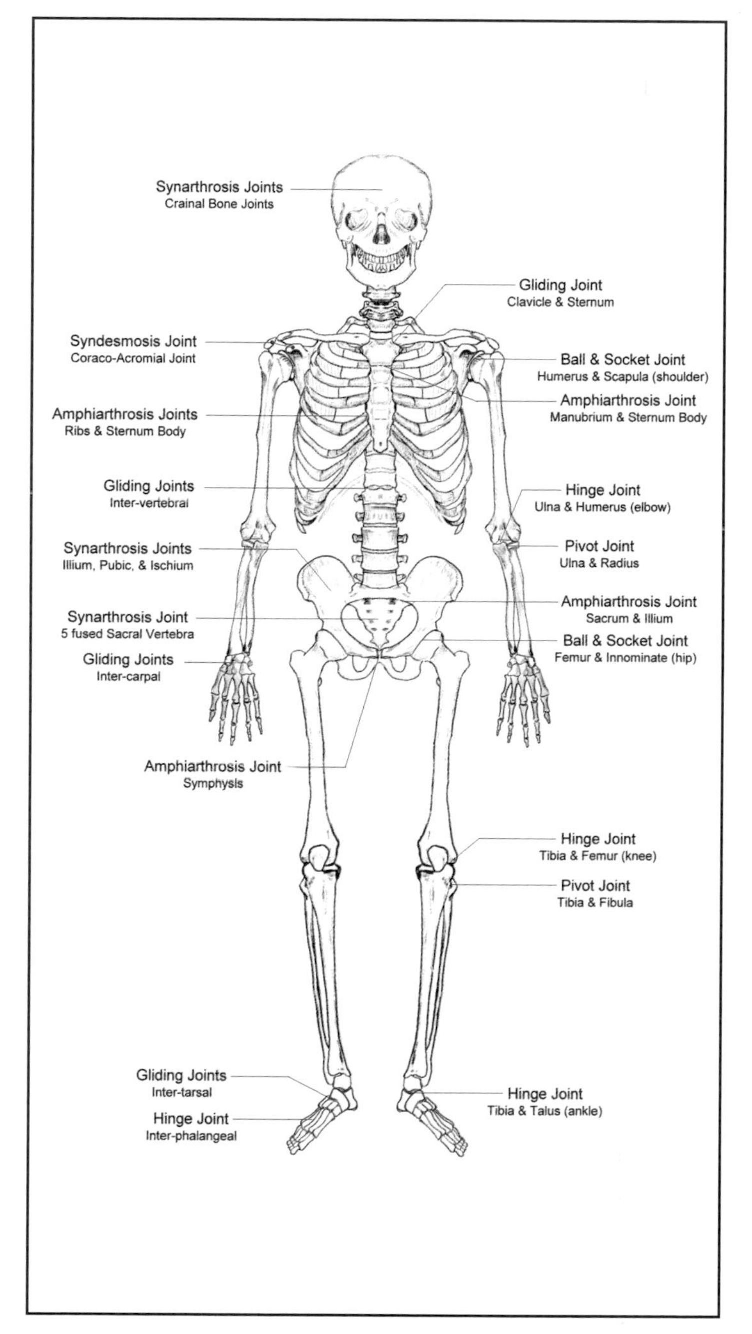

Figure 3-1. Anterior view of joint types

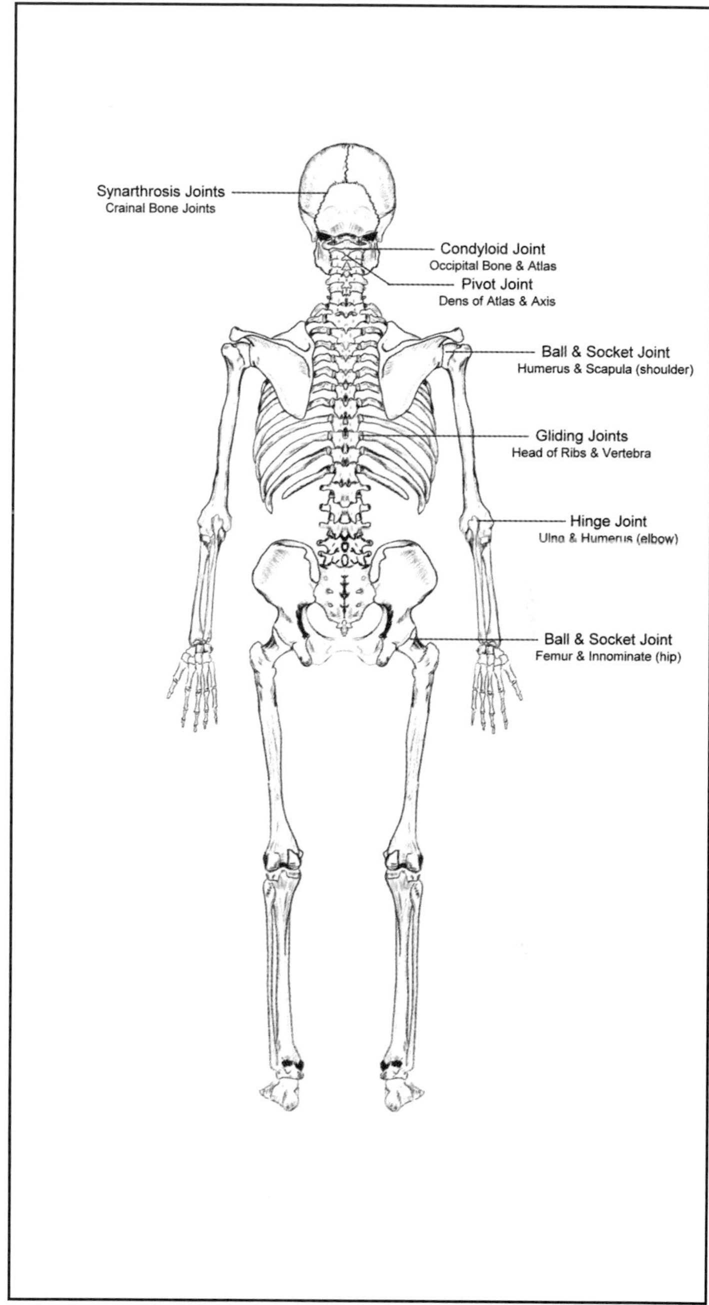

Figure 3-2. Posterior view of joint types

Figure 3-3. Joints that flex, extend, and hyperextend

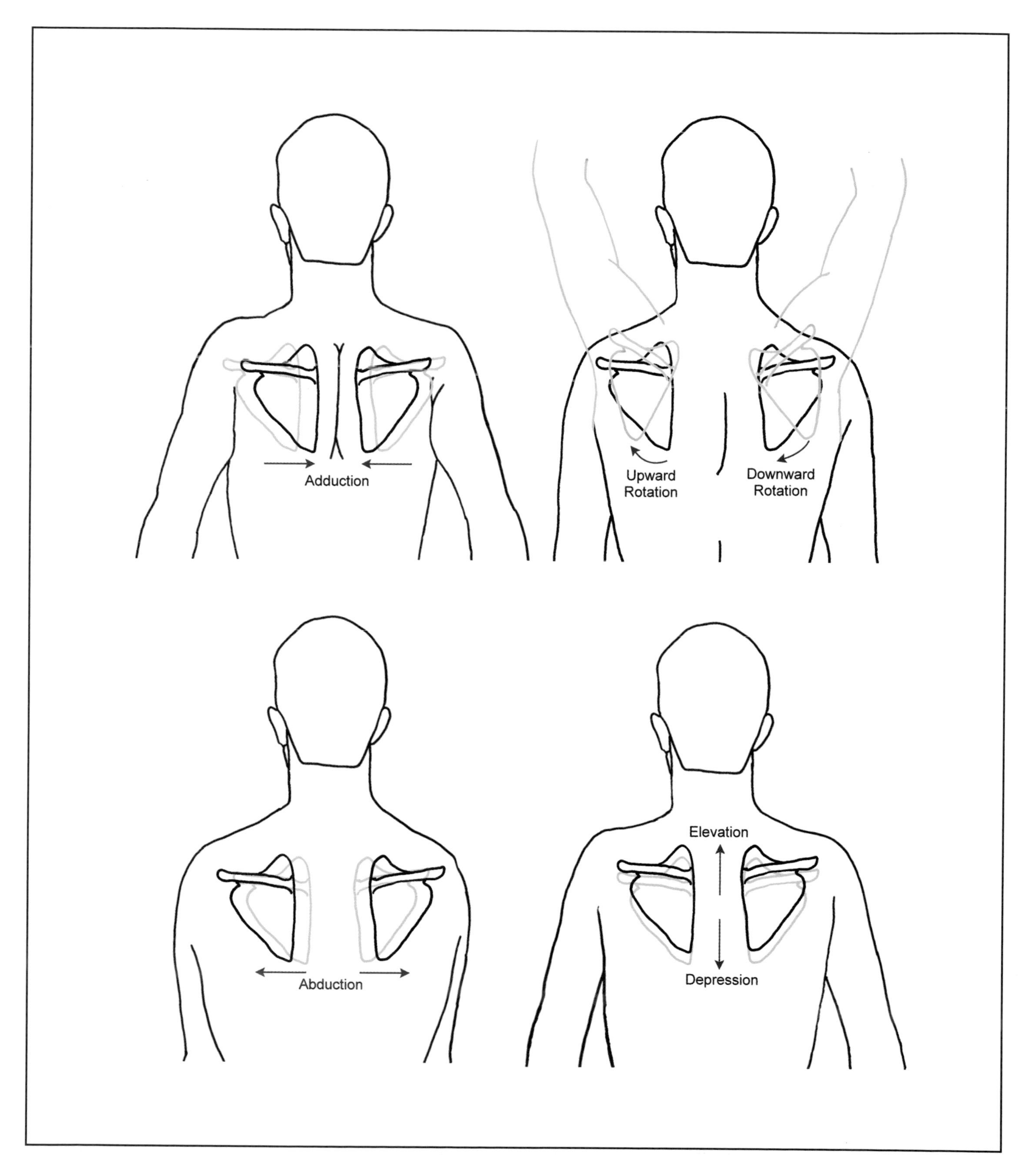

Figure 3-4. Scapula actions

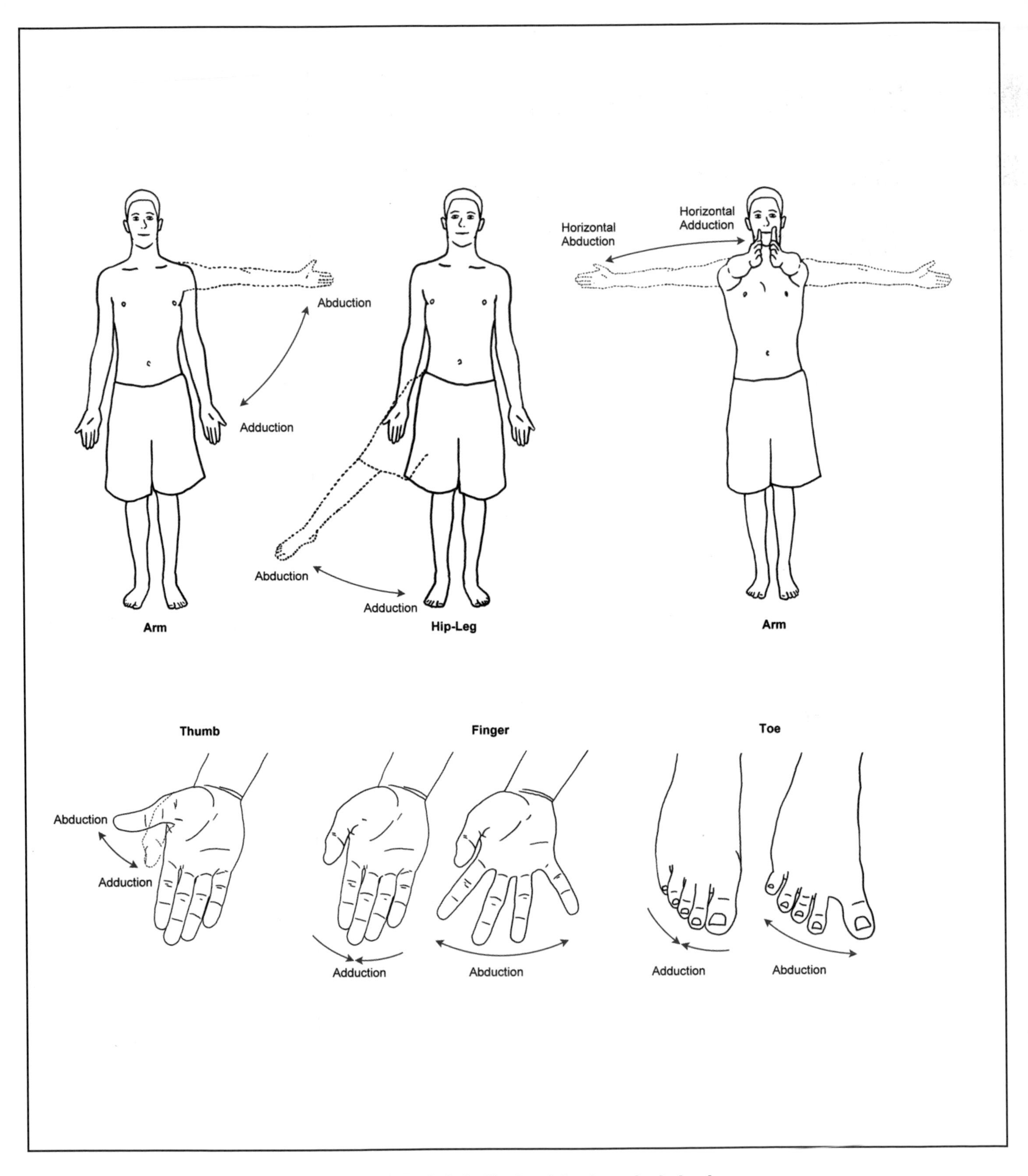

Figure 3-5. Joints that adduct and abduct

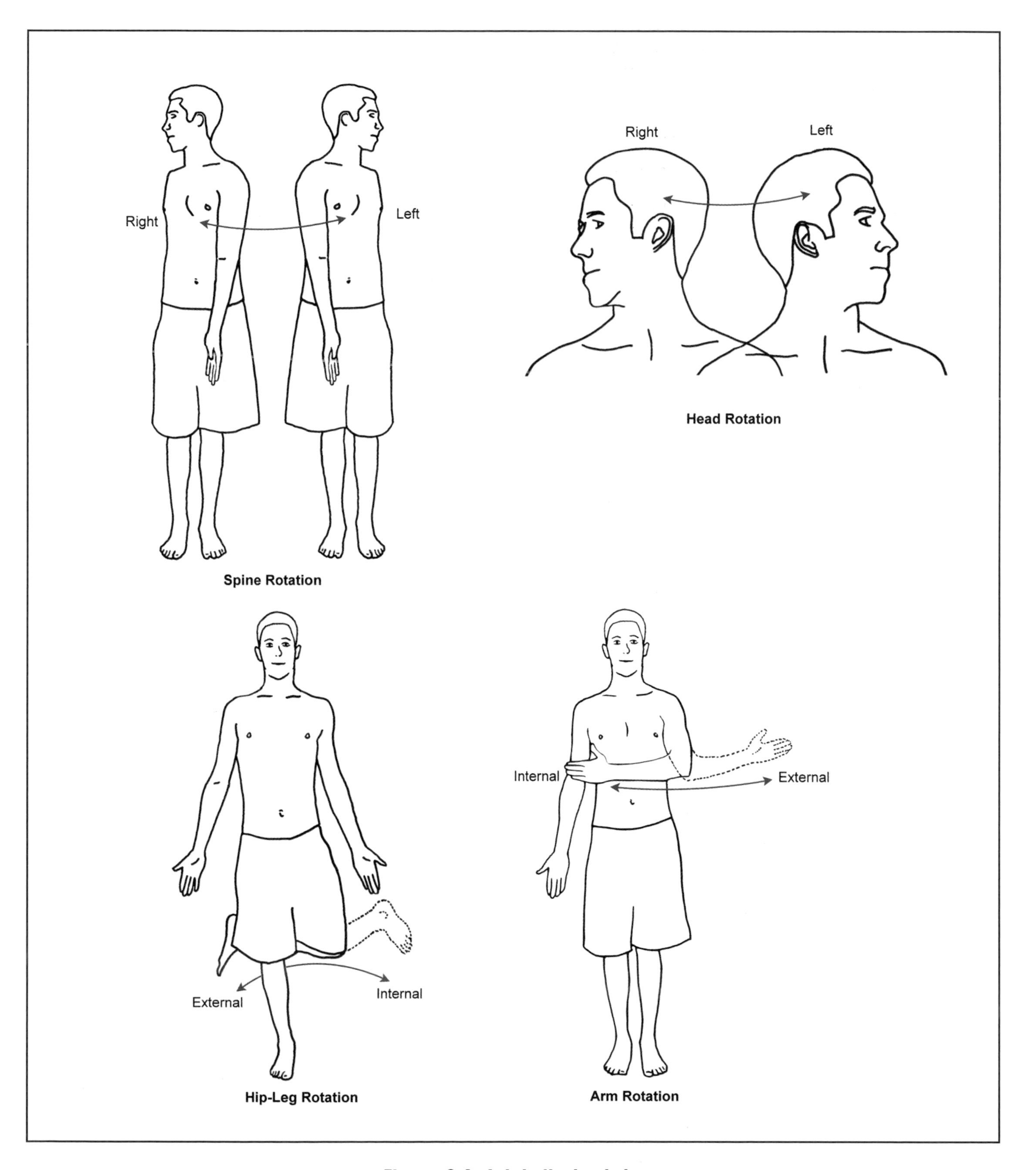

Figure 3-6. Joints that rotate

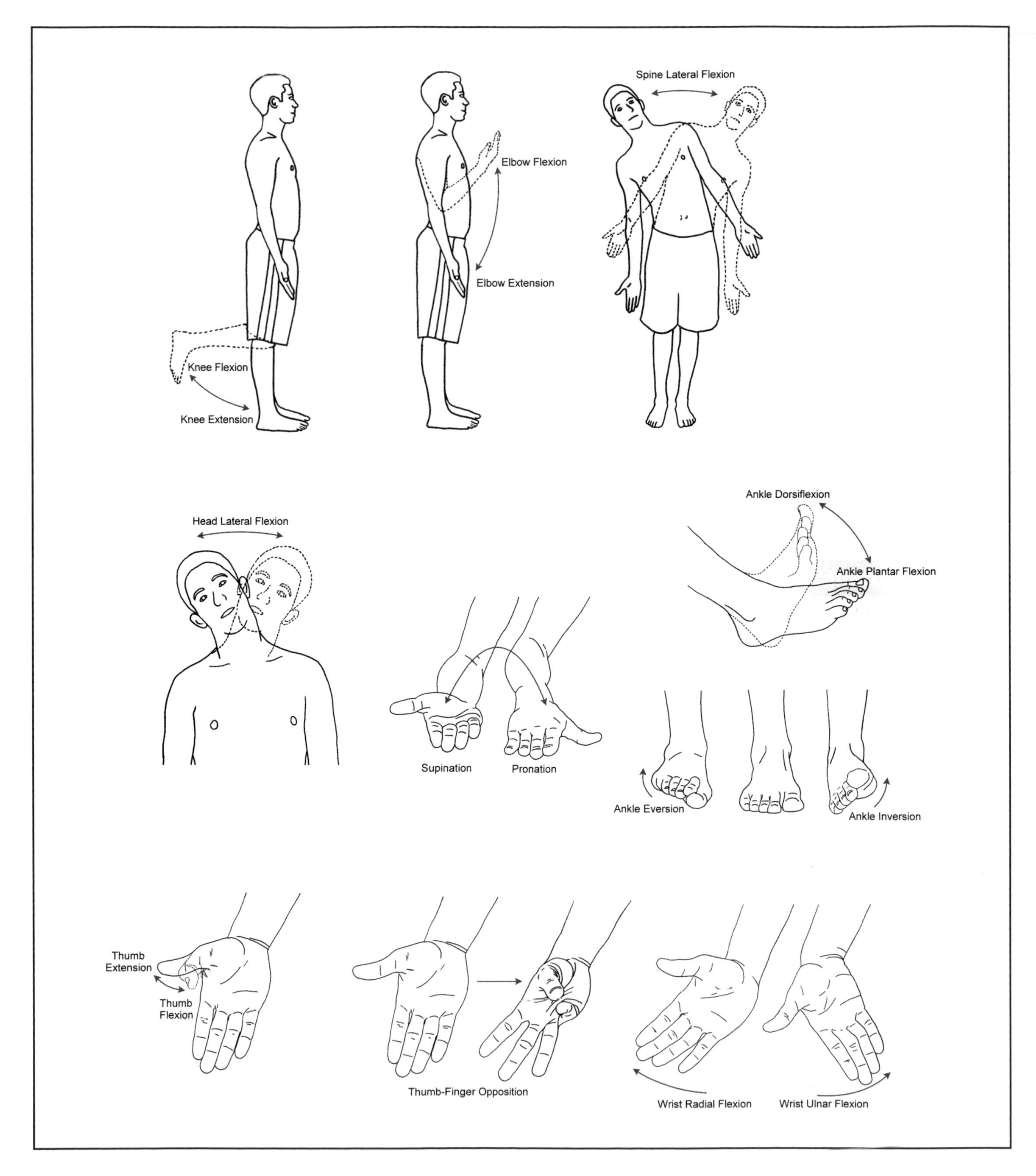

Figure 3-7. Miscellaneous body actions

Chapter 4 • Muscle Anatomy

This chapter deals with the body's muscular system. Since this book is geared toward physical activity and the analysis of movement, the muscles presented and discussed are primarily locomotor muscles (i.e., the muscles involved in locomotion and physical activity). These muscles are reviewed and discussed in detail. However, because fitness professionals should have a basic knowledge of and an appreciation for all the muscles in the body, the body's muscles are listed and briefly discussed in the next section of this chapter. In this section, muscle names and the number of each are listed by specific muscle (with a reference to the specific page on which details of the particular muscle appear) and organized by sections of the body, e.g. muscles of the head, muscles of the neck, muscles of the thorax, and so on.

The last section of this chapter is the main focal point of this chapter. In this section, each locomotor muscle is presented in alphabetical order with its origins and insertions and attachments identified, a full-colored illustration of the muscle, and other details about the muscle.

Muscle Classification

Because biologists are great classifiers, most parts of the body are classified. Nerves are classified, bones are classified, and muscles are classified. Classification is necessary because the body has several kinds of muscle, not just one. The early gross anatomist, (i.e., the individual who studied the body with the naked eye) classified muscles according to where they were located in the body. In the process, he came up with three basic classes of muscles by location.

Those muscles that attached to the skeleton he termed skeletal muscles. These muscles and the bones to which they were attached formed a lever system that provided locomotion and movement. However, when he dissected the abdomen (the viscera), he found other muscles. The stomach is a hollow organ with six big flat muscles that help digest food that enters the stomach. After leaving the stomach, the food travels down thirty-two feet of intestine, which has longitudinal and circular muscles in its walls that provide the peristaltic action that moves the food through the intestine. Because these muscles do not attach to the skeleton but are present in the viscera, he named them visceral muscles.

Much later, he discovered that blood vessels also have muscles in their walls. Although these muscles are not in the viscera, he classified them as viscera muscles. Finally, when the gross anatomist examined the heart, he found that it too is a muscle. Because the heart is a very unique muscle - different than any other muscle in the body, it forms a class of muscles all by itself, the cardiac muscle. Therefore, the first attempt to classify muscles involved grouping muscles by their location. In this process, all the muscles of the body were put into one of three classifications - skeletal, visceral, and cardiac.

With the advent of the microscope, a new branch of biology began called histology. In this process, the anatomist attempted to determine what the body looked like under a microscope. Microscopes enables the anatomist to closely examine cells, whether they are nerve cells, brain cells, bones cells or muscle cells. Subsequently, histologists developed a classification of muscles that was based on what the muscle cell looked like, a process that produced a way of grouping muscles called a histological classification.

When the skeletal muscle was examined under the microscope, the histologist saw that it had light and dark segments that ran down the length of the cells. Because these muscles were segmented or striated, he classified them as striated muscles. It just so happened that all skeletal muscles are striated. However, when he examined visceral muscles under the microscope, the striations were missing - visceral muscles were non-striated. Accordingly, he classified these muscles as smooth muscles. All visceral muscles are smooth muscles. When he studied heart muscle under the microscope, it was again different from either of the other two types of muscles. It had striations. These striations weren't regular, like those of the skeletal muscles, they were criss-crossed in a network pattern. As a result, he classified heart muscle as a syncytium (multinucleated) or a branch-striated muscle.

In other words, in addition to classifying muscles according to their location, the advent of microscopes enabled muscles to be classified according to what they look like under the microscope. Similar to their classification by location, all the muscles of the body can be put into one of three (additional) classifications - striated, smooth, and branch striated.

Another method of classifying muscles involves nerve control. Muscles, like all tissues of the body, are controlled by the nervous system. What kind of a nerve innervates a particular muscle will affect whether the muscle can be controlled volitionally (at will) or whether it is controlled automatically. Muscles are also classified according to how they are controlled by the nervous system. If a muscle can be controlled volitionally, the muscle is called a voluntary muscle. Most skeletal muscles are voluntary, although a few are both voluntary and involuntary.

Muscles that individuals have no control over are called involuntary muscles. Most of the body's visceral muscles are involuntary, although there are a few that are both voluntary and involuntary. When food is introduced to the stomach, the stomach muscles contract and start digesting food whether you want them to or not. As such, stomach muscles are involuntary muscles. Likewise, the cardiac muscle is an involuntary muscle. The muscles that can be both are called voluntary/involuntary (or mixed) muscles. The muscles involved in respiration are an example of mixed muscles. You can either breathe automatically, or hold your breath. Table 4-1 summarizes the three methods of classifying muscles and the three categories in each method.

Location	Microscopic Appearance	Nerve Control
Skeletal	Striated	Voluntary
Visceral	Smooth	Involuntary
Cardiac	Syncytium or Branch Striated	Voluntary/Involuntary

Table 4-1. Three methods of classifying muscles

Muscle Properties

Muscles have the following four properties or characteristics that make them unique:

- Contractibility – the ability to shorten when innervated, causing movement. This factor is a unique property of muscles.
- Extensibility – the ability to be stretched beyond their normal resting length. This property is what enables flexible movement to occur.
- Elasticity – the ability to rebound to their resting length after being stretched.
- Tonicity – a state of hardness of the muscle. When a muscle is exercised, more muscle fibers become active, and the muscle becomes firmer. This factor is called muscle tone. While the other properties of muscles can be readily measured, it is difficult to quantify the level of tonicity.

Types of Contraction

Contraction is a unique property of muscle tissue and involves the ability to shorten when innervated by a motor nerve. However, because more than one kind of contraction exists, muscle contraction is also classified by type. In anatomy, two kinds of muscle contraction exists – isotonic contraction and isometric contraction. If a muscle shortens and causes movement when it is innervated, this is an isotonic contraction. For example, if you move a 25-pound dumbbell from a table by curling it, your elbow flexors are isotonically contracting. On the other hand, if a muscle is prevented from shortening when it is innervated, this is an isometric contraction. For example, in the previous instance, if the dumbbell you are trying to curl weighs 400lbs (as opposed to 25 lbs), and you try to curl it, but it doesn't move, your elbow flexors are contracting, but not shortening, since no movement is occurring. In this example, your elbow flexors are isometrically contracting.

When the fields of kinesiology, physical education, and sports medicine were initially being established, many individuals in these fields discovered that they needed to analyze muscular movements. Because they found the anatomical classification somewhat limiting in this regard, they created a new classification of muscular contractions. When a muscle was innervated and that muscle shortened and movement took place, they renamed isotonic contraction as "concentric contraction." When the muscle was innervated and movement did not take place, that type of contraction was called a "static contraction" instead of an isometric contraction.

The type of movement (term) that was needed that the anatomist didn't have was what happens when a muscle moves against gravity, but the muscle lengthens during the movement. Using the previous example, if the 25-pound dumbbell on a table is actually moved when you attempt to curl it, your elbow flexors are isotonically or concentrically contracting. If you now slowly return the dumbbell to the table, then the same muscles that curled the dumbbell are allowing it to be slowly returned to the table. Your elbow flexors are lengthening, but are doing the work. Gravity wants to pull the dumbbell quickly to the table. A kinesiologist calls this type of contraction an "eccentric contraction." Specifically, the muscle is working, but lengthening. If an anatomist were asked about this, he would say that your elbow flexors are actually slowly relaxing. Such an analysis is true, but someone analyzing muscular movement needs a descriptive term to use. In this instance, that term is "eccentric contraction."

A mix of the aforementioned terms are employed in this book. When a muscle is working, but lengthening, this is referred to as an eccentric contraction. When the muscle is working, but not changing length, this is called an isometric contraction. When the muscle is working and shortening, this is either designated as an concentric contraction, or no designation is used at all, since this is the typical muscle action.

Description	Human Anatomy Terms	Kinesiological Terms
Muscle length shortens	Isotonic	Concentric
Muscle length doesn't change	Isometric	Static
Muscle length increases		Eccentric

Table 4-2. Types of muscle contractions

Muscles in the Human Body

Tables 4-3, 4-4, and 4-5 itemize all the skeletal muscles in the body. Many of the muscles (like the facial muscles) will not be dealt with in detail, but are listed for reference. A number in parenthesis after a muscle group indicates the number of muscles in the group. These muscles frequently have page numbers in brackets following their name. You should refer to the appropriate page for more information on that muscle. The muscles have been divided into two main groups: axial muscles (muscles that are in the body) and appendicular muscles (muscles found in the arms and legs). They are then broken into groups pertaining to the body area in which they are located.

Muscles of the head:
- Cranial muscles (7)
- Facial muscles (24)
- Muscles of the eye (10)
- Muscles of the tongue (8)
- Muscles of the soft palate (6)
- Muscles of the pharynx (4)

Muscles of the neck:
- Muscles of the larynx (22)
- Superficial lateral muscles of the neck:
  - sternocleidomastoid (p 116)
  - rectus capitis lateralis (p 115)
- Anterior neck muscles (7):
  - rectus capitis anterior (p 114)
  - longus capitis (p 114)
- Posterior neck muscles (15):
  - rectus capitis posterior (major) (p 118)
  - rectus capitis posterior (minor) (p 118)
  - splenius cervicis (p 119)
  - splenius capitus (p 119)
  - obliquus capitis inferior (p 117)
  - obliquus capitis superior (p 117)

Muscles of the thorax:
- Diaphragm
- External intercostals (11) (p 151)
- Internal intercostals (11) (p 151)
- Serratus posterior (inferior) (p 154)
- Serratus posterior (superior) (p 155)
- Scalenus medius (p 153)
- Transversus thoracis (p 156)
- Scalenus posterior (p 153)
- Subcostales (p 156)
- Levator costarum (p 152)
- Scalenus anterior (p 152)

The abdominal muscles:
- External oblique abdominal (p 40)
- Internal oblique abdominal (p 41)
- Rectus abdominis (p 44)
- Transversus abdominis (p 45)
- Quadratus lumborum (p 43)
- Pyramidalis (p 42)

The spinal muscles:
- Sacrospinalis (erector spinae):
  - longissimus cervicis (p 146)
  - longissimus thoracis (p 147)
  - longissimus capitis (p 145)
  - iliocostalis cervicis (p 142)
  - iliocostalis thoracis (p 144)
  - iliocostalis lumborum (p 143)
  - spinalis capitis (p 148)
  - spinalis cervicis (p 149)
  - spinalis thoracis (p 150)
- Deep muscles:
  - longus colli (superior oblique) (p 138)
  - longus colli (vertical) (p 139)
  - longus colli (inferior oblique) (p 138)
  - semispinalis cervicis (p 140)
  - semispinalis thoracis (p 141)
  - interspinalis (p 137)
  - rotatores (p 140)
  - intertransversarii (p 137)
  - multifidus (p 139)

Table 4-3. Axial muscles

Muscles of the upper extremity:

- Muscles that act primarily on the shoulder girdle:
  - pectoralis minor (p 132)
  - serratus anterior (p 134)
  - rhomboids (major and minor) (p 133)
  - levator scapulae (p 131)
  - trapezius (p 136)
  - subclavius (p 135)
- Muscles that act primarily on the shoulder joint:
  - pectoralis major (p 126)
  - coracobrachialis (p 120)
  - deltoid anterior (p 121)
  - deltoid middle (p 122)
  - deltoid posterior (p 123)
  - latissimus dorsi (p 125)
  - teres major (p 129)
  - teres minor (p 130)
  - supraspinatus (p 128)
  - infraspinatus (p 124)
  - subscapularis (p 127)
- Muscles that act primarily on the elbow and radioulnar joints:
  - biceps brachii (p 59)
  - brachialis (p 60)
  - triceps brachii (p 65)
  - anconeus (p 58)
  - pronator teres (p 63)
  - brachioradialis (p 61)
  - supinator (p 64)
  - pronator quadratus (p 62)
- Muscles that act primarily on the wrist and fingers:
  - superficial flexors:
    - √ flexor carpi radialis (p 168)
    - √ palmaris longus (p 171)
    - √ flexor digitorum superficilias (p 170)
    - √ flexor carpi ulnaris (p 169)
  - deep flexors:
    - √ flexor digitorum profundus (p 161)
    - √ flexor pollicis longus (p 162)
  - superficial extensors:
    - √ extensor carpi radialis longus (p 164)
    - √ extensor carpi radialis brevis (p 163)
    - √ extensor digitorum (p 167)
    - √ extensor digiti minimi (p 166)
    - √ extensor carpi ulnaris (p 165)
  - deep extensors:
    - √ abductor pollicis longus (p 157)
    - √ extensor pollicis longus (p 160)
    - √ extensor pollicis brevis (p 159)
    - √ extensor indicis (p 158)
- Intrinsic muscles of the hand that act on the fingers:
  - opponens digiti minimi (p 84)
  - abductor digiti minimi (p 77)
  - adductor pollicis (p 79)
  - opponens pollicis (p 85)
  - flexor pollicis brevis (p 82)
  - abductor pollicis brevis (p 78)
  - palmar interossei (p 86)
  - dorsal interossei (4) (p 80)
  - flexor digiti minimi brevis (p 81)
  - lumbricals (4) (p 83)

Table 4-4. Upper-extremity appendicular muscles

Muscles of the lower extremity:

- Muscles that act primarily on the hip joint:
  - ilio-psoas:
    - √ iliacus (p 93)
    - √ psoas major (p 95)
    - √ psoas minor (p 96)
  - pectineus (p 94)
  - tensor fascia latae (p 97)
  - adductor longus (p 88)
  - adductor brevis (p 87)
  - adductor magnus (p 89)
  - gluteus maximus (p 90)
  - gluteus medius (p 91)
  - gluteus minimus (p 92)
- Muscles that primarily rotate the hip:
  - piriformis (p 104)
  - gemellus superior (p 101)
  - gemellus inferior (p 100)
  - obturator internus (p 103)
  - obturator externus (p 102)
  - quadratus femoris (p 105)
- Muscles that act on the hip and knee:
  - sartorius (p 99)
  - gracilis (p 98)
- Muscles that act primarily on the knee joint:
  - quadriceps
    - √ rectus femoris (p 110)
    - √ vastus lateralis (p 112)
    - √ vastus medialis (p 113)
    - √ vastus intermedius (p 111)
  - hamstrings:
    - √ biceps femoris (p 107)
    - √ semitendinosus (p 109)
    - √ semimembranosus (p 108)
  - popliteus (p 106)
- Muscles that act primarily on the ankle, and toes:
  - anterior group:
    - √ extensor digitorum longus (p 46)
    - √ extensor hallucis longus (p 47)
    - √ peroneus tertius (p 48)
    - √ tibialis anterior (p 49)
  - lateral group:
    - √ peroneus brevis (p 50)
    - √ peroneus longus (p 51)
  - posterior group
    - √ gastrocnemius (p 54)
    - √ soleus (p 56)
    - √ plantaris (p 55)
    - √ flexor digitorum longus (p 52)
    - √ flexor hallucis longus (p 53)
    - √ tibialis posterior (p 57)
- Intrinsic muscles of the foot that act on the toes:
  - dorsal interossei (4) (p 69)
  - abductor hallucis (p 67)
  - extensor digitorum brevis (p 70)
  - plantar interossei (3) (p 75)
  - flexor digiti minimi (quinti) brevis (p 71)
  - adductor hallucis (p 68)
  - flexor hallucis brevis (p 73)
  - lumbricals (4) (p 74)
  - quadratus plantae (p 76)
  - flexor digitorum brevis (p 72)
  - abductor digiti minimi (quinti) (p 66)

Table 4-5. Lower-extremity appendicular muscles

Muscle Details and Illustrations

The following pages offer the essential details of every skeletal muscle, grouped by body part and presented in alphabetical order.

NOTE: Sometimes a muscle is classified as two distinct muscles—one referred to as a major, the other as a minor. The difference, however, is academic in that the two parts have the same action and are usually learned as just one muscle. For example, early anatomists divided the psoas muscle into two separate muscles—the psoas major (upper fibers), and the psoas minor (lower fibers), but the muscle works as a whole. Likewise, the rhomboid major and rhomboid minor act as one muscle and are usually learned as one muscle. On the other hand, on occasion, the terms "major" and "minor" reflect two totally different muscles and must be treated as such, e.g., pectoralis major and minor, or teres major and minor.

It is also important to note that some muscles, due to their extensive origin, have opposite actions. In these instances, anatomists look at the actions of the anterior and posterior fibers that are located in a particular part of a muscle to define that muscle, e.g. the adductor magnus. Some of these parts are different enough to be considered and studied as two or three muscles, even though they are one muscle, e.g., the deltoid muscle (refer to pages 121-123). The deltoid is one muscle but can be learned as three muscles: anterior deltoid, middle deltoid, and posterior deltoid. The aforementioned three terms separate muscles are also referred by some textbooks as the clavicular deltoid, acromional deltoid, and scapula deltoid—terms that correspond to their origin. Although books deal with the deltoid as one muscle and refer to it as the major abductor of the arm, it may be beneficial to learn it as three muscles, because although the deltoid is an abductor of the arm, the anterior and posterior deltoids are arm adductors. As such, it is helpful to study and learn the deltoid as three separate muscles.

External Oblique Abdominal

Description:
These muscles, on each side of the abdomen, aid in rotating the spine when working independently. When working together, they assist the rectus abdominis in spine flexion and abdomen compression.

Joint Crossings:
Spine

Rank/Body Action:

1 Spine flexion (both)
2 Spine rotation (to the opposite side - each)
3 Abdomen compression (both)
4 Spine lateral flexion (slight)

Blood Supply:
- Lower intercostal arteries
- Subcostal arteries
- Lumbar arteries

Insertion:
- Linea alba
- Anterior portion of the crest of the Ilium and inguinal ligament

Nerves:
- Intercostal nerves T8-T12
- Iliohypogastric nerves T12, L1
- Ilioinguinal nerve L1

Origin:
External surfaces of the lower eight ribs

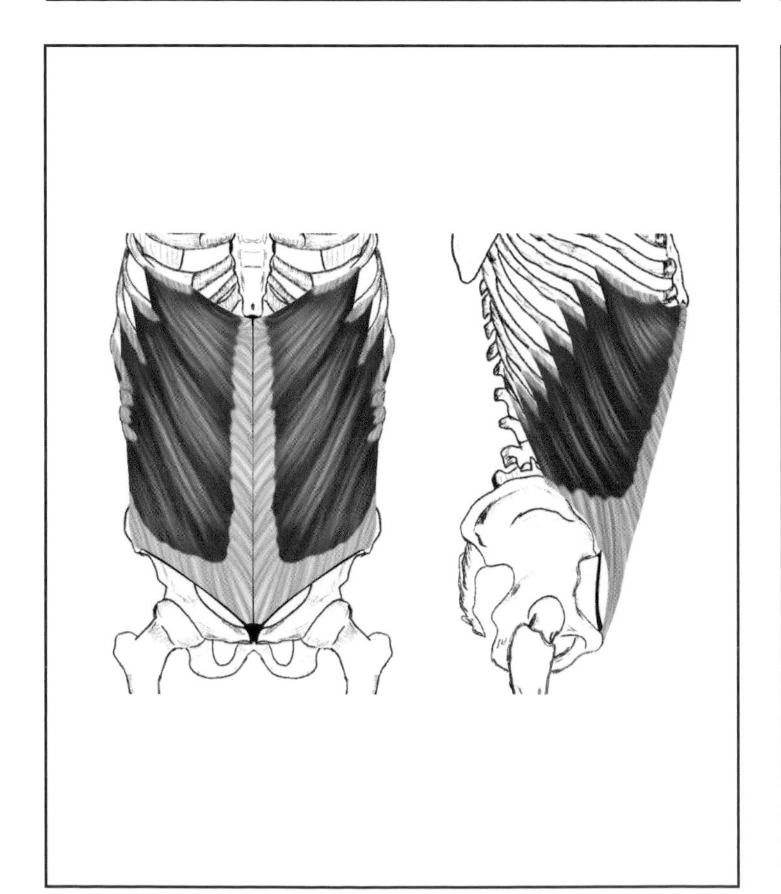

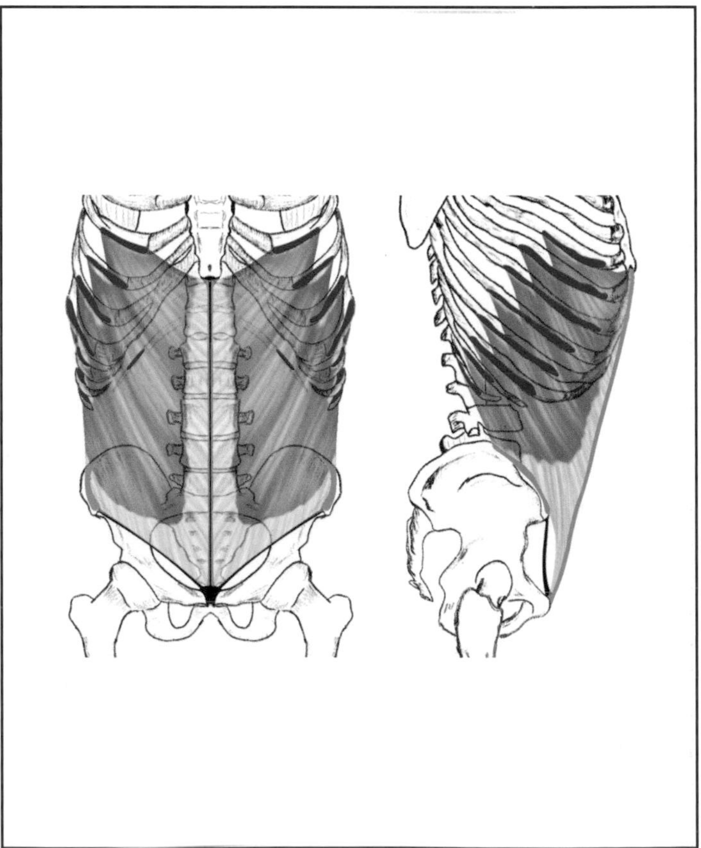

Internal Oblique Abdominal

Description:
The fibers of this muscle run diagonally in the opposite direction to that of the external obliques, and help rotate and flex the spine as well as compress the abdomen.

Joint Crossings:
Spine

Rank/Body Action:
1 Spine flexion (both)
2 Spine rotation (to the same side - each)
3 Abdomen compression (both)

Blood Supply:
- Lower intercostal arteries
- Subcostal arteries
- Lumbar arteries

Insertion:
- Lower three ribs
- Linea alba
- Superior ramus of the pubis

Nerves:
- Intercostal nerves T8-T12
- Iliohypogastric nerves T12, L1
- Ilioinguinal nerve L1

Origin:
- Lumbodorsal fascia
- Anterior portion of the crest of the Ilium
- Lateral one-half of the inguinal ligament

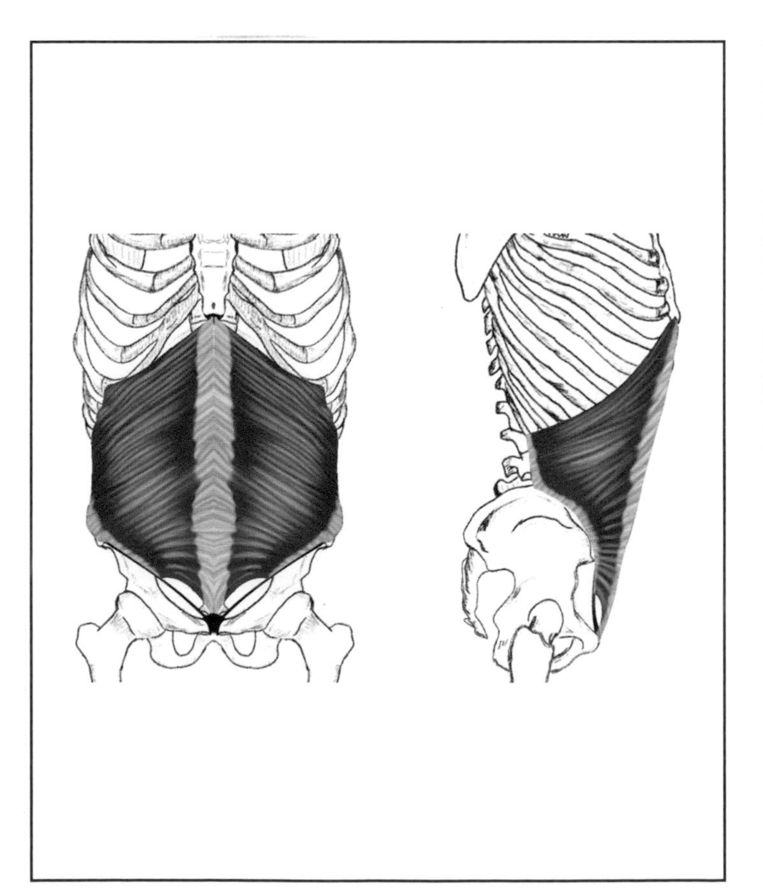

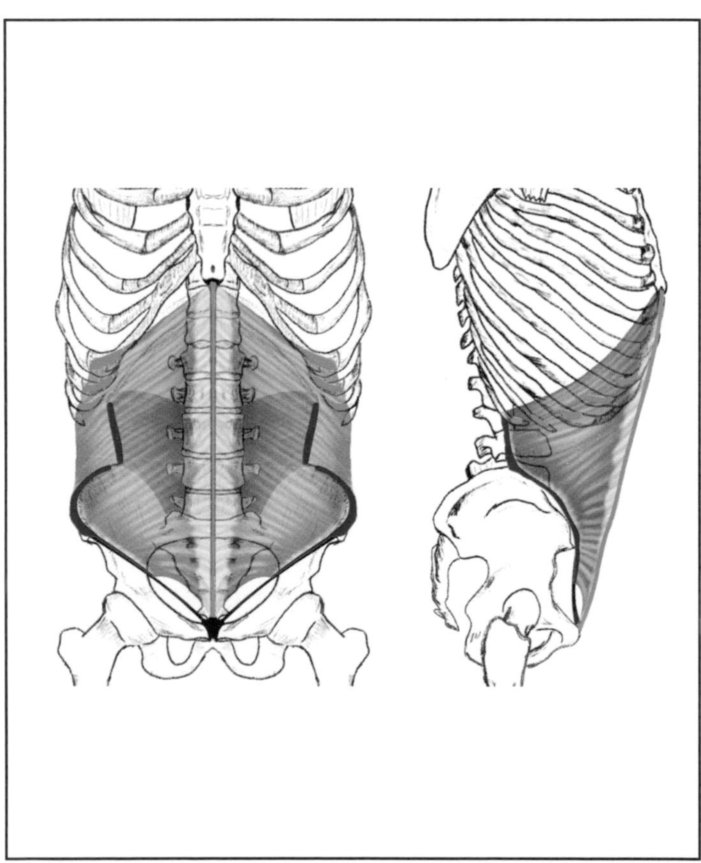

Pyramidalis

Description:
A small muscle of the lower anterior part of the abdomen that is in the same sheath with the rectus abdominis and functions to pull on the linea alba.

Joint Crossings:
N/A

Body Action:
Pulls on the linea alba

Blood Supply:
Superior and inferior epigastric arteries

Insertion:
Lower portion of the linea alba

Nerves:
Branch of the subcostal nerve T12

Origin:
Anterior surface of pubis

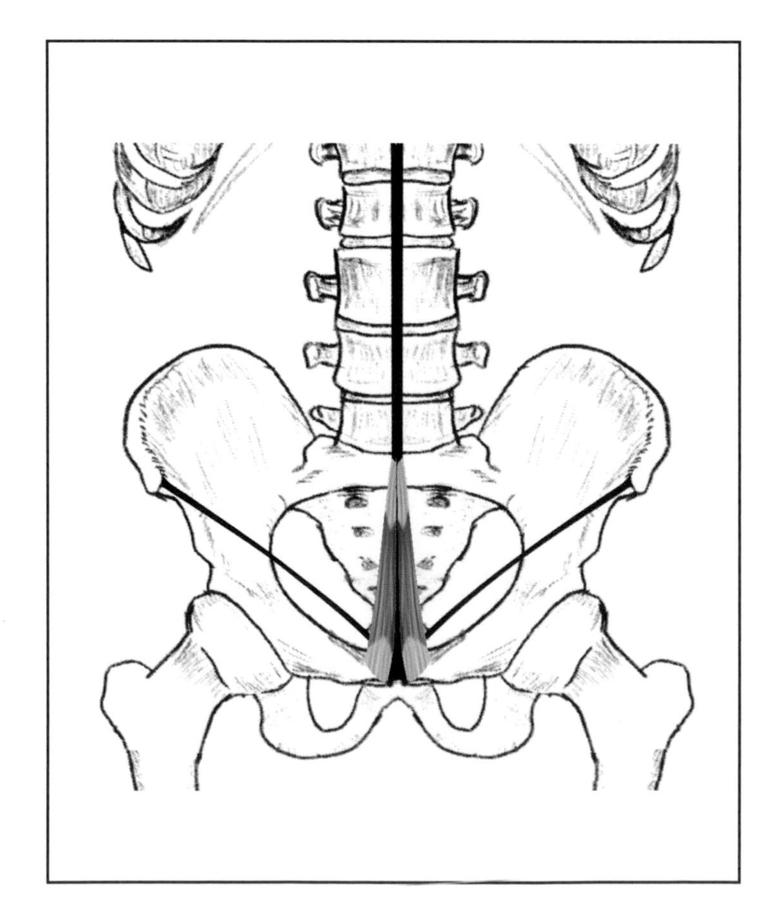

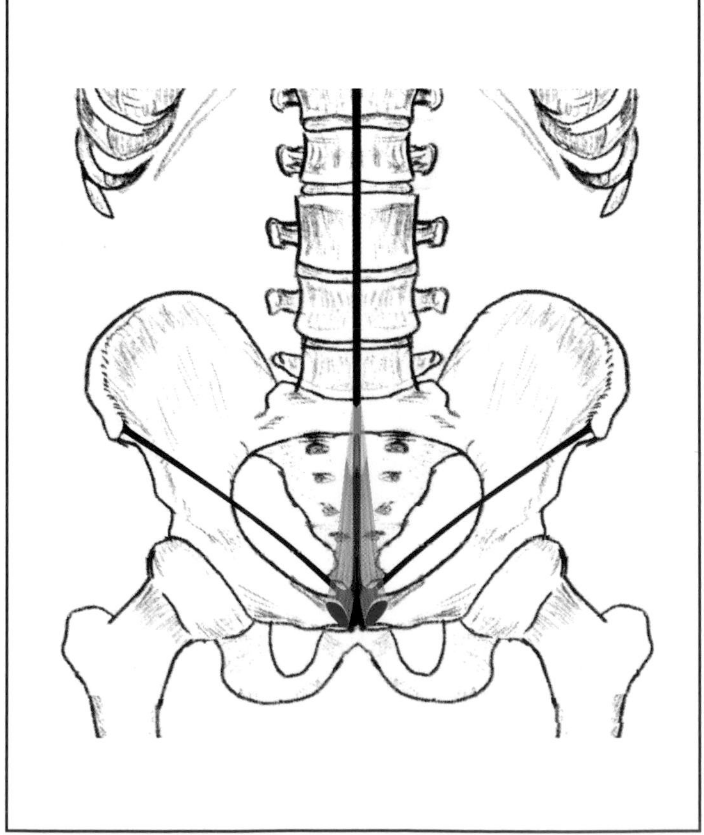

Quadratus Lumborum

Description:
A quadrilateral-shaped muscle of the back that flexes the spine laterally, and when working together with the other back muscles to extend the spine.

Joint Crossings:
Spine

Rank/Body Action:

1	Spine extension (both)
2	Spine hyperextension (both)
3	Spine lateral flexion (each)

Blood Supply:
- Lumbar artery
- Subcostal artery

Insertion:
- Twelfth rib
- Transverse processes of the five lumbar vertebrae

Nerves:
Branches of nerves T12, L1

Origin:
Posterior crest of the ilium

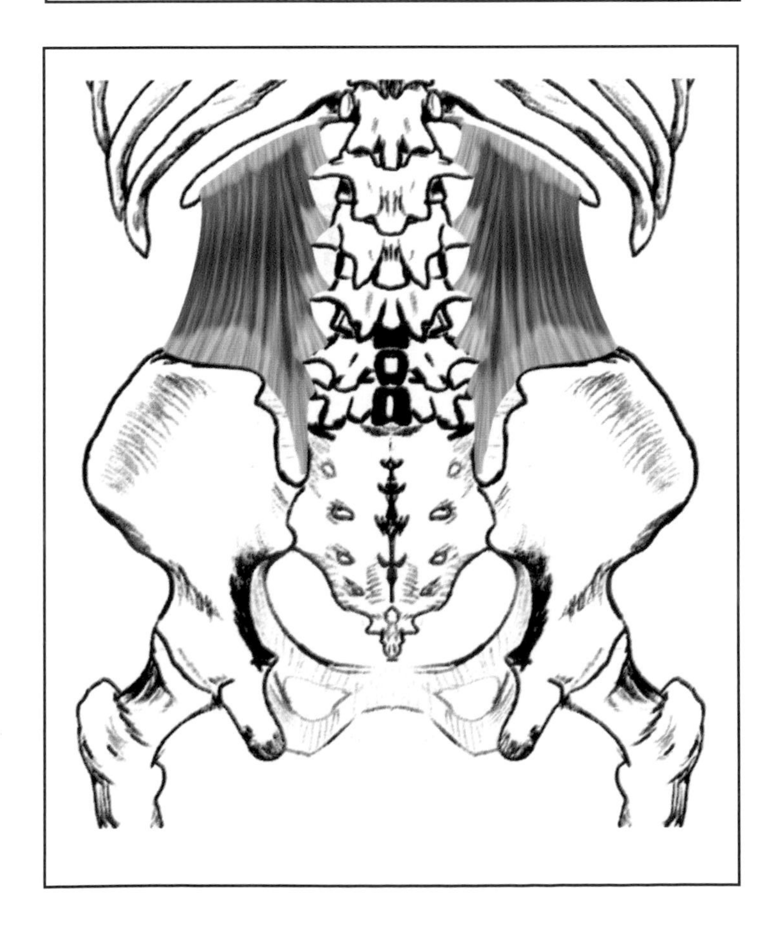

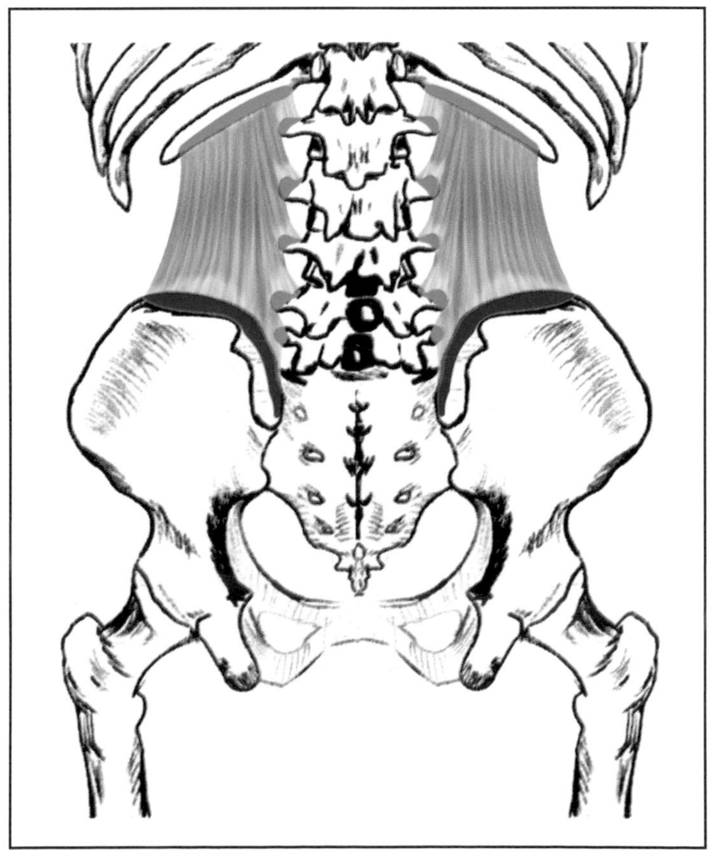

RECTUS ABDOMINIS

Description:
A long flat muscle on either side of the linea alba, extending along the whole length of the front of the abdomen, that acts to flex the spine, tense the anterior wall of the abdomen, and assist in compressing the contents of the abdomen.

Joint Crossings:
Spine

Rank/Body Action:

1	Spine flexion
2	Abdomen compression

Blood Supply:
Superior and inferior epigastric arteries

Insertion:
- Xiphoid process of the sternum
- Cartilages of the fifth, sixth, and seventh ribs

Nerves:
Seventh to twelfth intercostal nerves

Origin:
Superior ramus of the pubis (crest of the pubic bone)

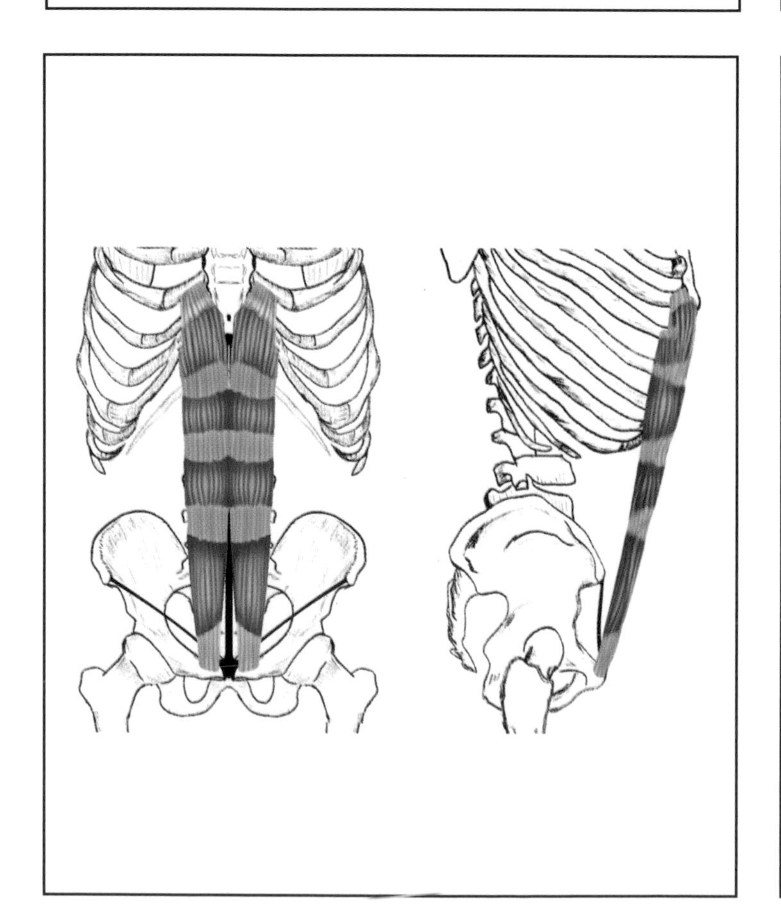

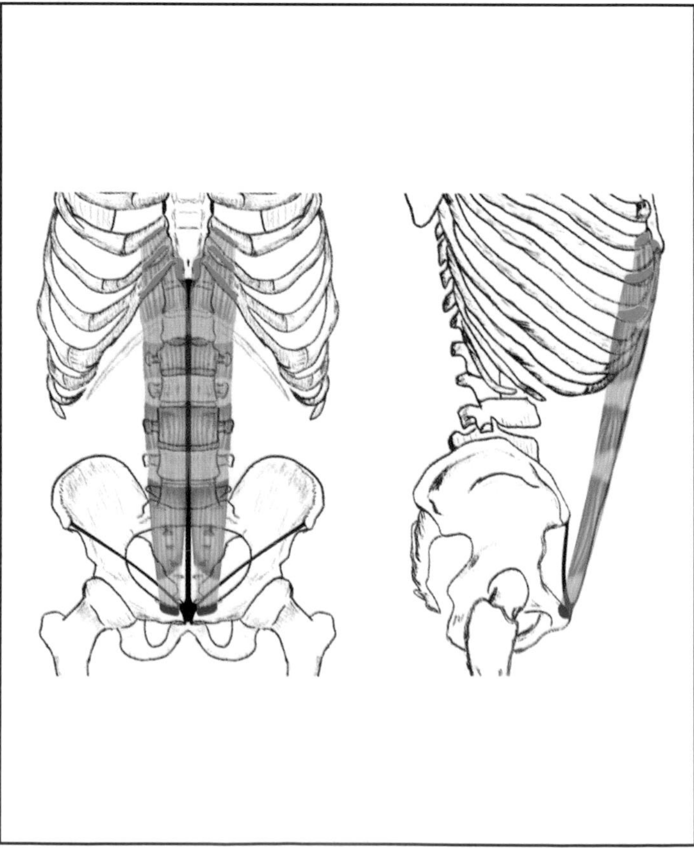

Transversus Abdominis

Description:
A flat muscle with transverse fibers that forms the innermost layer of the wall of the abdomen and ends in a broad aponeurosis. It acts to compress the abdominal viscera and assists in the expulsion of the contents of various abdominal organs (as in urination, defecation, vomiting and parturition).

Joint Crossings:
N/A

Body Action:
Abdomen compression

Blood Supply:
- Lower intercostal arteries
- Subcostal arteries
- Lumbar arteries

Insertion:
- Linea alba
- Superior ramus of the pubis

Nerves:
- Intercostal nerves T7-T12
- Iliohypogastric nerves T12, L1
- Ilioinguinal nerve L1

Origin:
- Lateral one-half of the inguinal ligament
- Crest of the ilium
- Lumbodorsal fascia
- Inside surfaces of the lower six ribs

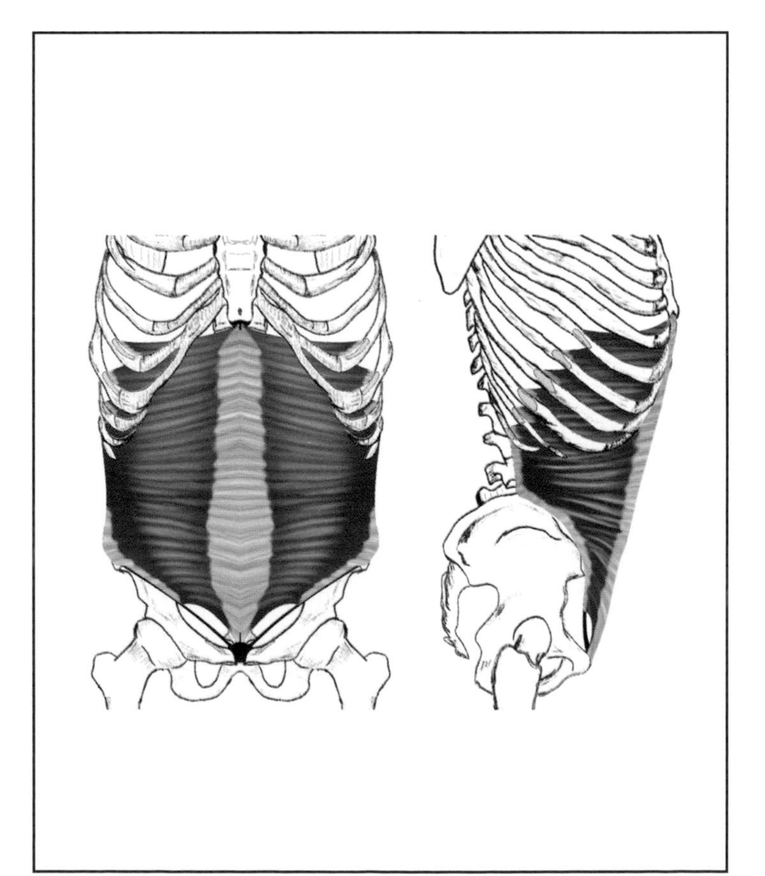

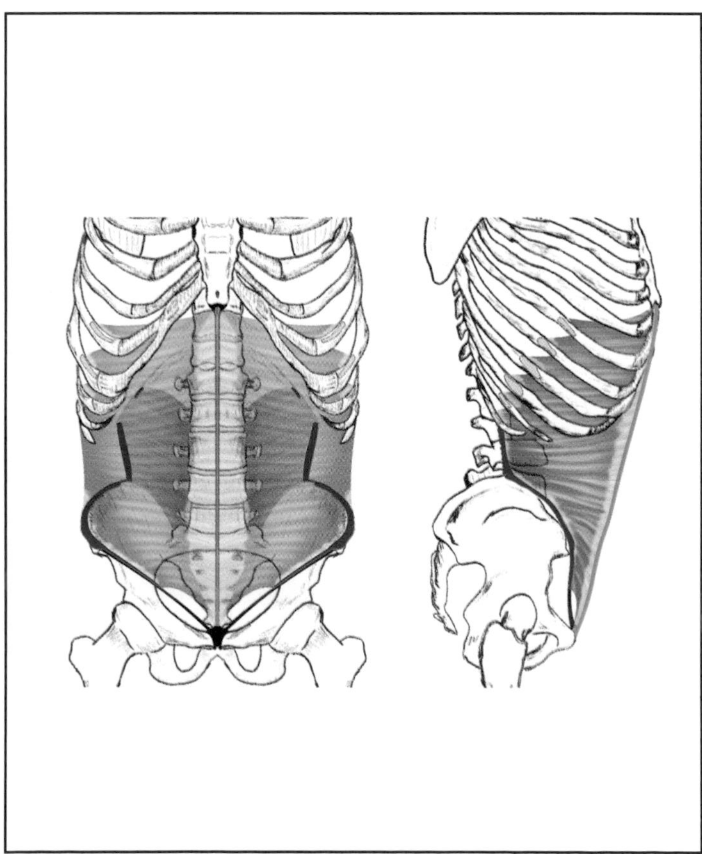

EXTENSOR DIGITORUM LONGUS

Description:
A muscle in which fibers extend obliquely from either side of a central tendon on the outer front of each leg. It extends the four small toes and dorsiflexes and everts the ankle.

Joint Crossings:
Two

Rank/Body Action:

1	Ankle dorsiflexion
2	Toe extension (2nd - 5th toes)
3	Toe hyperextension
4	Ankle eversion

Blood Supply:
Anterior tibial artery

Insertion:
Dorsal surface of the second and third phalange of the 2nd - 5th toes

Nerves:
Deep peroneal nerve L4, 5, S1

Origin:
- Outside surface of the lateral condyle of the tibia
- Head and anterior surface upper three-fourths of the fibula

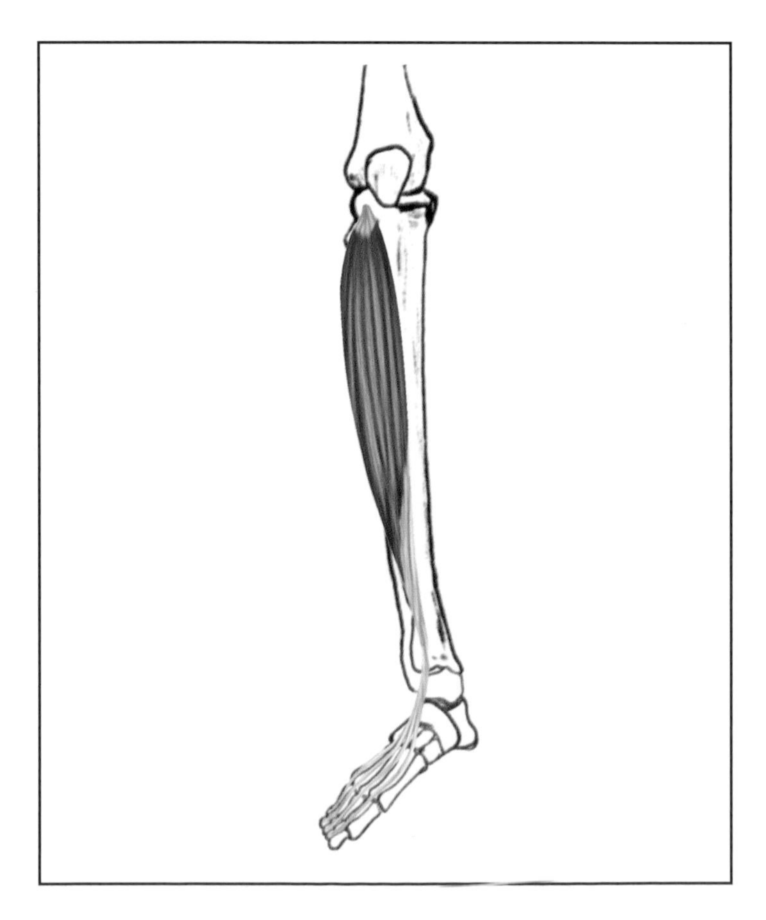

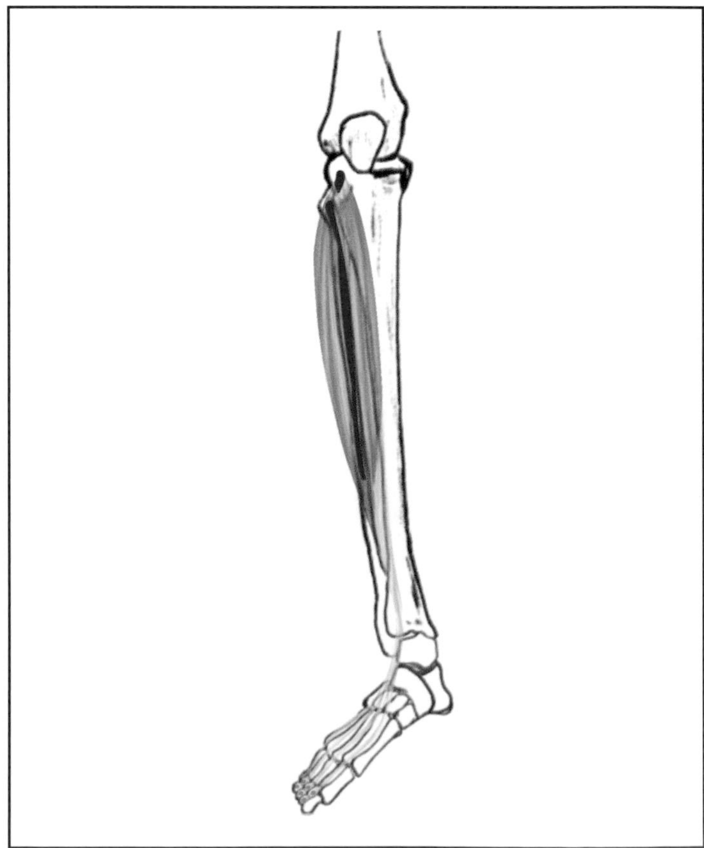

EXTENSOR HALLUCIS LONGUS

Description:
A long thin muscle situated on the fibula that extends the big toe and dorsiflexes and inverts the ankle.

Joint Crossings:
Two

Rank/Body Action:
1 Toe extension (big toe)
2 Toe hyperextension (big toe)
3 Ankle dorsiflexion
4 Ankle inversion

Blood Supply:
Anterior tibial artery

Insertion:
Base of the second phalange of the big toe (dorsal surface)

Nerves:
Deep peroneal nerve L4, 5, S1

Origin:
Anterior surface of the middle portion one-fourth of the fibula

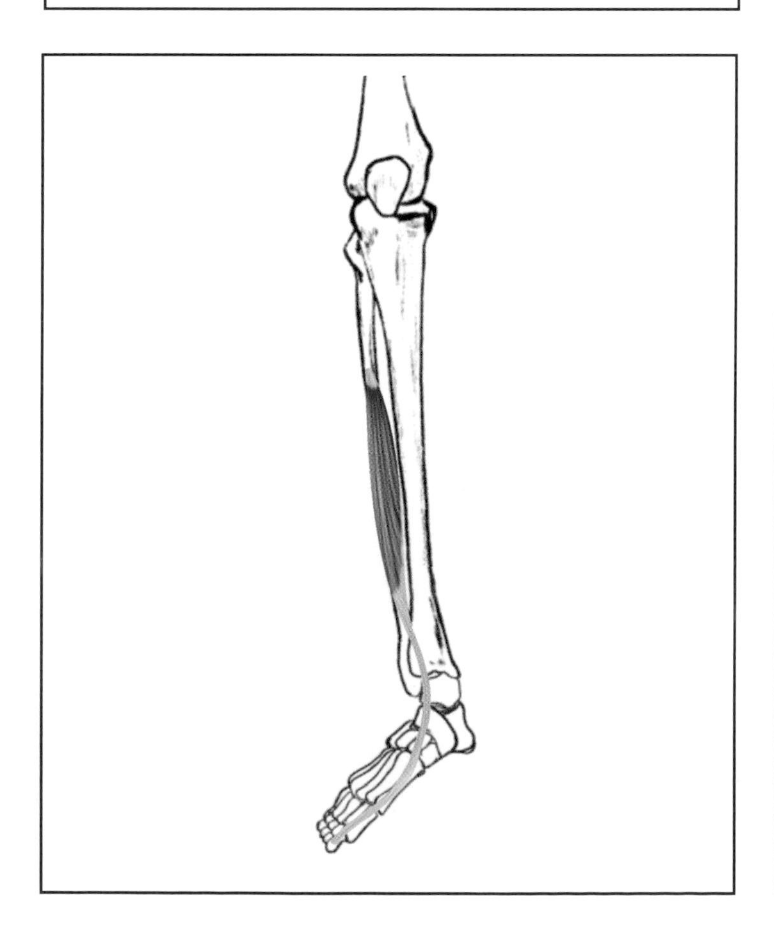

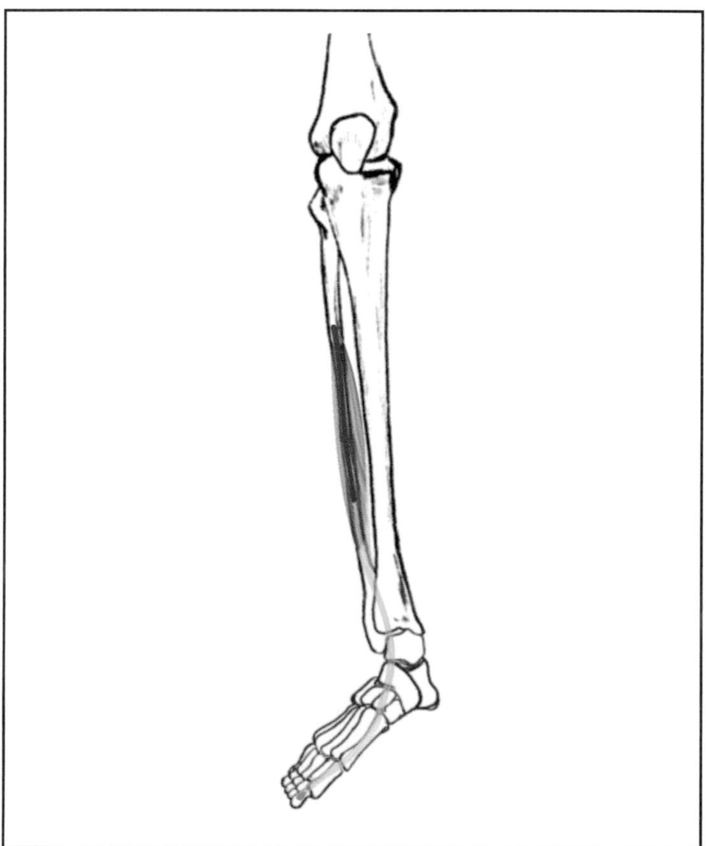

PERONEUS TERTIUS

Description:
A muscle that is sometimes referred to as a branch of the extensor digitorum longus. Absent in some individuals, it dorsiflexes the ankle and assists in everting it.

Joint Crossings:
One

Rank/Body Action:
1 Ankle dorsiflexion
2 Ankle eversion

Blood Supply:
Anterior tibial artery

Insertion:
Base of the fifth metatarsal

Nerves:
Deep peroneal nerve L4, 5, S1

Origin:
Anterior/lateral surface of the lower one-third of the fibula

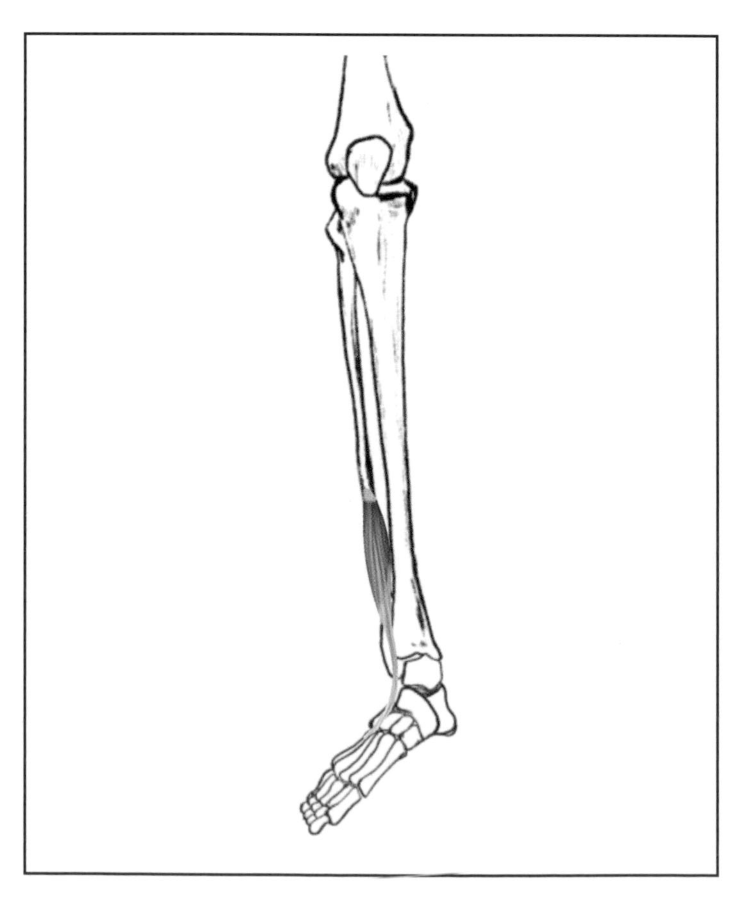

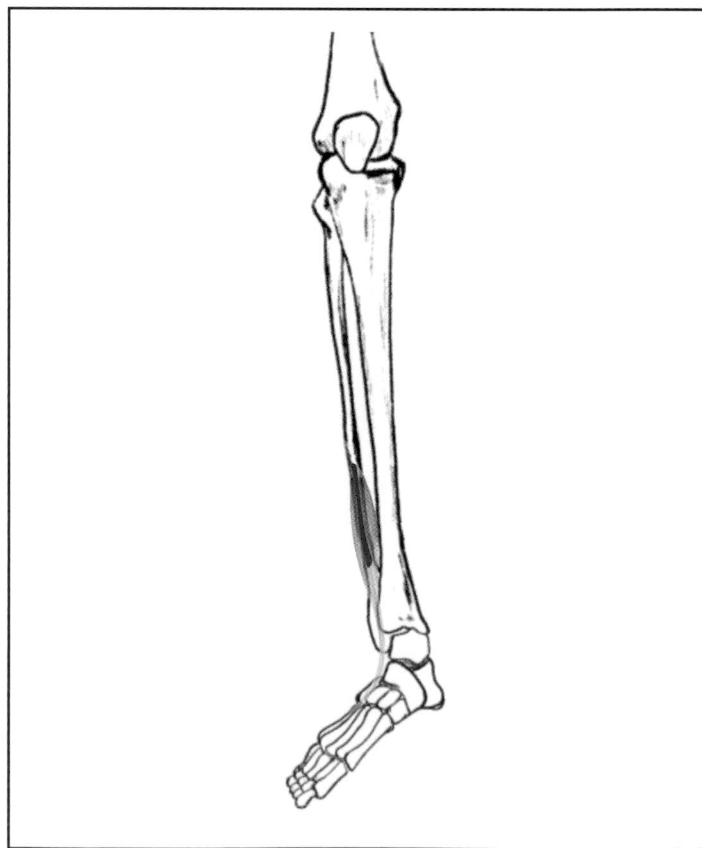

TIBIALIS ANTERIOR

Description:
One of the muscles of the lower leg that act to dorsiflex the foot and to invert it. Also called the tibialis anticus.

Joint Crossings:
One

Rank/Body Action:
1 Ankle dorsiflexion
2 Ankle inversion

Blood Supply:
Anterior tibial artery

Insertion:
Crosses the top of the foot to the plantar surface of the first cuneiform and base of the first metatarsal

Nerves:
Deep peroneal nerve L4, 5, S1

Origin:
- Outside surface of the lateral condyle of the tibia
- Two-thirds of the anterior shaft of the tibia

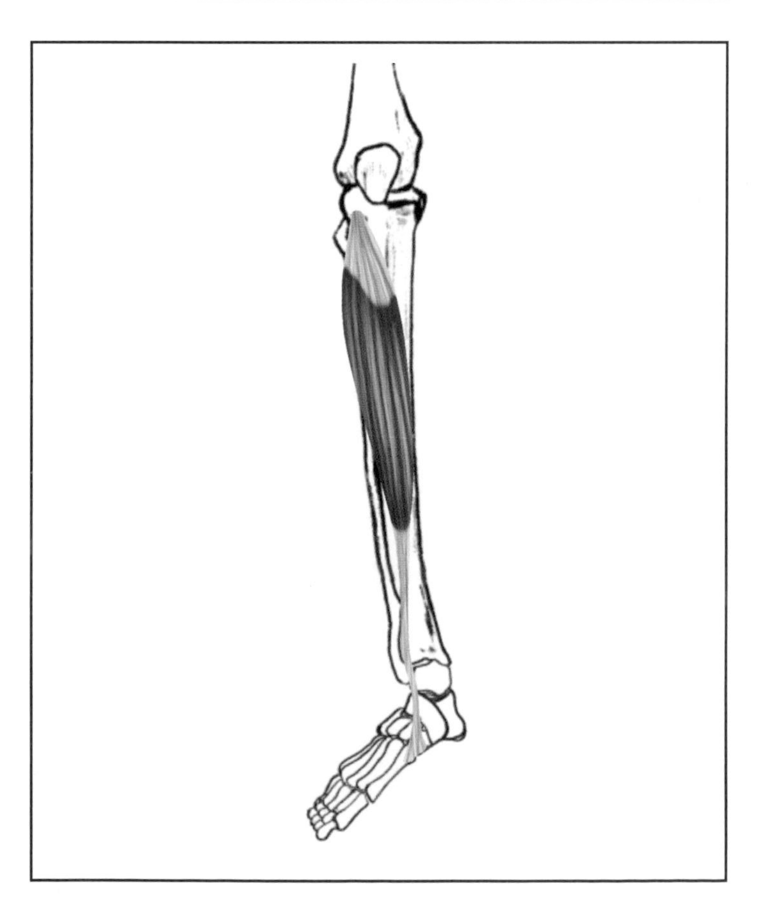

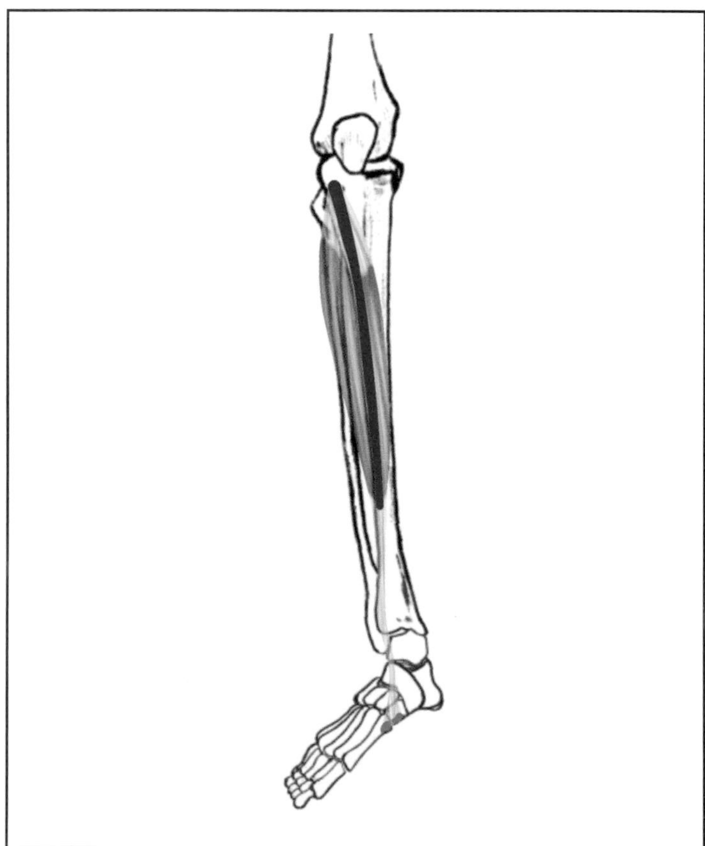

PERONEUS BREVIS

Description:
A lower leg muscle that assists in everting and plantar flexing the ankle.

Joint Crossings:
One

Rank/Body Action:
1 Ankle eversion
2 Ankle plantar flexion

Blood Supply:
Muscular branches of the peroneal artery

Insertion:
Around the outside ankle to the base of the fifth metatarsal

Nerves:
Superficial peroneal nerve L4, 5, S1

Origin:
Lower lateral two-thirds surface of the fibula

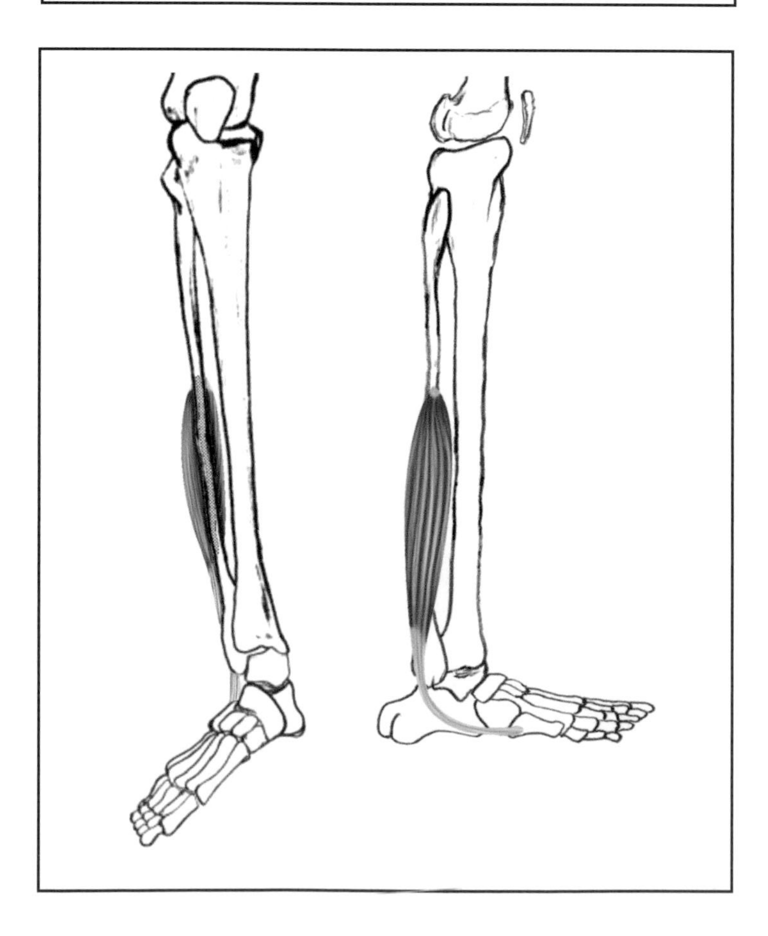

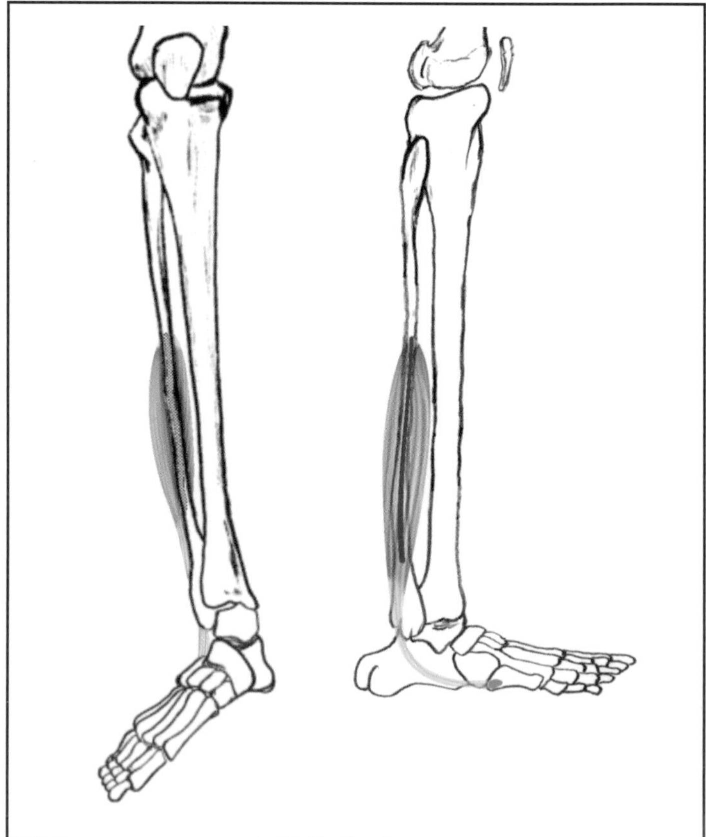

Peroneus Longus

Description:
A lower leg muscle that aids in everting and plantar flexing the ankle.

Joint Crossings:
One

Rank/Body Action:

1 Ankle eversion
2 Ankle plantar flexion

Blood Supply:
Muscular branches of the peroneal artery

Insertion:
Around the outside ankle, crossing under the foot to the first cuneiform and base of the first metatarsal

Nerves:
Superficial peroneal nerve L4, 5, S1

Origin:

- Lateral surface of the head of the tibia
- Lateral surface of the head and upper lateral two-thirds surface of the fibula

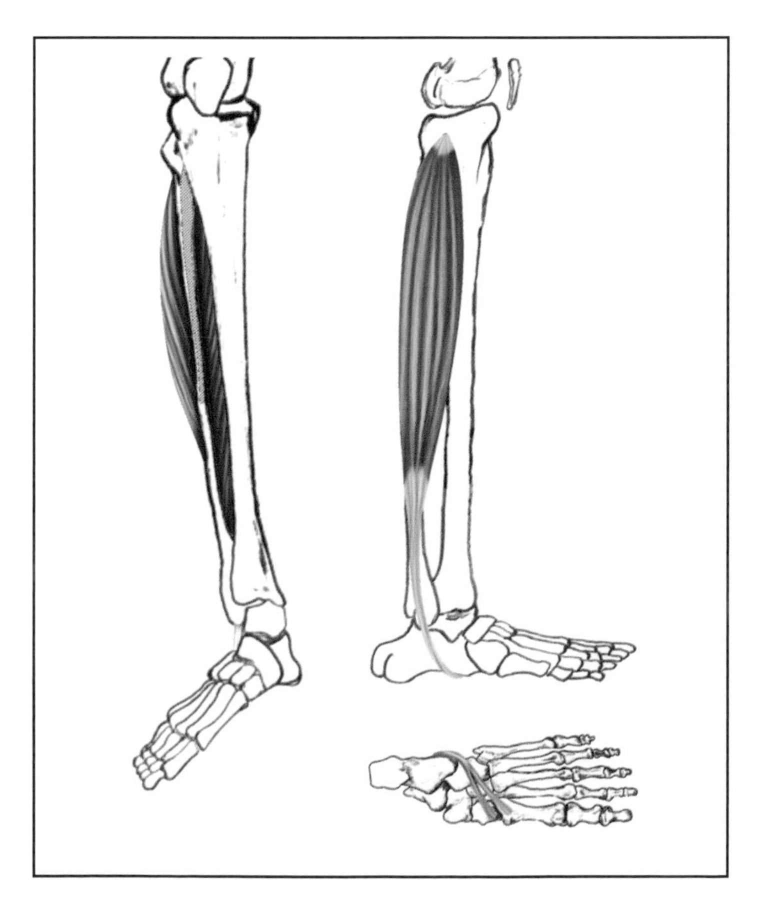

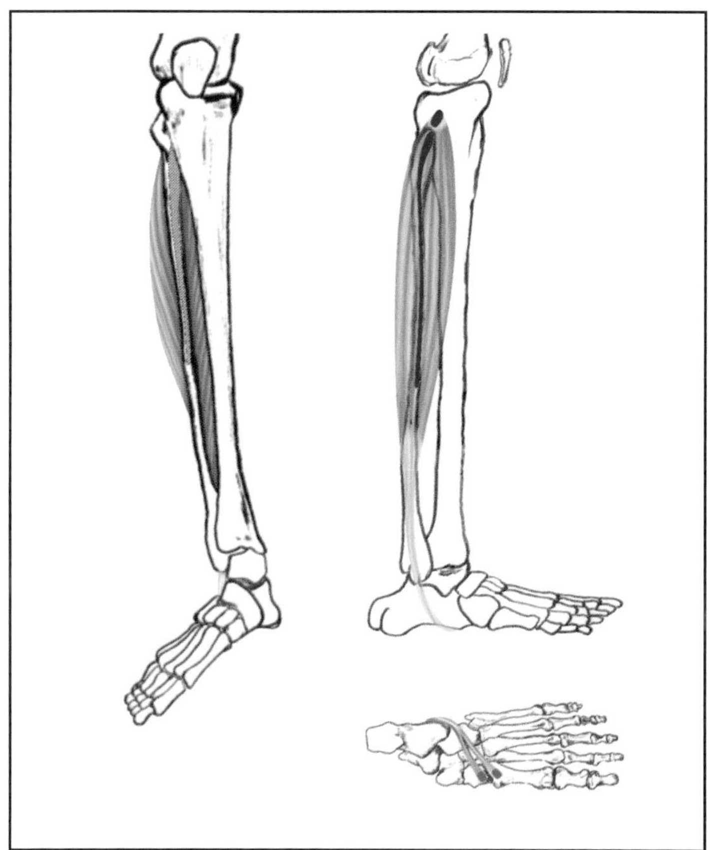

Flexor Digitorum Longus

Description:
A muscle of the outside of the lower leg that flexes the last phalanx of each of the four small toes, and helps plantar flex and invert the ankle.

Blood Supply:
- Peroneal artery
- Posterior tibial artery

Insertion:
Around the medial ankle to the plantar surface of the base of the third phalange of the 2nd-5th toes

Joint Crossings:
Two

Nerves:
Tibial nerve L5, S1

Rank/Body Action:
1 Toe flexion (2nd-5th toes)
2 Ankle plantar flexion
3 Ankle inversion

Origin:
- Posterior surface of the middle one-third portion of the tibia, between the soleus and tibialis posterior

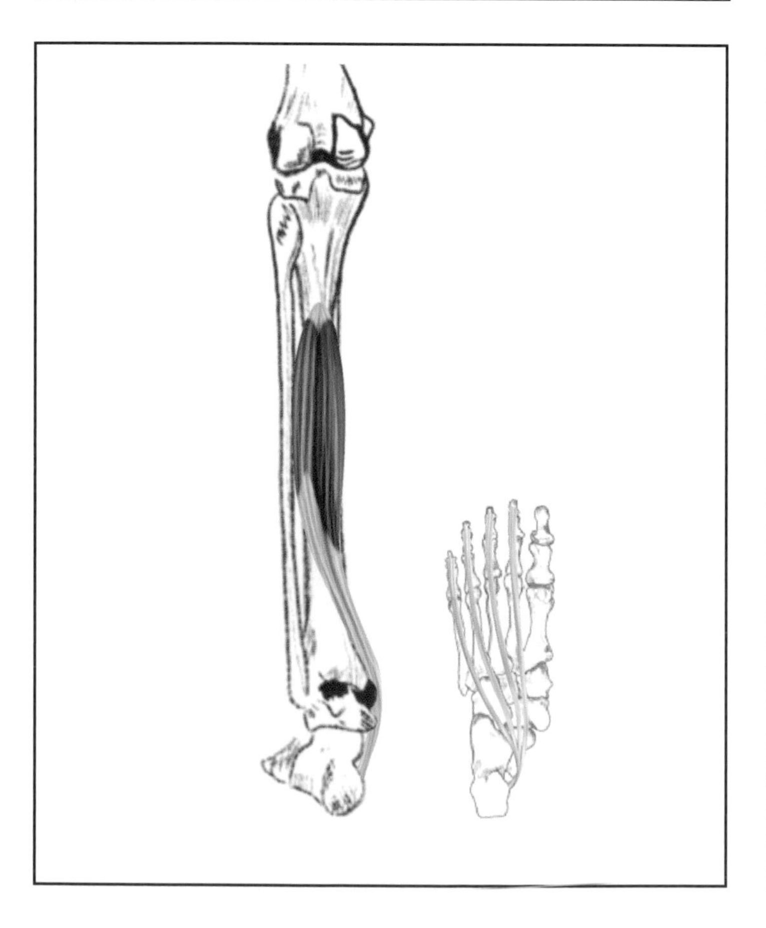

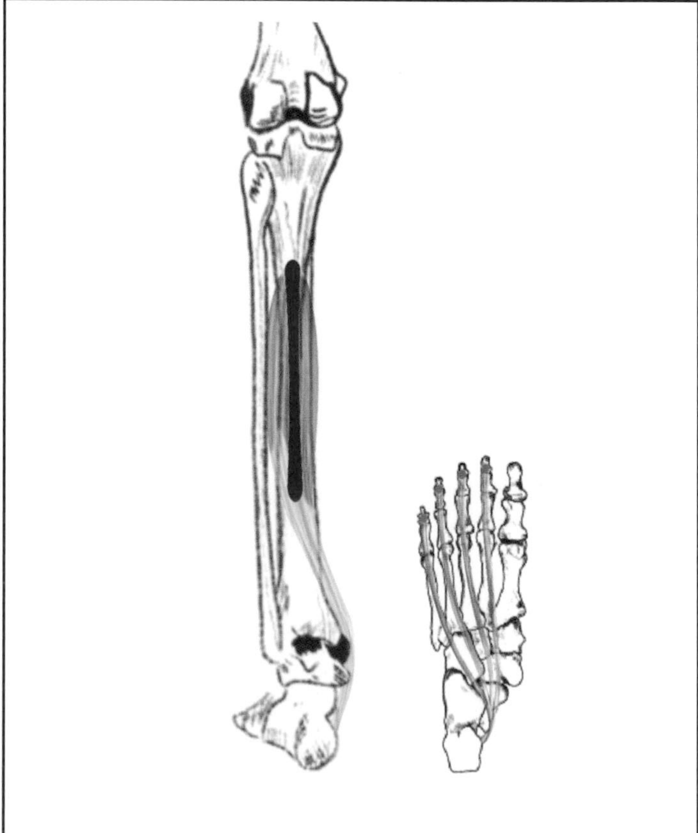

Flexor Hallucis Longus

Description:
A long, deep muscle of the outside of the lower leg that flexes the terminal phalange of the big toe.

Joint Crossings:
Two

Rank/Body Action:
1 Toe flexion (big toe)
2 Ankle inversion
3 Ankle plantar flexion

Blood Supply:
- Peroneal artery
- Posterior tibial artery

Insertion:
Around the inside ankle to the plantar surface of the base of the terminal phalange of the big toe

Nerves:
Tibial nerve L5, S1, 2

Origin:
Posterior surface of the lower two-thirds of the fibula

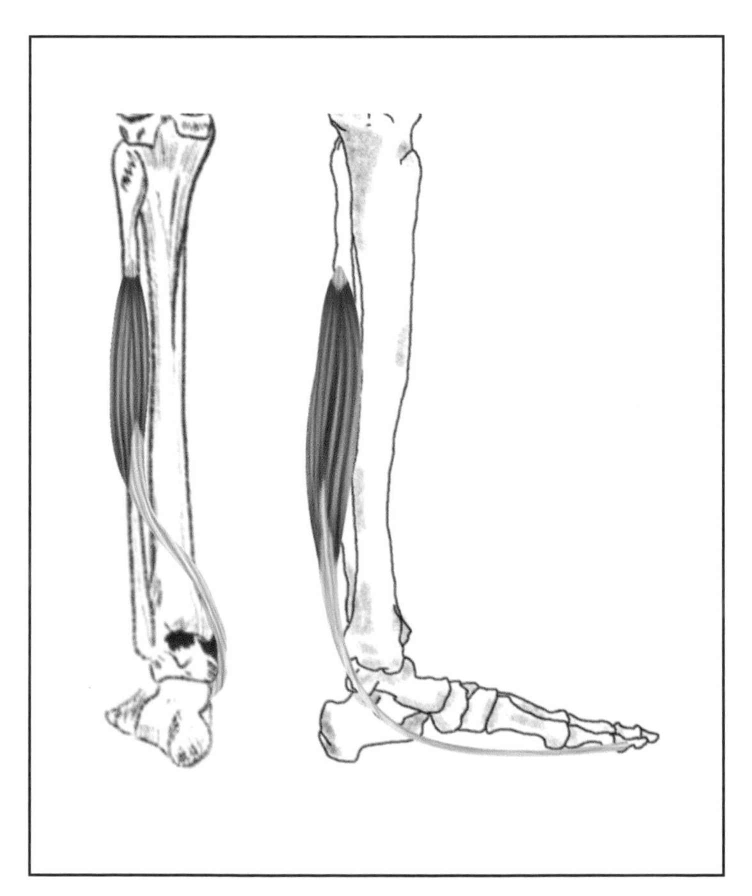

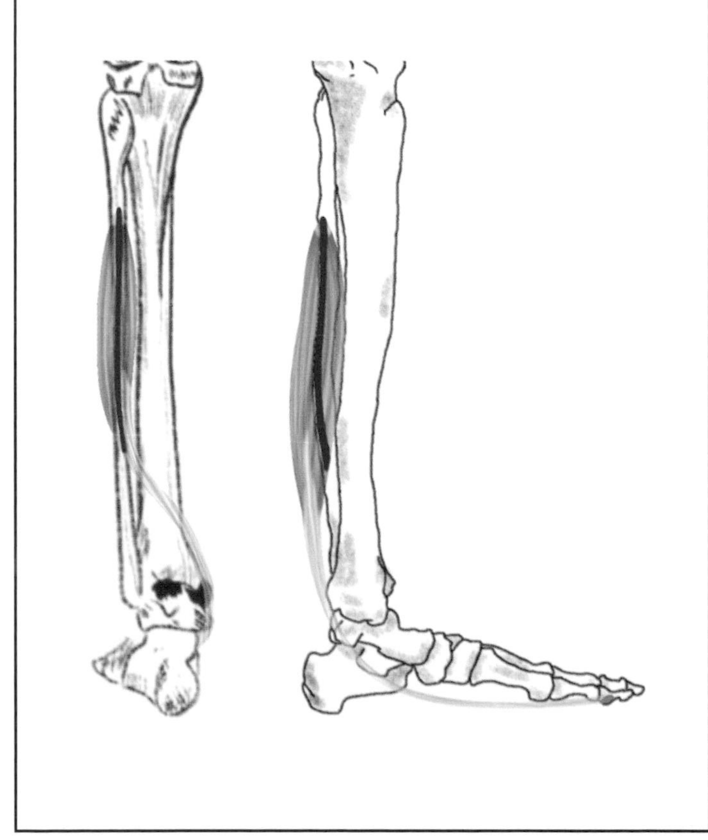

GASTROCNEMIUS

Description:
The largest and most superficial muscle of the lower leg, heavily used to propel the body up and forward when walking, running, and jumping.

Joint Crossings:
Two

Rank/Body Action:
1 Ankle plantar flexion (strong)
2 Knee flexion (slight)

Blood Supply:
- Sural branches of popliteal artery
- Muscular branches of peroneal artery
- Posterior tibial artery

Insertion:
Tuberosity of the calcaneus via the achilles tendon

Nerves:
Tibial nerve S1, 2

Origin:
Posterior surface of medial and lateral epicondyles of the femur

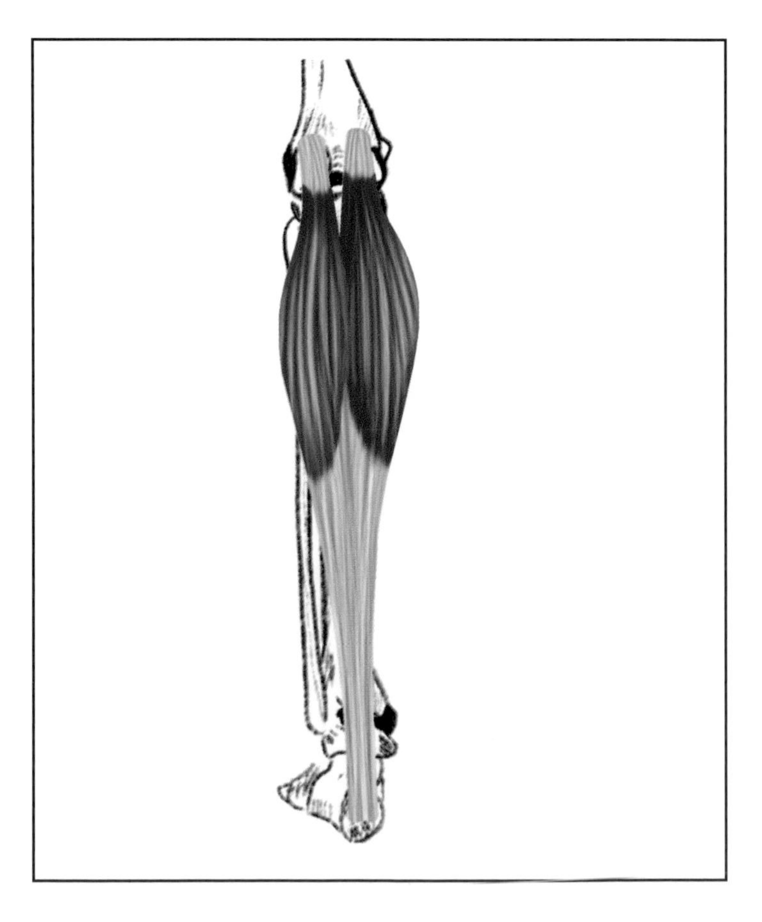

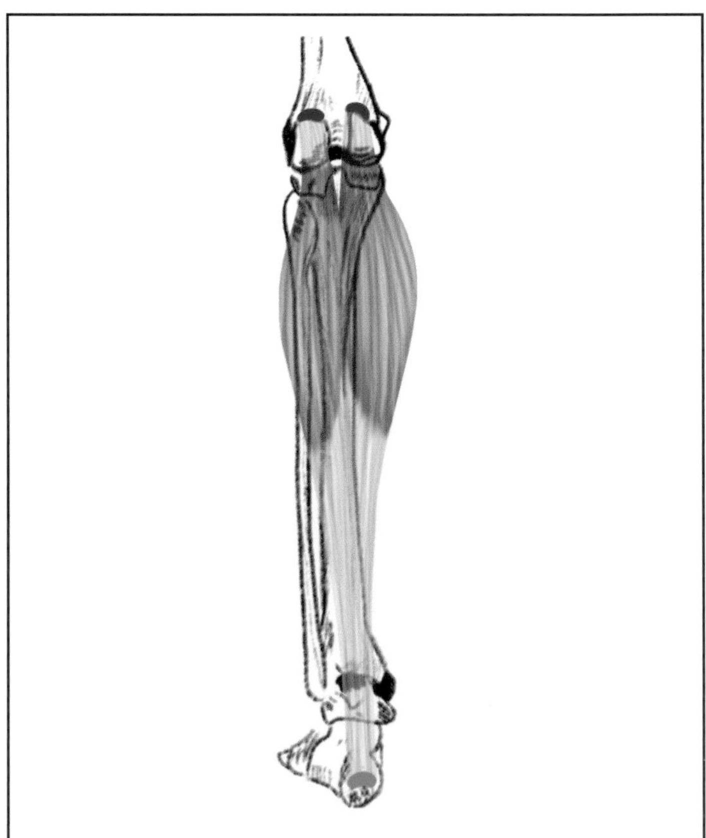

PLANTARIS

Description:
A small muscle of the lower leg that weakly flexes the knee and plantar flexes the ankle. Assists the gastrocnemius and soleus in plantar flexion. Missing in some people.

Blood Supply:
- Sural branches of popliteal artery
- Muscular branches of peroneal artery
- Posterior tibial artery

Joint Crossings:
Two

Insertion:
Tuberosity of the calcaneus via the achilles tendon

Rank/Body Action:
1 Ankle plantar flexion
2 Knee flexion (slight)

Nerves:
Tibial nerve S1, 2

Origin:
Posterior surface of the lateral epicondyle of the femur (above the gastrocnemius)

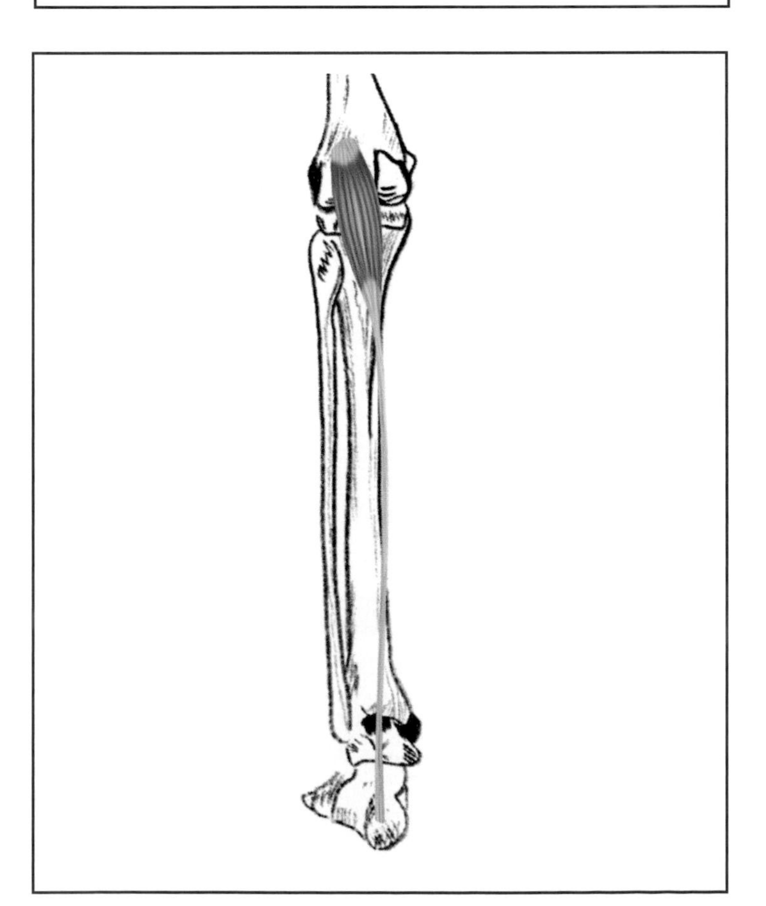

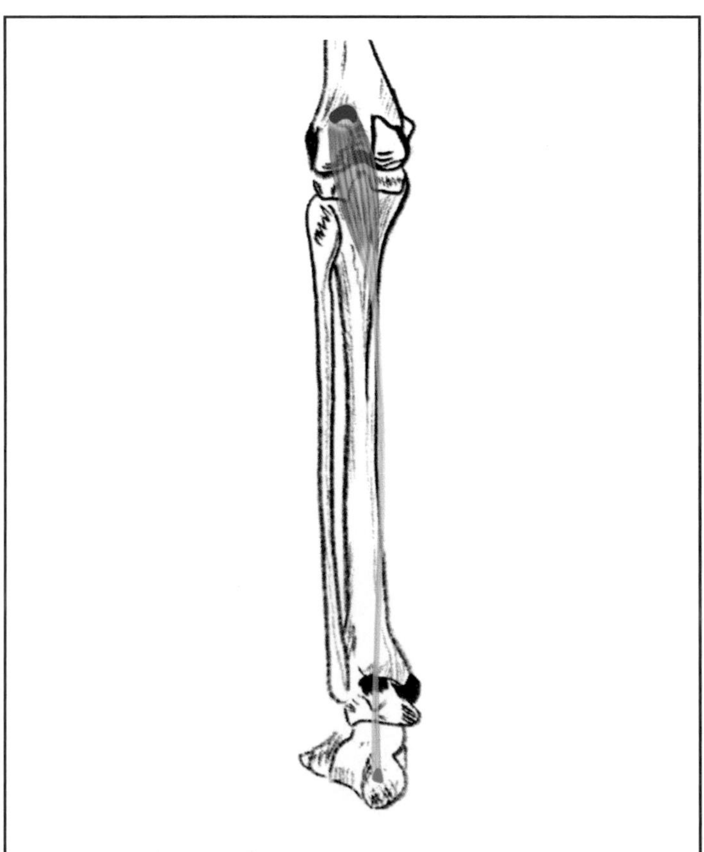

Soleus

Description:
A broad flat muscle of the lower leg that lies under the gastrocnemius which plantar flexes the ankle.

Joint Crossings:
One

Body Action:
Ankle plantar flexion (strong)

Blood Supply:
- Sural branches of popliteal artery
- Muscular branches of peroneal artery
- Posterior tibial artery

Insertion:
Tuberosity of the calcaneus via the achilles tendon

Nerves:
Tibial nerve S1, 2

Origin:
- Upper 1/3 posterior shaft of the fibula and the posterior surface of the head of the fibula
- Popliteal line
- Middle 1/3 of the medial border of the tibia

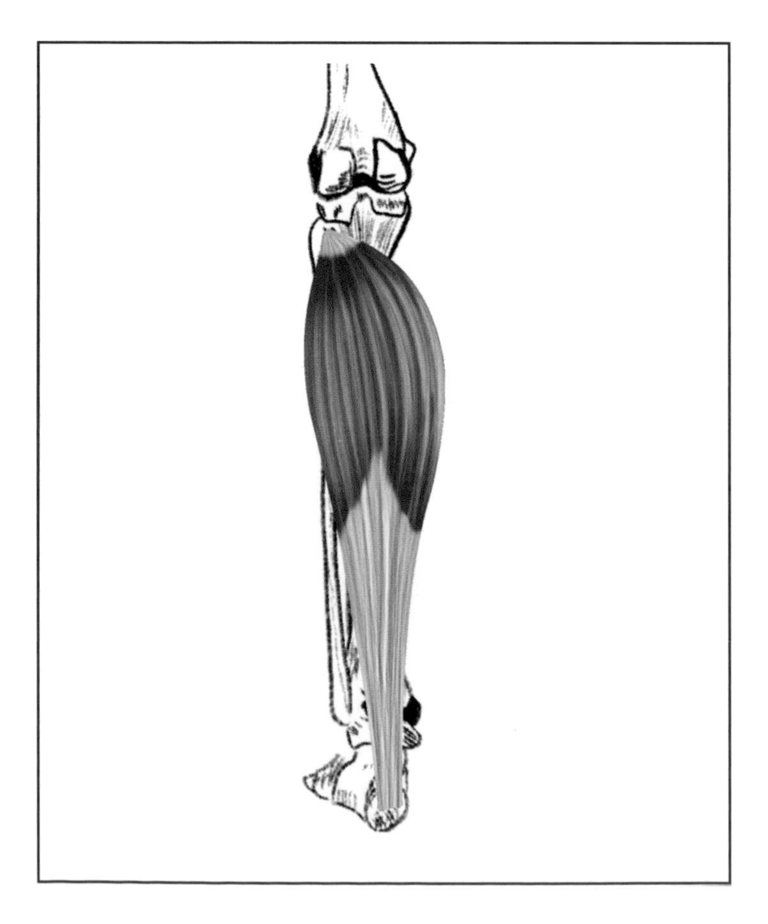

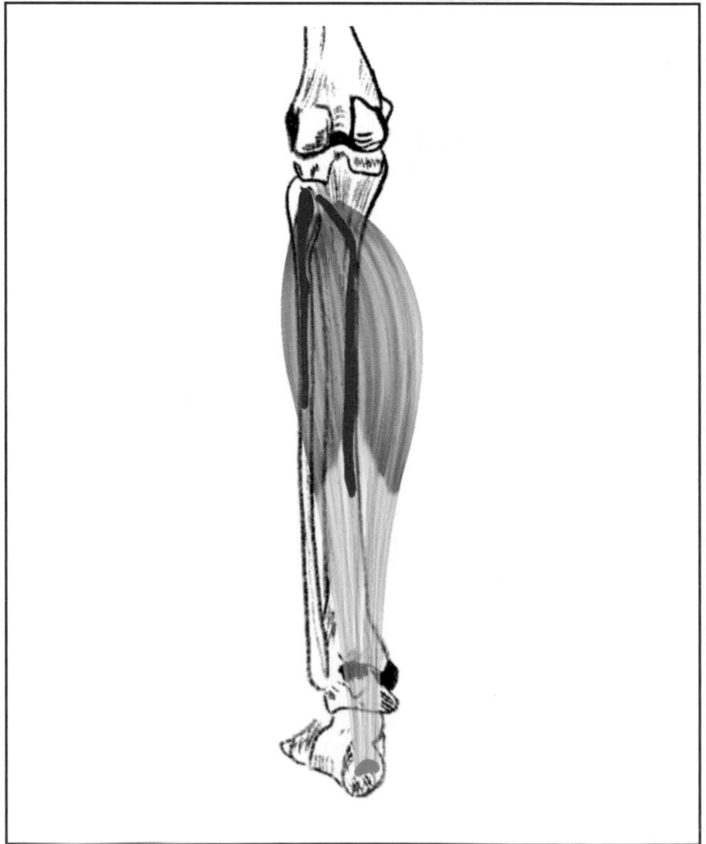

TIBIALIS POSTERIOR

Description:
A deep muscle of the posterior lower leg that plantar flexes and inverts the ankle. Also called the tibialis posticus.

Joint Crossings:
One

Rank/Body Action:
1 Ankle plantar flexion
2 Ankle inversion

Blood Supply:
- Peroneal artery
- Posterior tibial artery

Insertion:
Around the medial ankle to the plantar surface of the navicular, first cuneiform, cuboid, and the junction of the following four bones – the second and third cuneiforms, and the second and third metatarsals

Nerves:
Tibial nerve L5, S1

Origin:
- Middle one-third of the posterior lateral surface of the tibia
- Middle one-third of the posterior medial surface of the fibula

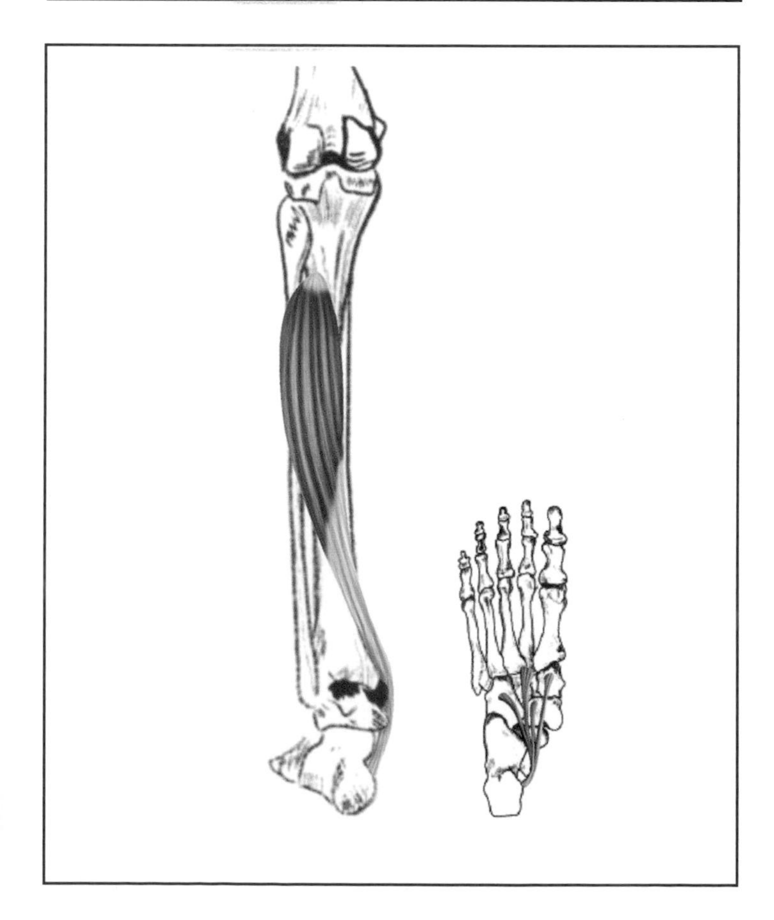

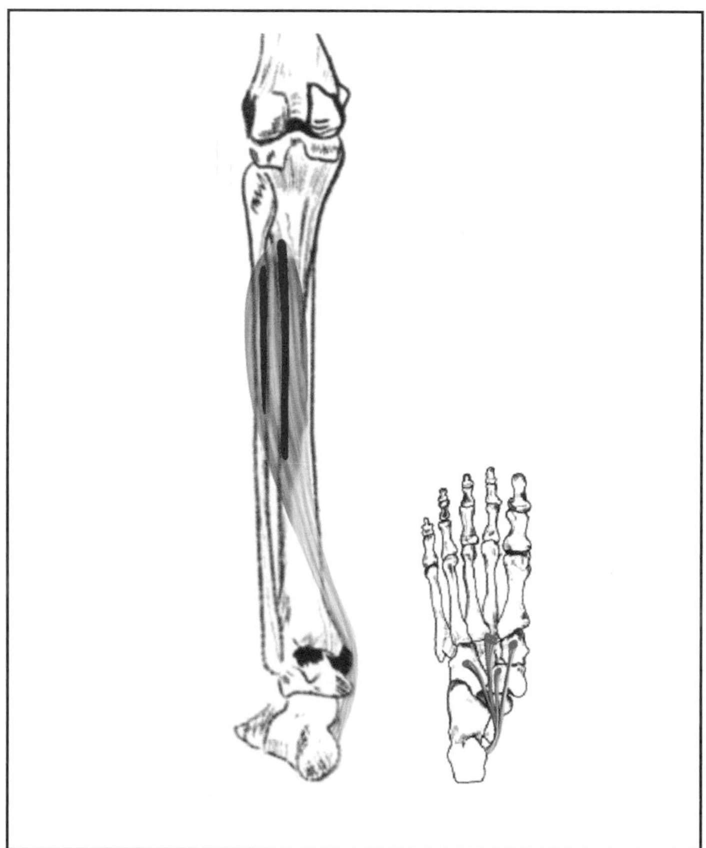

Anconeus

Description:
A small triangular extensor muscle that is superficially situated behind and below the elbow joint that extends the elbow.

Joint Crossings:
One

Body Action:
Elbow extension

Blood Supply:
Middle collateral artery from the profunda brachii artery

Insertion:
Posterior/lateral surface of the olecranon process and posterior surface of the ulna

Nerves:
Radial nerve C7, 8

Origin:
Posterior surface of the lateral epicondyle of the humerus

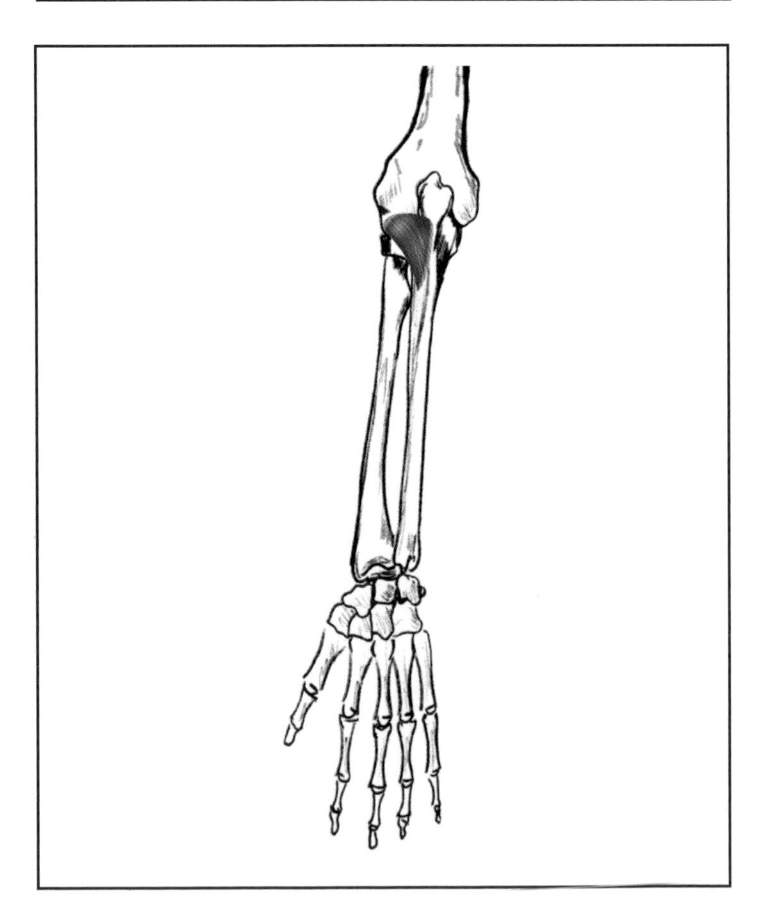

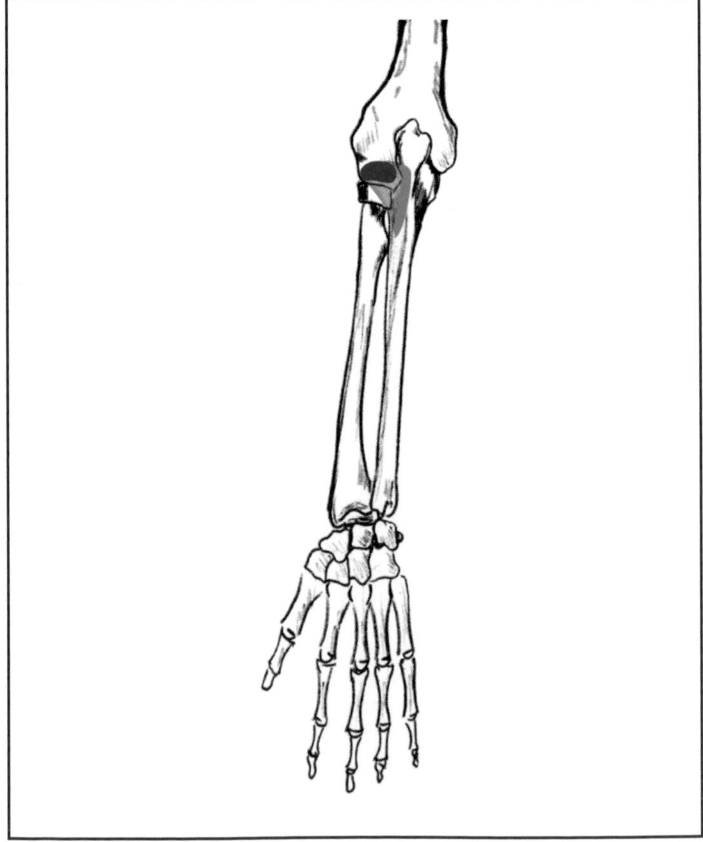

Biceps Brachii

Description:
The most superficial muscle of the upper, anterior arm. Thought of as a major elbow flexor (curls), it is also a strong supinator of the radio-ulnar joint, as well as a flexor of the arm.

Joint Crossings:
Three

Rank/Body Action:
1 Elbow flexion
2 Forearm supination
3 Arm flexion

Blood Supply:
Muscular branches of brachial artery

Insertion:
Radial tuberosity of the radius

Nerves:
Musculocutaneous nerve C5, 6

Origin:
- Long head: supraglenoid tubercle of the glenoid fossa of the scapula
- Short head: coracoid process of the scapula

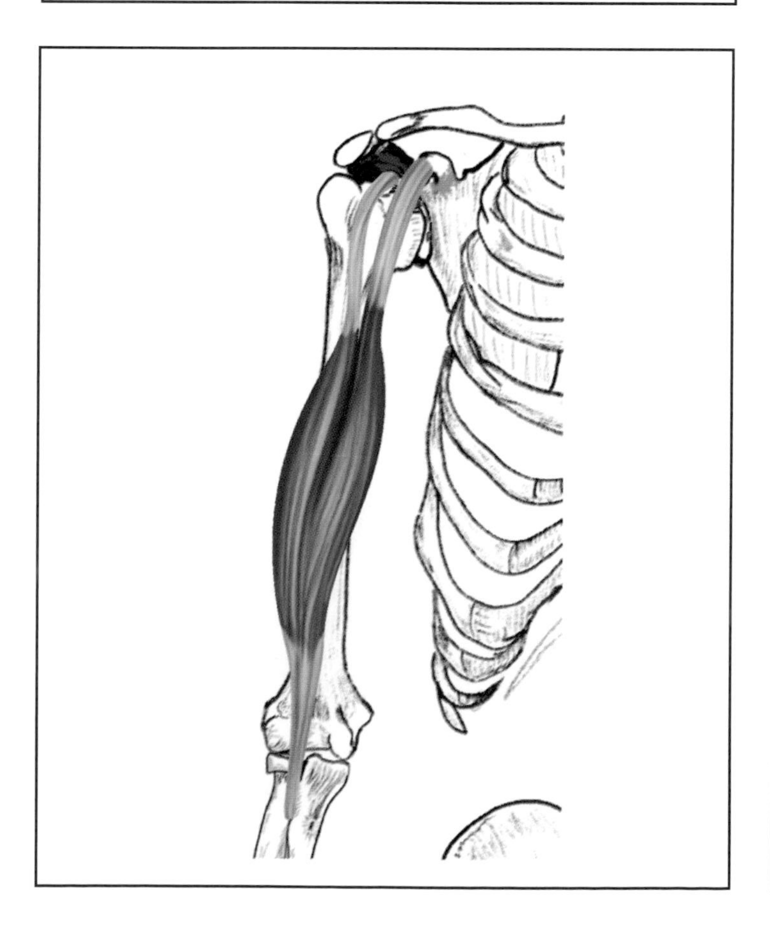

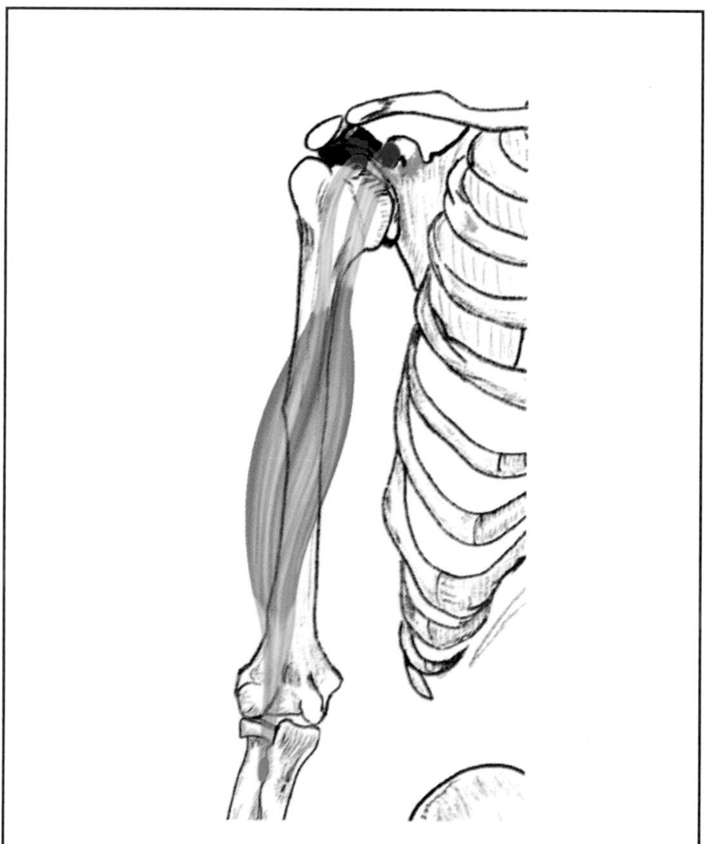

BRACHIALIS

Description:
An anterior, upper arm muscle lying under the biceps. Its only action is as a true flexor of the elbow.

Joint Crossings:
One

Body Action:
Elbow flexion

Blood Supply:
- Muscular branches of brachial artery
- Radial recurrent artery

Insertion:
- Coronoid process of the ulna
- The ulnar tuberosity

Nerves:
Musculocutaneous nerve C5, 6

Origin:
Lower portion of the anterior surface of the humerus

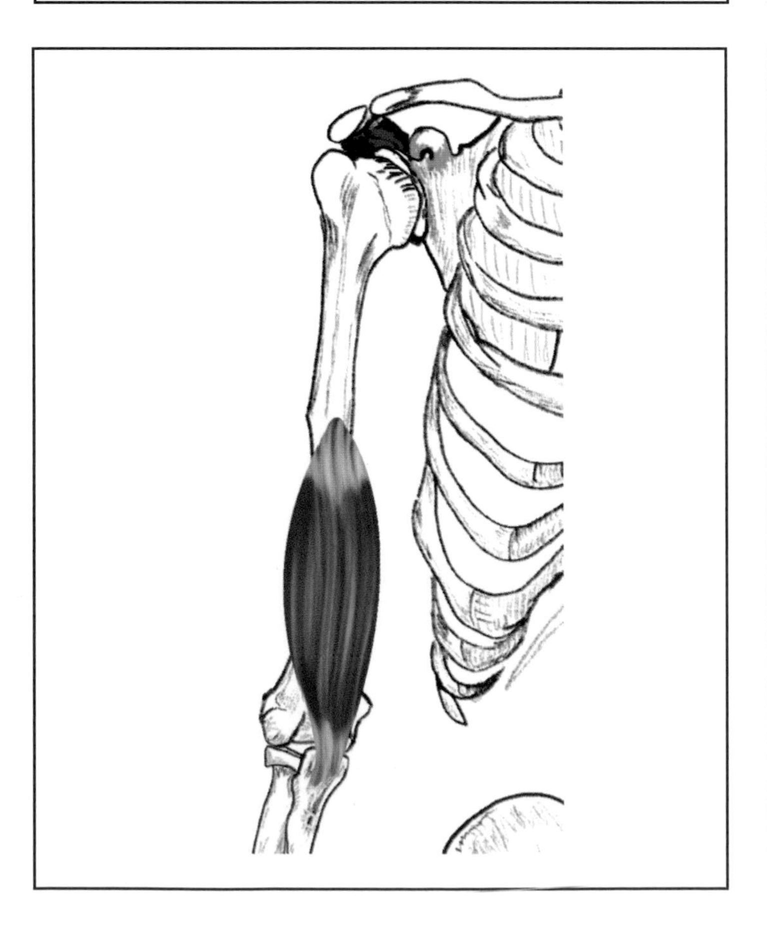

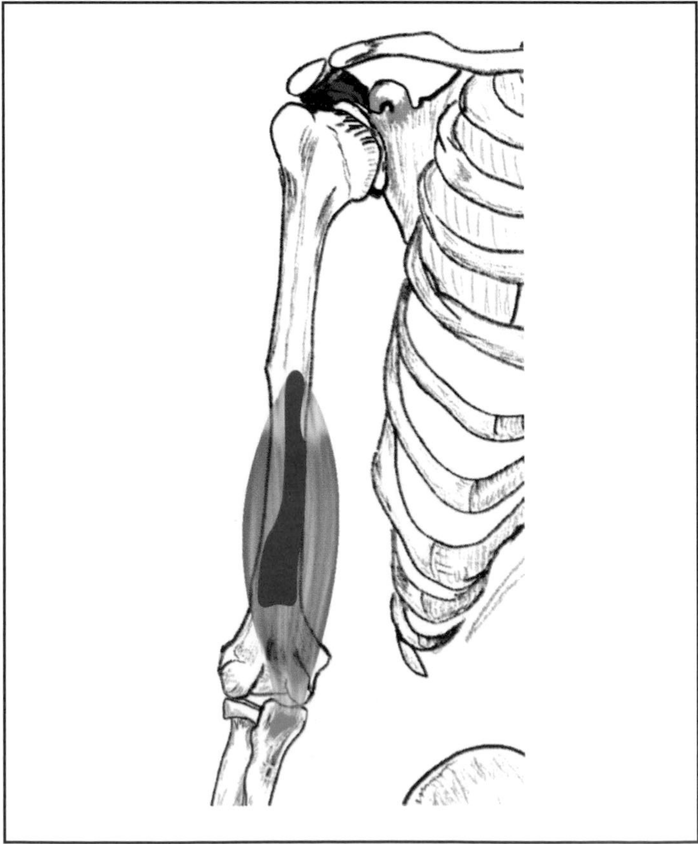

BRACHIORADIALIS

Description:
A long, narrow muscle of the forearm that flexes the elbow and supinates the radio-ulna joint. Elbow flexion is strongest when the forearm is already supinated.

Joint Crossings:
Two

Rank/Body Action:
1 Elbow flexion
2 Forearm supination

Blood Supply:
Radial recurrent artery

Insertion:
Styloid process of the radius

Nerves:
Radial nerve C5, 6

Origin:
Upper two-thirds of the lateral supracondylar ridge of the humerus (about the lower lateral one-third of the humerus)

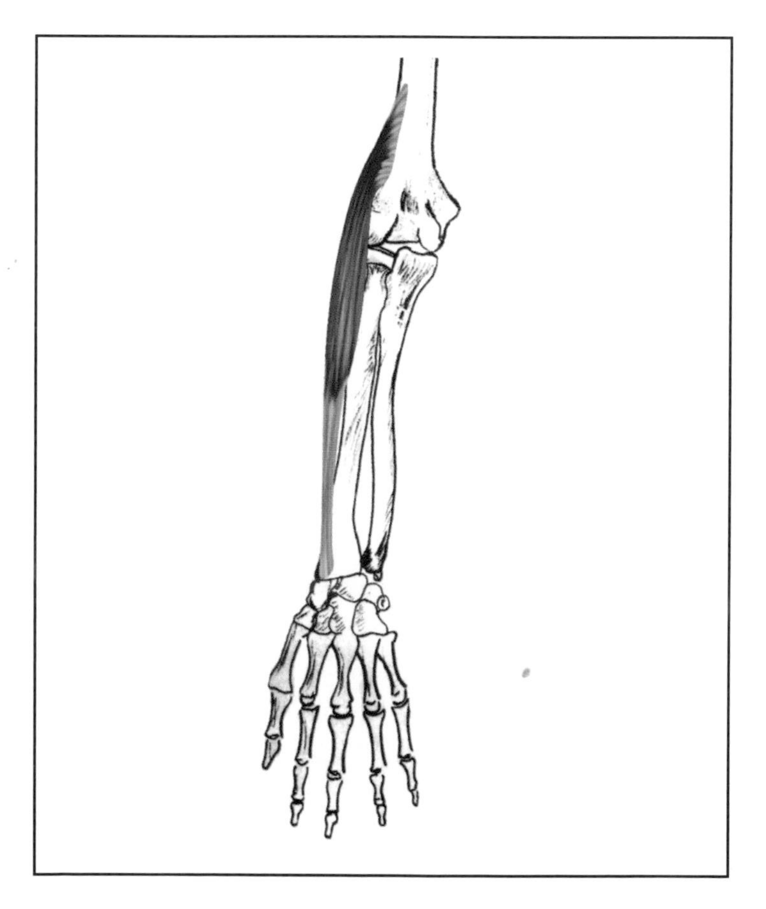

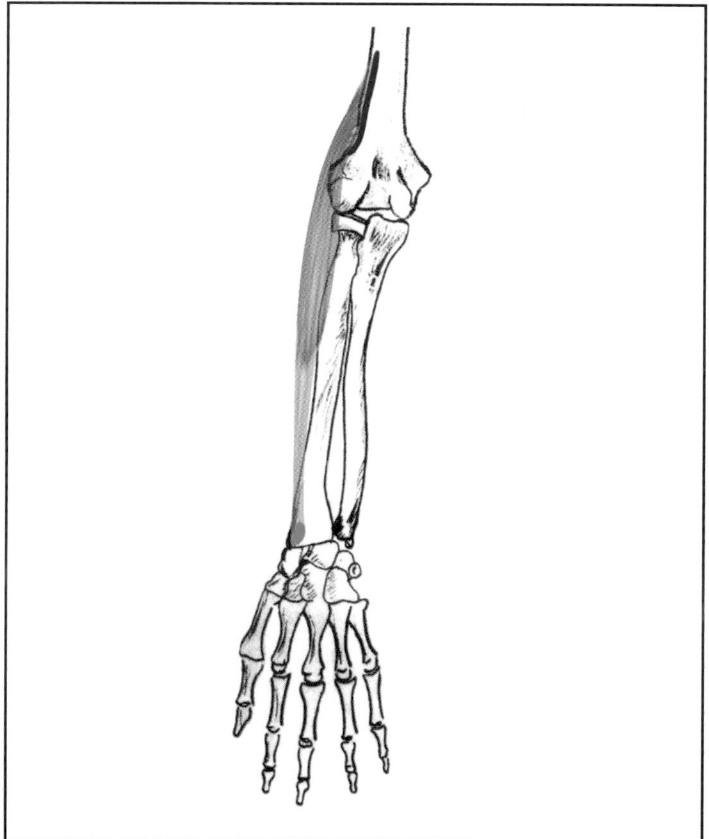

Pronator Quadratus

Description:
A deep muscle of the lower forearm, passing transversely from the ulna to the radius and serving to pronate the forearm.

Joint Crossings:
One

Body Action:
Forearm pronation

Blood Supply:
- Anterior interosseous artery
- Muscular branches of the radial artery

Insertion:
Anterior surface of the lower one-fourth of radius

Nerves:
Palmar interosseous branch of median nerve C6, 7

Origin:
Anterior surface of the lower one-fourth of ulna

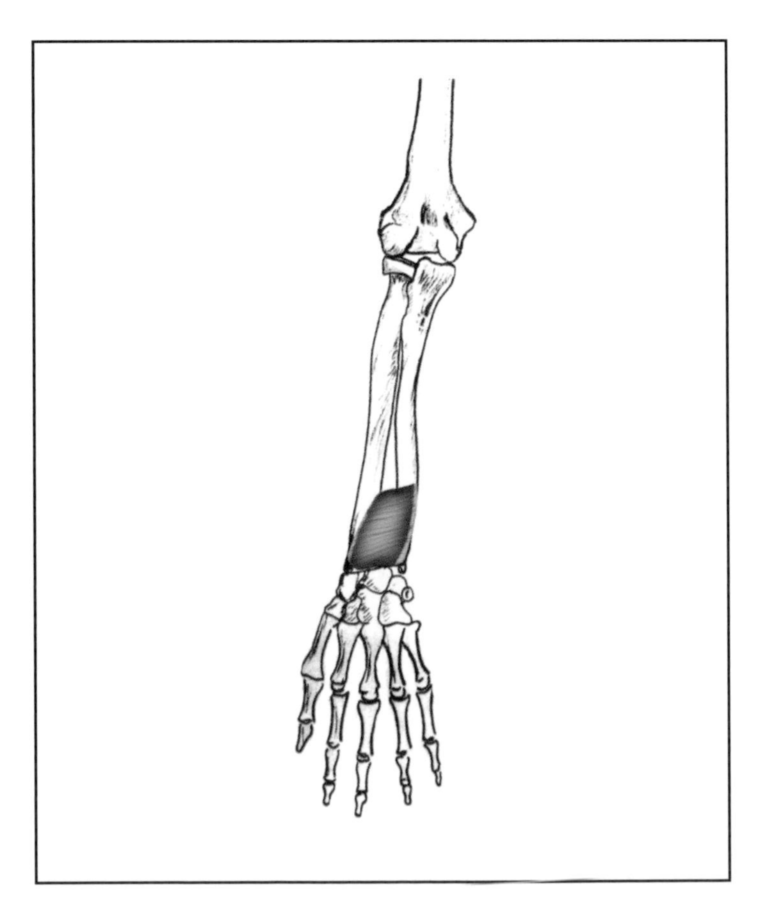

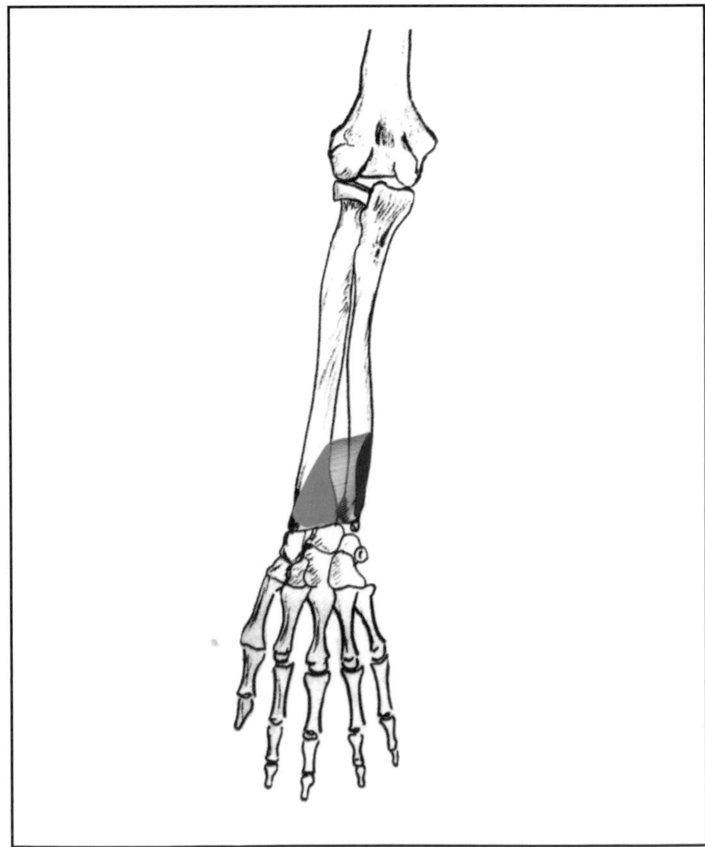

Pronator Teres

Description:
A muscle of the forearm that serves to pronate the forearm and flex the elbow. Elbow flexion is weaker when the forearm is supinated.

Joint Crossings:
Two

Rank/Body Action:
1 Forearm pronation
2 Elbow flexion

Blood Supply:
- Muscular branches of ulnar artery
- Muscular branches of radial artery

Insertion:
Middle one-third of the lateral surface of the radius

Nerves:
Median nerve C6, 7

Origin:
- Medial epicondyle of the humerus
- Coronoid process of the ulna

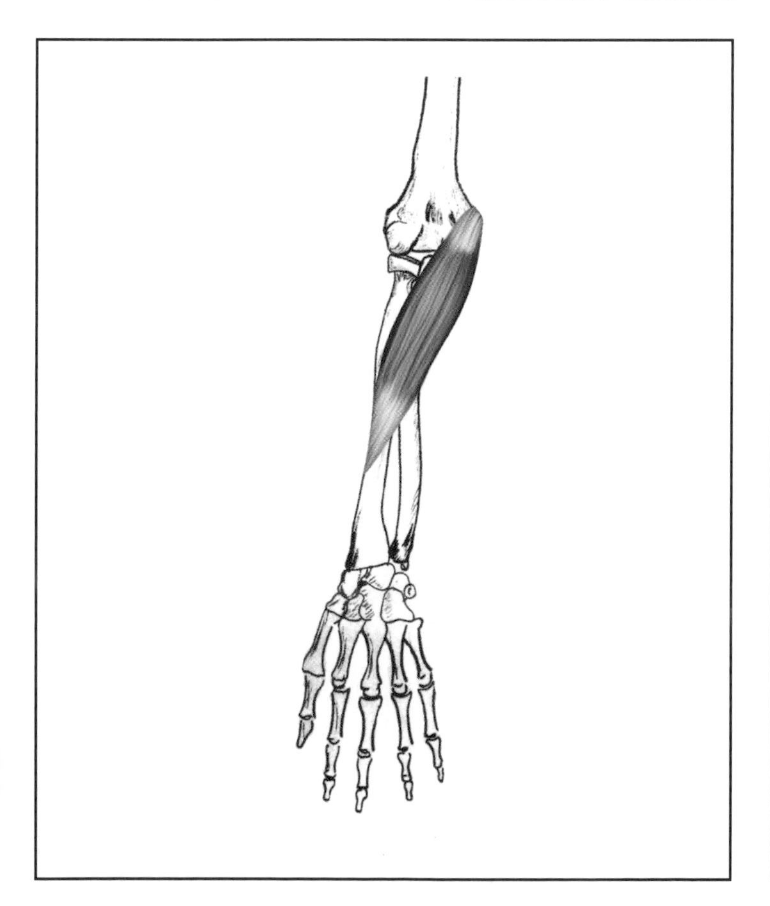

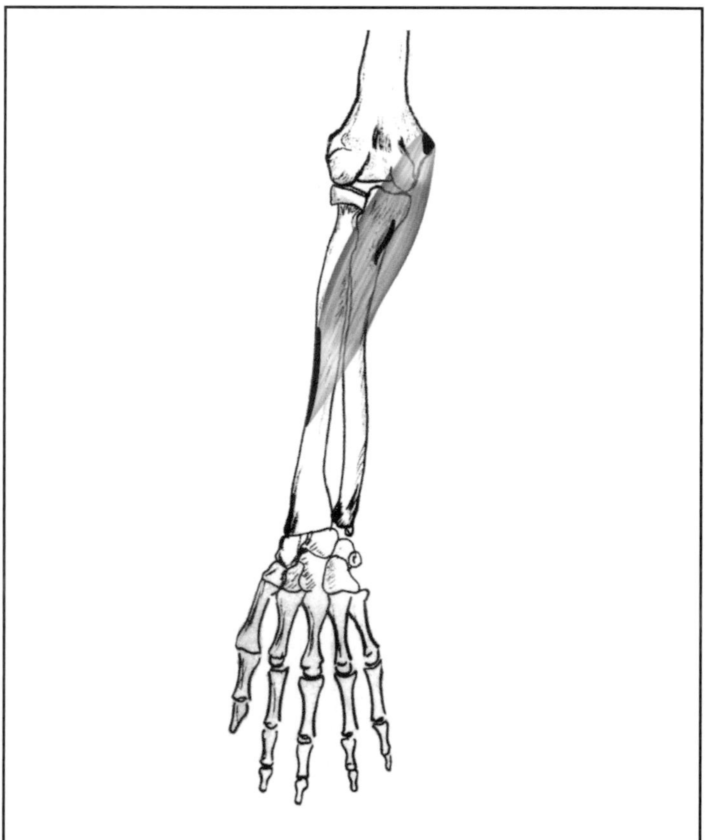

Supinator

Description:
A muscle that produces the motion of supination, rotating the wrist to make the thumb move outward.

Joint Crossings:
Two

Rank/Body Action:
1 Forearm supination
2 Elbow flexion

Blood Supply:
Radial recurrent artery

Insertion:
Lateral surface of the upper one-third of the radius

Nerves:
Deep branch of radial nerve C6

Origin:
- Lateral/posterior epicondyle of the humerus
- Supinator crest of the ulna

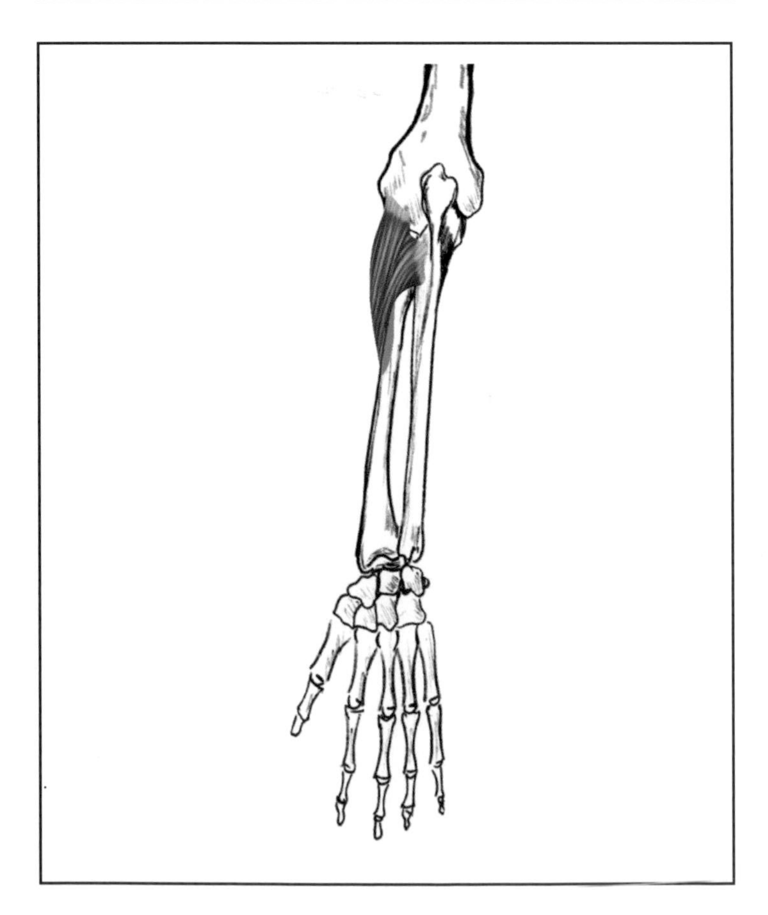

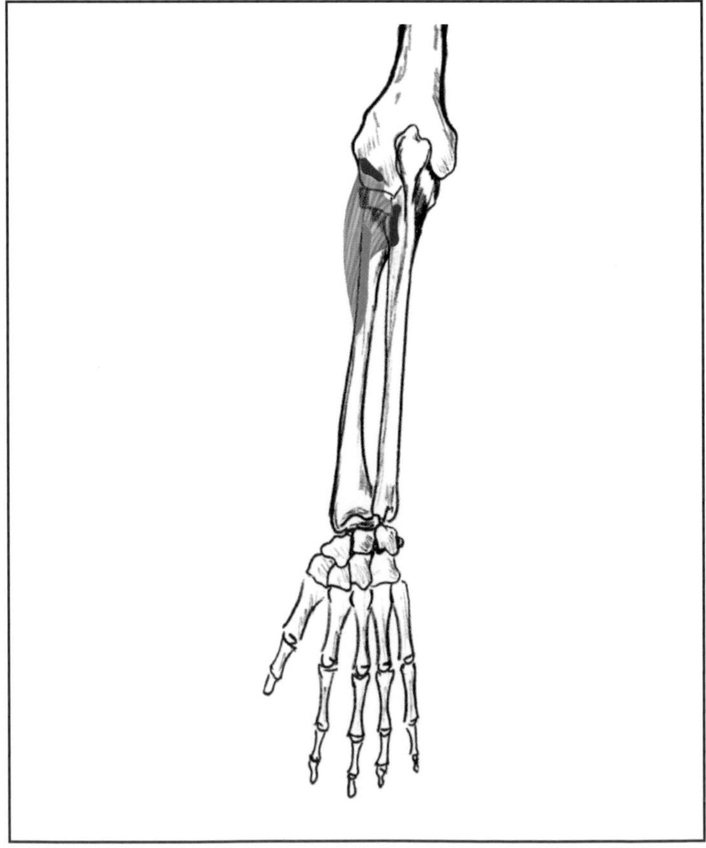

Triceps Brachii

Description:
A major upper, posterior arm muscle that strongly extends the elbow. Because of its attachment to the scapula, it also assists with arm extension and adduction.

Joint Crossings:
Two

Rank/Body Action:
1 Elbow extension
2 Arm extension (long head)
3 Arm hyperextension (long head)
4 Arm adduction

Blood Supply:
- Branches of the brachial artery
- Superior ulnar collateral artery
- Profunda brachii artery

Insertion:
Olecranon process of ulna

Nerves:
Radial nerve C7, 8

Origin:
- Long head: infraglenoid tubercle of the scapula
- Lateral head: upper, 1/3 lateral-posterior surface of the humerus
- Medial head: lower 3/4 posterior surface of the humerus

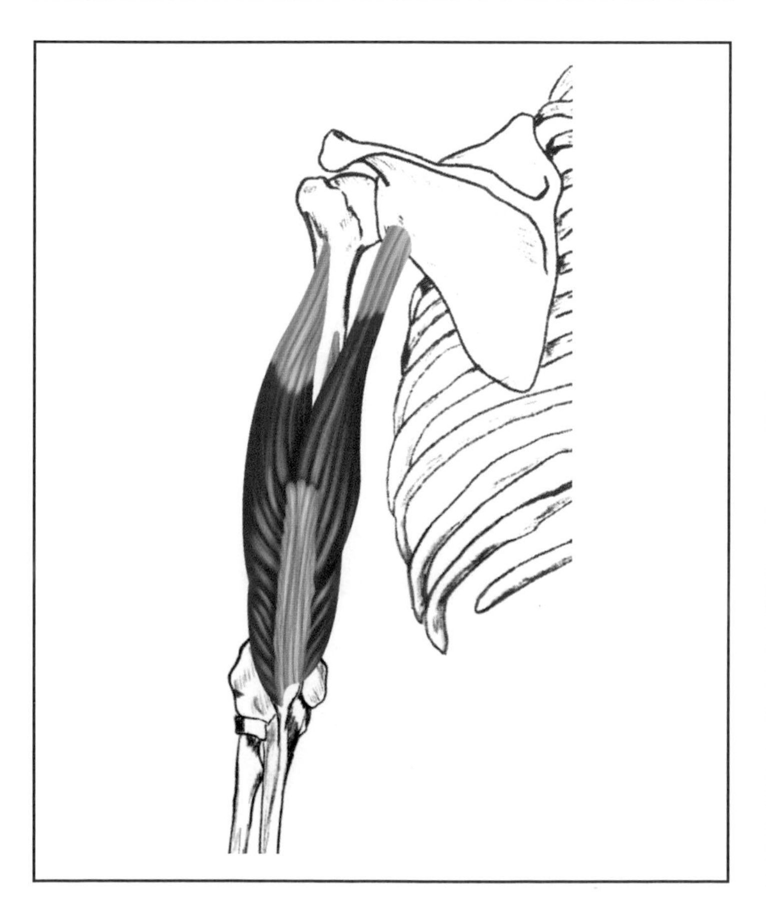

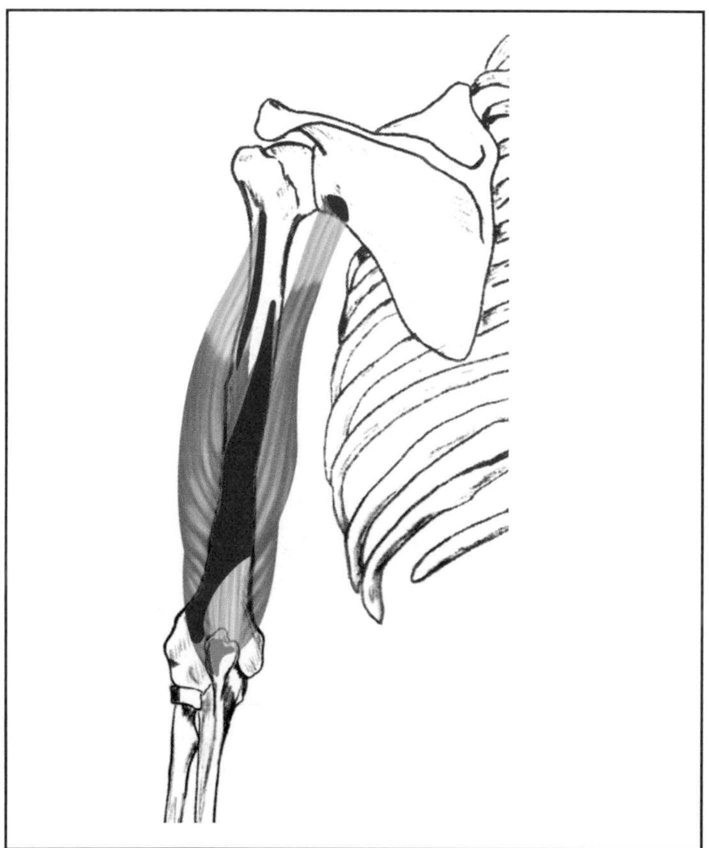

Abductor Digiti Minimi (Quinti)

Description:
A muscle that lies along the lateral side of the foot and abducts the little toe. Also known as the abductor digiti quinti.

Joint Crossings:
One

Body Action:
Toe abduction (little toe)

Blood Supply:
Lateral plantar artery

Insertion:
Lateral side of the base of the first phalange of the little toe

Nerves:
Lateral plantar nerve S1, 2

Origin:
- Lateral and medial processes of the tuberosity of calcaneus
- The plantar aponeurosis and the septum between the lateral and medial processes

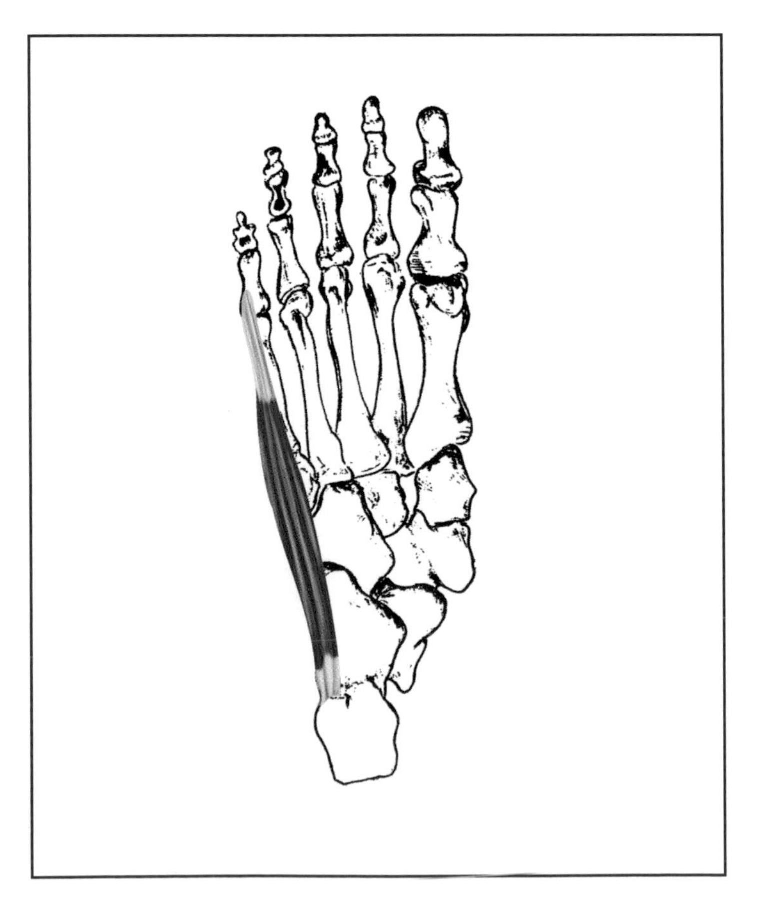

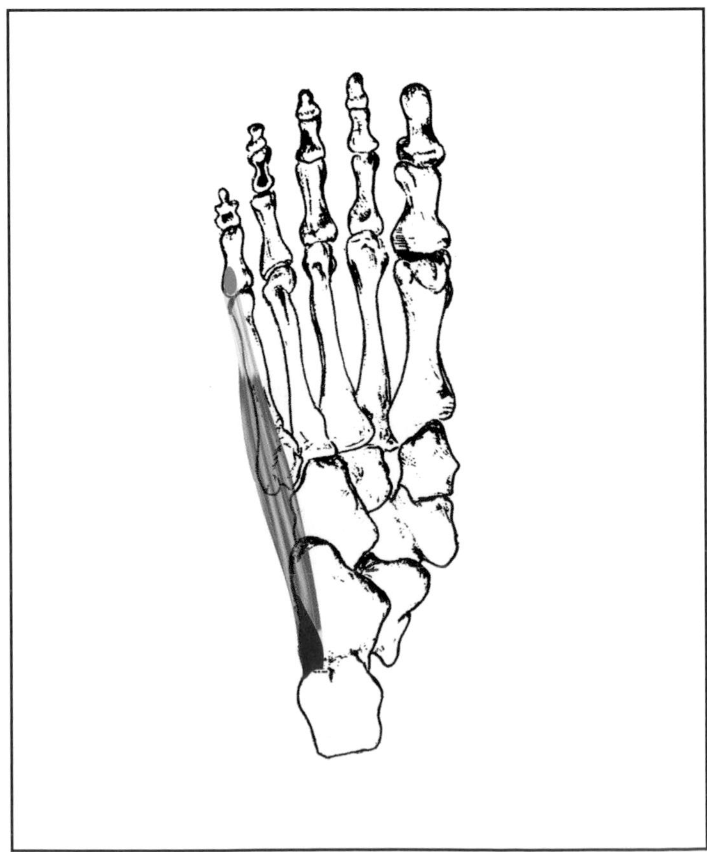

Abductor Hallucis

Description:
This muscle lies along the medial border of the foot and abducts the big toe.

Joint Crossings:
One

Rank/Body Action:
1 Toe abduction (big toe)
2 Toe flexion (big toe)

Blood Supply:
Medial plantar artery

Insertion:
Tibial side of the base of the first phalange of the big toe

Nerves:
Medial plantar nerve L4, 5

Origin:
- Medial process of the tuberosity of the calcaneum
- A strong ligament-the flexor retinaculum
- A strong ligament-the plantar aponeurosis

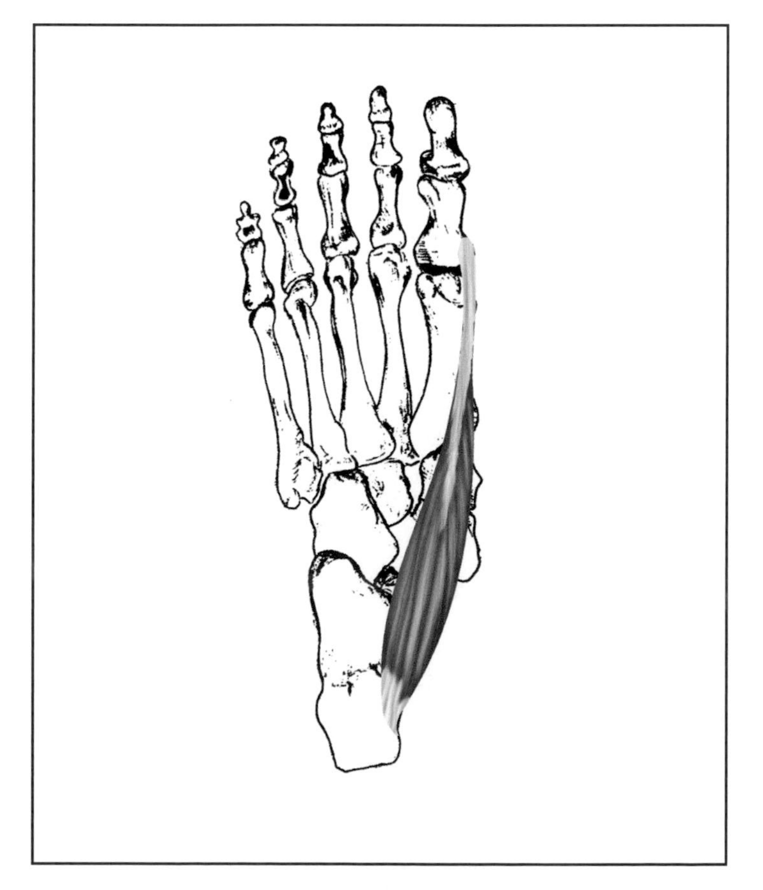

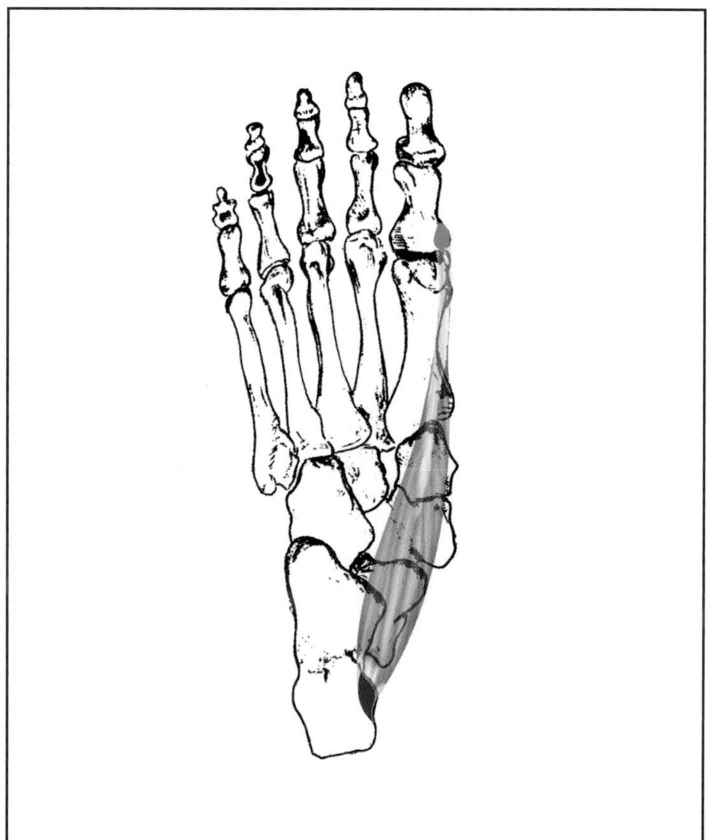

Adductor Hallucis

Description:
A foot muscle that adducts the big toe. Also known as the adductor obliquus hallucis.

Joint Crossings:
One

Body Action:
Toe adduction (big toe)

Blood Supply:
Lateral plantar artery

Insertion:
Lateral side of base of the first phalange of the big toe

Nerves:
Lateral plantar nerve S1, 2

Origin:
- Oblique head: base of the second and fourth metatarsals and the tendon of the peroneus longus
- Transverse head: plantar metatarsophlangeal ligaments of the 3rd-5th toes

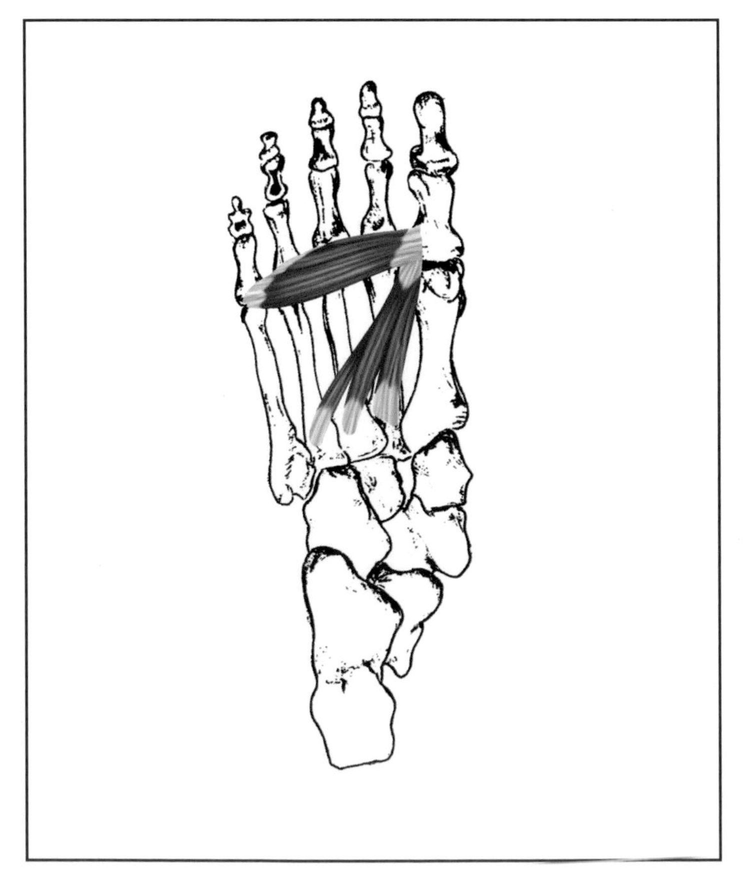

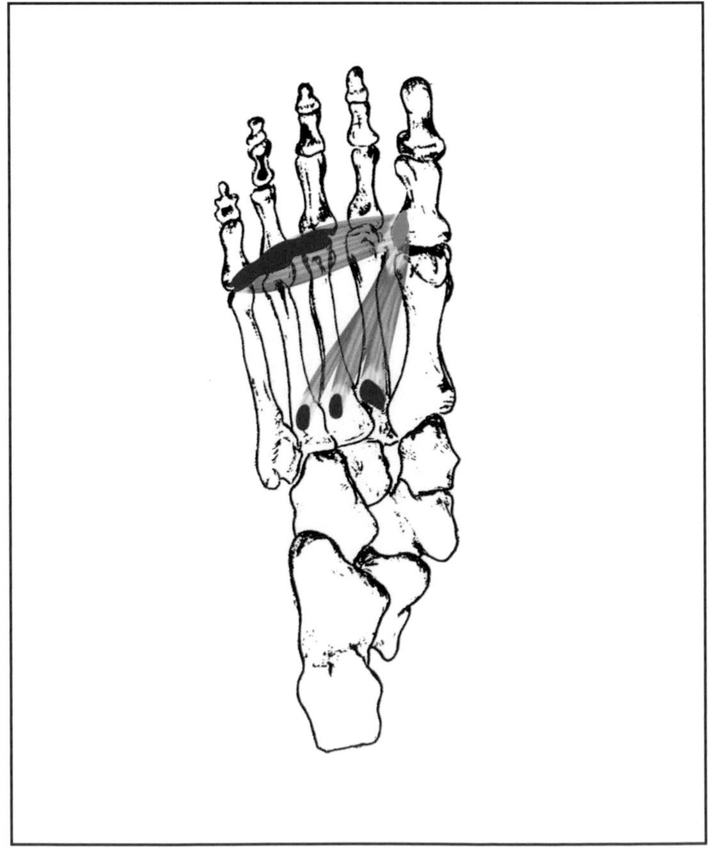

Dorsal Interossei (Four Muscles)

Description:
A group of four muscles that abduct and flex the 2nd, 3rd, and 4th toes.

Joint Crossings:
One

Rank/Body Action:
1 Toe flexion
2 Toe abduction

Blood Supply:
Lateral plantar artery

Insertion:
- First heads – the medial side of the second phalange of the 2nd, 3rd, and 4th toes
- Second heads – the lateral sides of the second phalange of the 2nd, 3rd, and 4th toes

Nerves:
Lateral plantar nerve S1, 2

Origin:
Two heads from the adjacent sides of the metatarsals

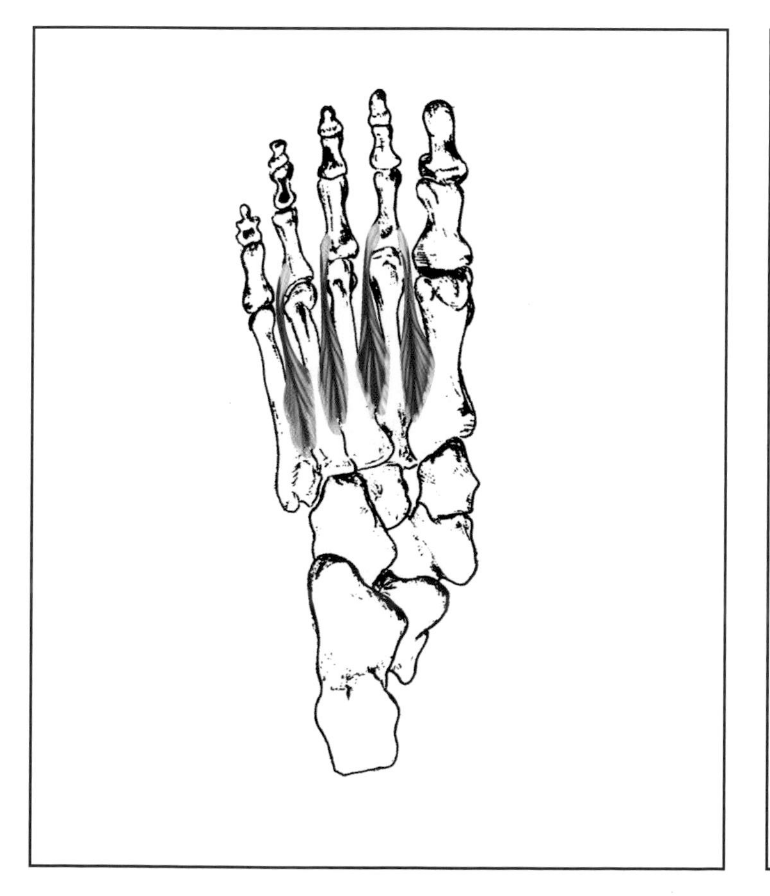

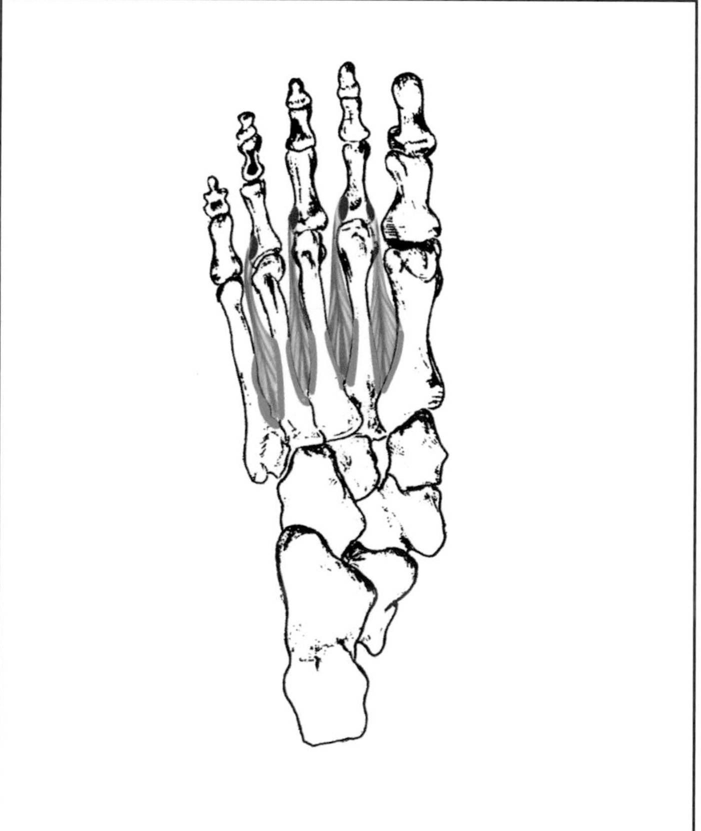

Extensor Digitorum Brevis

Description:
A broad, thin muscle on the dorsal surface of the foot that extends the toes. The most medial tendon is frequently identified as a separate muscle – the extensor hallucis brevis.

Joint Crossings:
One

Rank/Body Action:
1 Toe extension (2nd-4th toes, assists big toe – metatarsophalangeal joints)
2 Toe hyperextension (2nd-4th toes, assists big toe – metatarsophalangeal joints)

Blood Supply:
Dorsalis pedis artery

Insertion:
- By four tendons to the base of the first phalange of the big toe
- Lateral sides of extensor digitorum longus tendons of the 2nd-4th toes

Nerves:
Deep peroneal nerve L5, S1

Origin:
- Anterior and lateral calcaneus
- Lateral talocalcaneal ligament

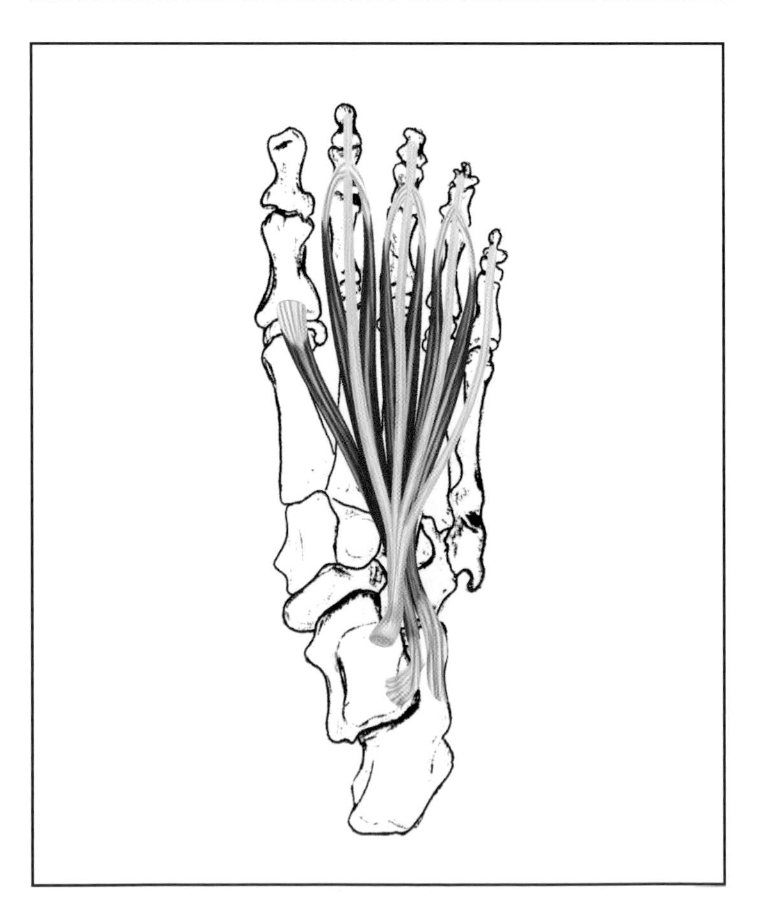

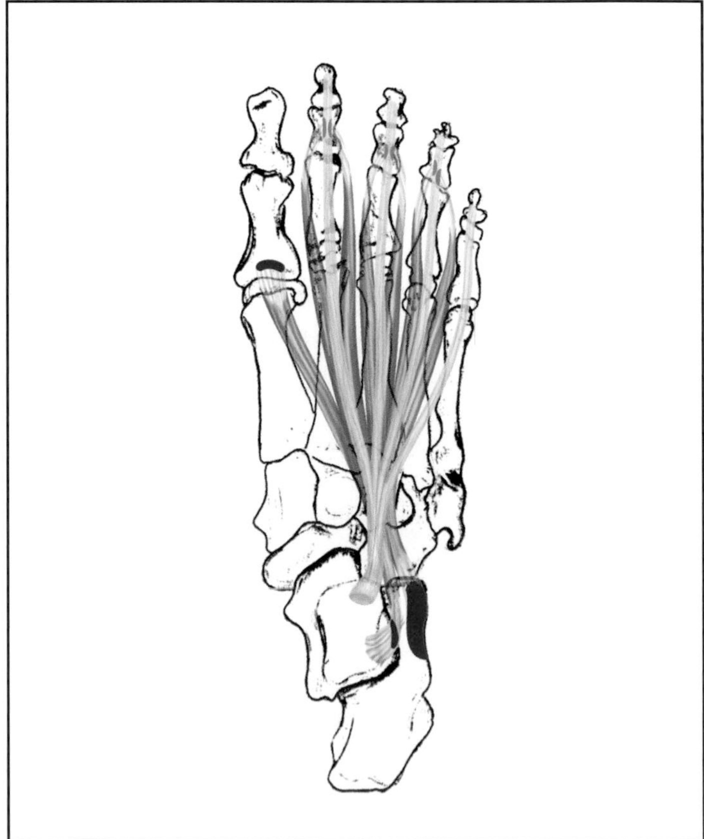

Flexor Digiti Minimi (Quinti) Brevis

Description:
A muscle of the sole of the foot that flexes the first phalange of the little toe.

Joint Crossings:
One

Body Action:
Toe flexion (little toe)

Blood Supply:
Lateral plantar artery

Insertion:
Lateral side of base of first phalange of the little toe

Nerves:
Lateral plantar nerve S2, 3

Origin:
- Base of the fifth metatarsal
- Sheath of the peroneus longus tendon

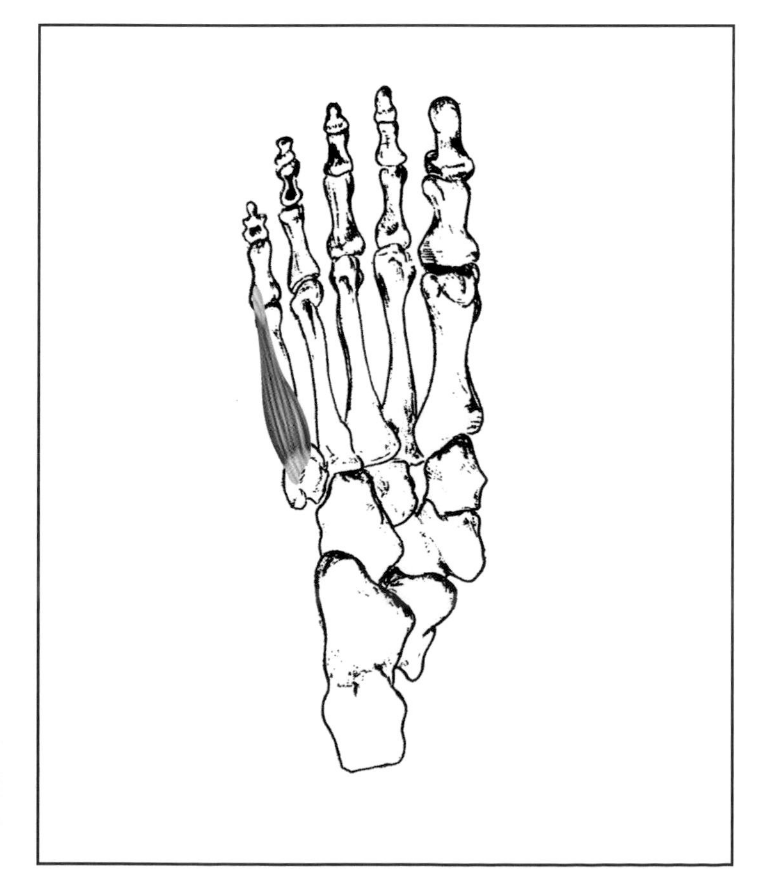

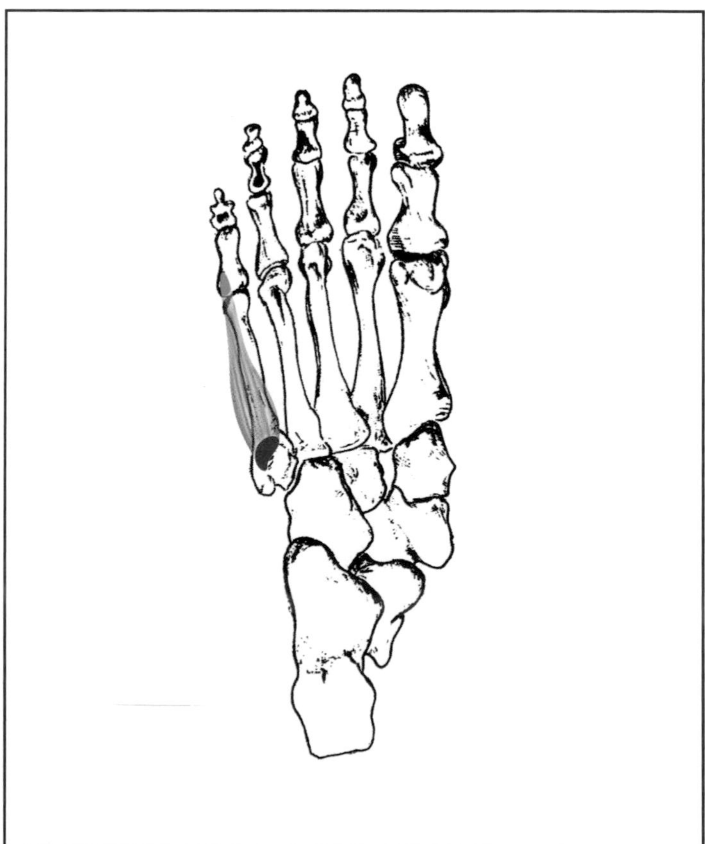

Flexor Digitorum Brevis

Description:
A muscle that lies in the middle of the plantar surface of the foot and is united with the plantar aponeurosis. It flexes the middle phalanx of each of the four small toes.

Blood Supply:
Medial plantar artery

Insertion:
A split tendon to the four toes which splits again to insert on each side of the second phalange of each toe

Joint Crossings:
One

Nerves:
Medial plantar nerve L4, 5

Body Action:
Toe flexion (2nd-5th toes)

Origin:
- Medial process of the tuberosity of calcaneus
- Central part of the plantar aponeurosis

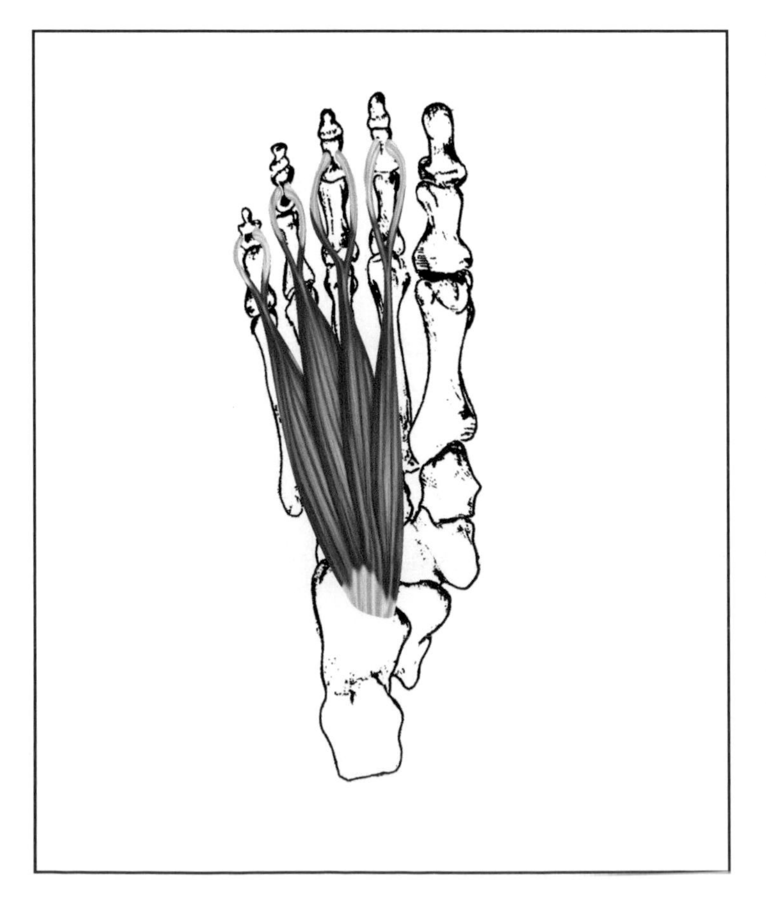

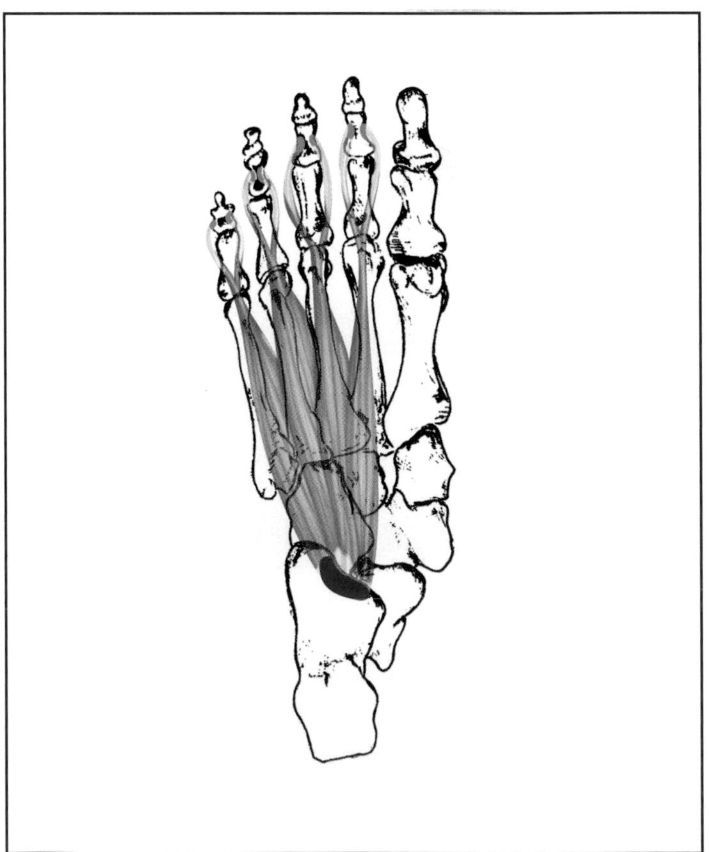

Flexor Hallucis Brevis

Description:
A short, two-bellied muscle on the plantar surface of the foot that flexes the first phalange of the big toe.

Joint Crossings:
One

Body Action:
Toe flexion (big toe)

Blood Supply:
Medial plantar artery

Insertion:
Medial and lateral portion of the first phalange of the big toe

Nerves:
Medial plantar nerve L4, 5, S1

Origin:
- Cuboid
- Lateral cuneiform

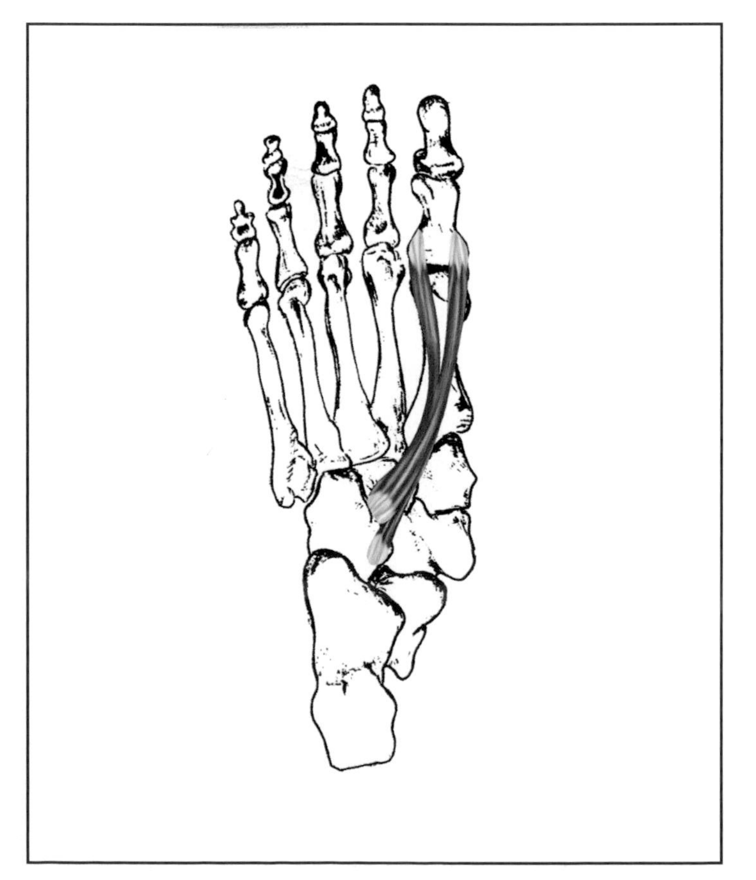

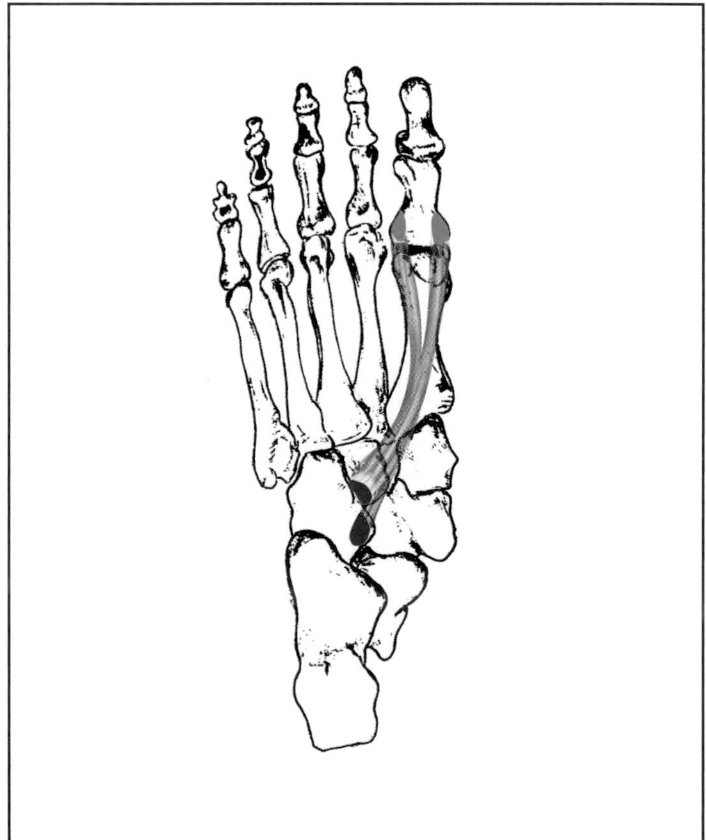

Lumbricals (Four Muscles)

Description:
Four small muscles of the foot, comparable to the lumbricales of the hand, that flex the proximal phalanges and extend the two terminal phalanges of each toe.

Joint Crossings:
One

Rank/Body Action:

1	Toe extension (two distal phalanges of four toes)
2	Toe hyperextension (two distal phalanges of four toes)
3	Toe flexion (proximal phalanges of four toes)

Blood Supply:
- First – medial plantar artery
- Second - fourth – lateral plantar artery

Insertion:
Dorsal surface of the tendons of the extensor digitorum longus

Nerves:
- First – medial plantar nerve L4, 5
- Second - fourth – lateral plantar nerve S1, 2

Origin:
On the tendons of the flexor digitorum longus, the four muscles are numbered from medial to lateral

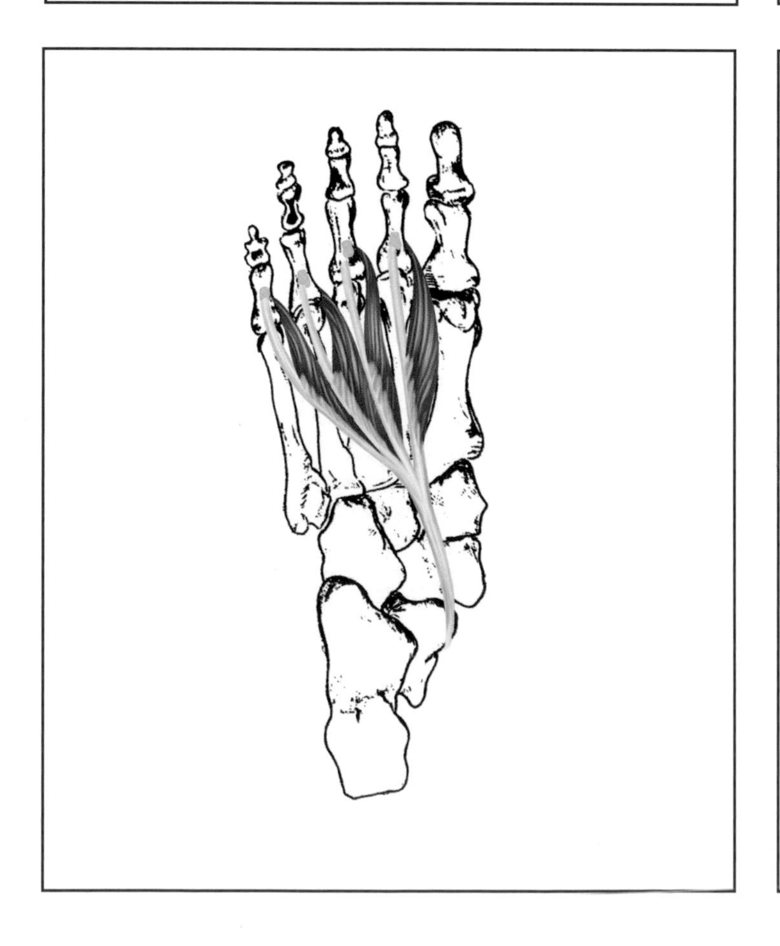

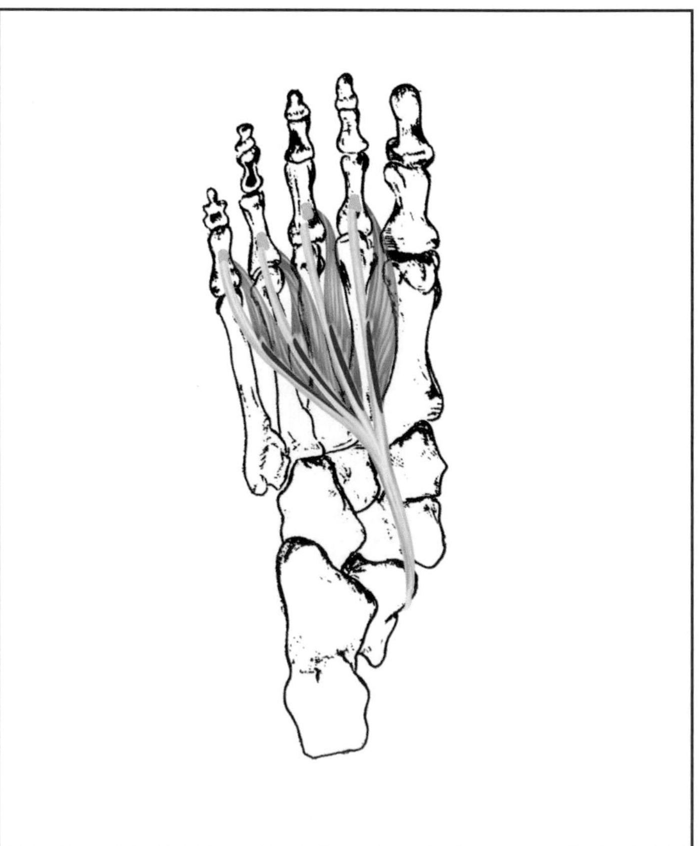

Plantar Interossei (Three Muscles)

Description:
Three muscles that lie beneath the metatarsal bones along the plantar side of the third, fourth, and fifth toes. Acts to flex the first phalanges and adduct the toes. Also called the plantaris interosseus.

Joint Crossings:
One

Rank/Body Action:
1 Toe extension (3rd-4th toes)
2 Toe hyperextension (3rd-4th toes)
3 Toe adduction (3rd-5th toes)

Blood Supply:
Lateral plantar artery

Insertion:
Medial sides of bases of first phalanges of the third, fourth, and fifth toes

Nerves:
Lateral plantar nerve S1, 2

Origin:
Medial sides of the bases of the third, fourth, and fifth metatarsals

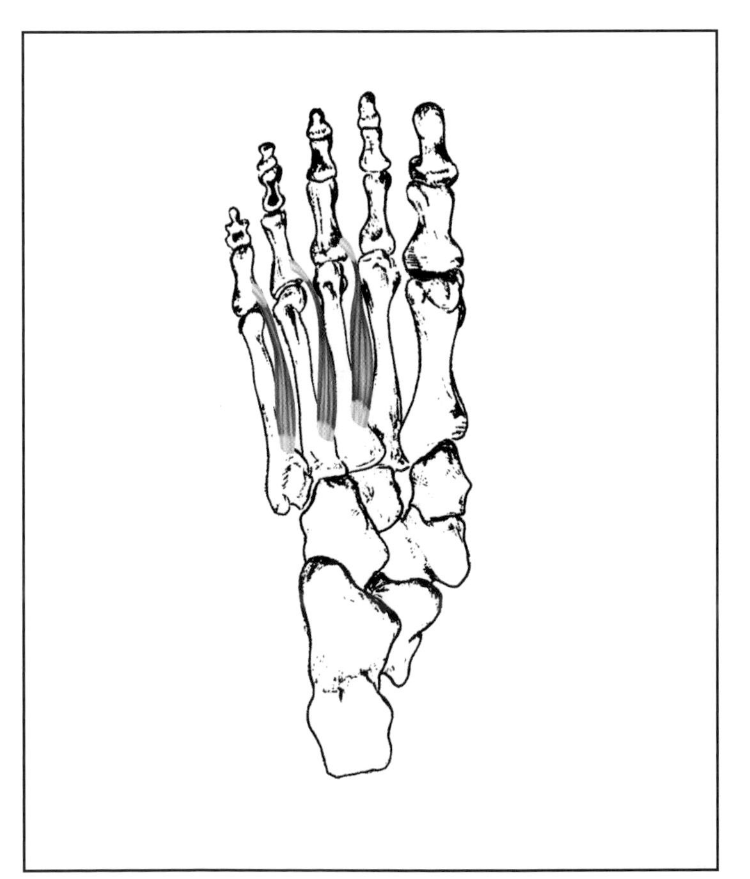

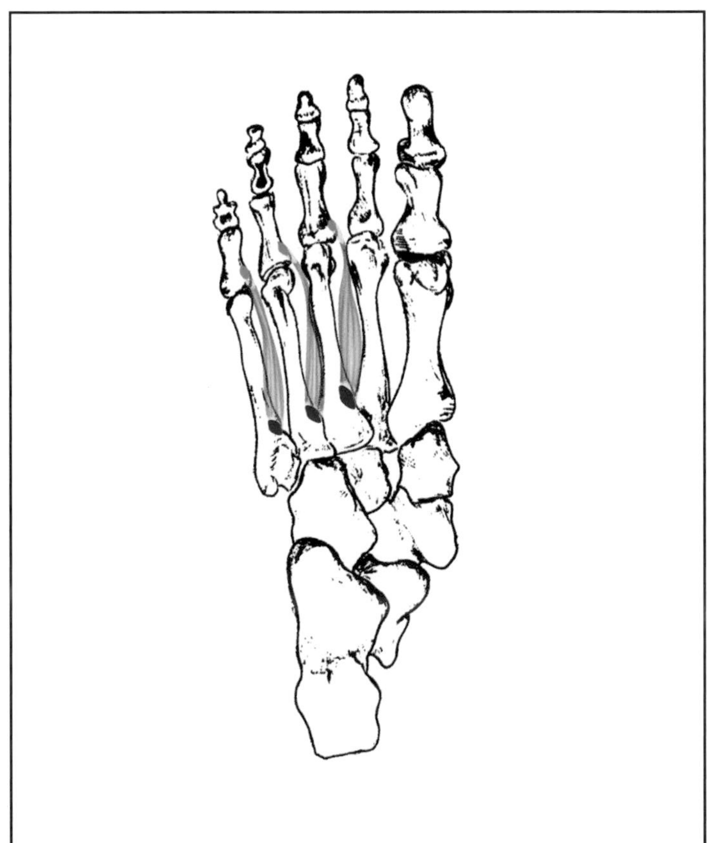

Quadratus Plantae

Description:
A muscle on the plantar surface of the foot that has two heads and aids in flexing the toes. Also known as the flexor accessorius.

Joint Crossings:
One

Body Action:
Toe flexion (2nd - 5th toes – distal interphalangeal joints)

Blood Supply:
Lateral plantar artery

Insertion:
Lateral margin of tendon of the flexor digitorum longus at about the metatarsophalangeal joint

Nerves:
Lateral plantar nerve S1, 2

Origin:
- Its origins are separated by the plantar ligament
- Medial head – medial/plantar surface of calcaneus
- Lateral head – lateral border of inferior surface of calcaneus

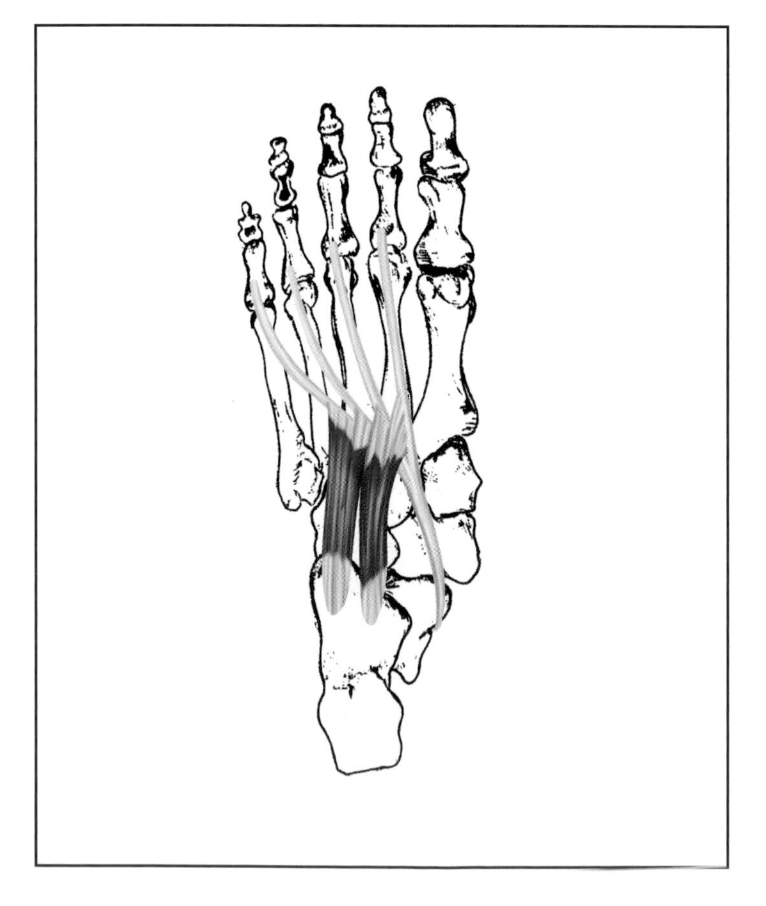

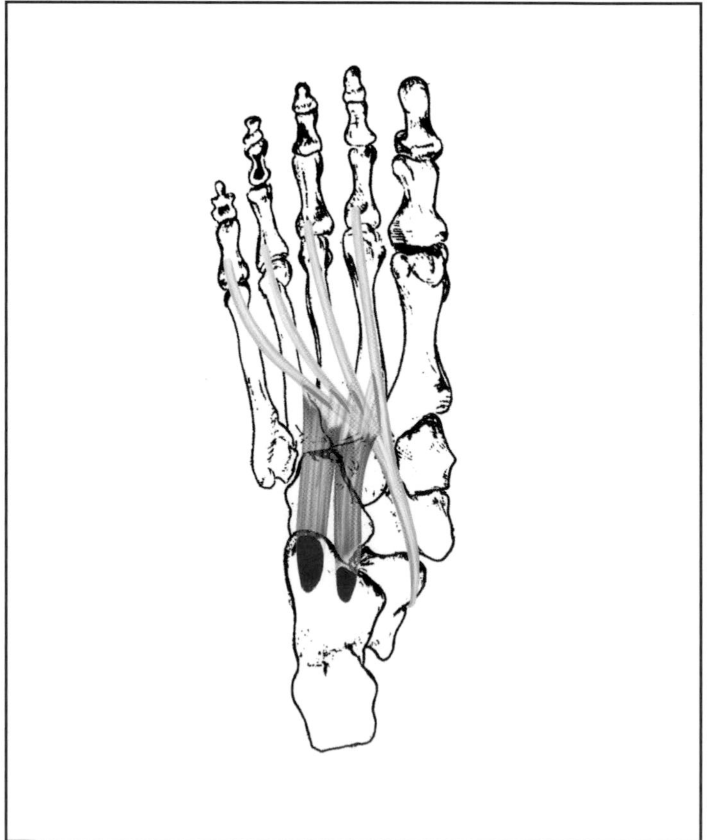

ABDUCTOR DIGITI MINIMI

Description:
A small muscle of the hand that abducts the little finger and flexes the phalange nearest the hand. Also called the abductor digiti quinti.

Joint Crossings:
One

Rank/Body Action:
1 Finger abduction (little finger)
2 Finger flexion (little finger – first phalange)

Blood Supply:
Deep palmar branches of ulnar artery

Insertion:
- The ulnar side of base of first phalange of the little finger
- The ulna border of the extensor digiti minimi

Nerves:
Deep branch of ulnar nerve C8, T1

Origin:
- Pisiform
- Tendon of Flexor Carpi Ulnaris

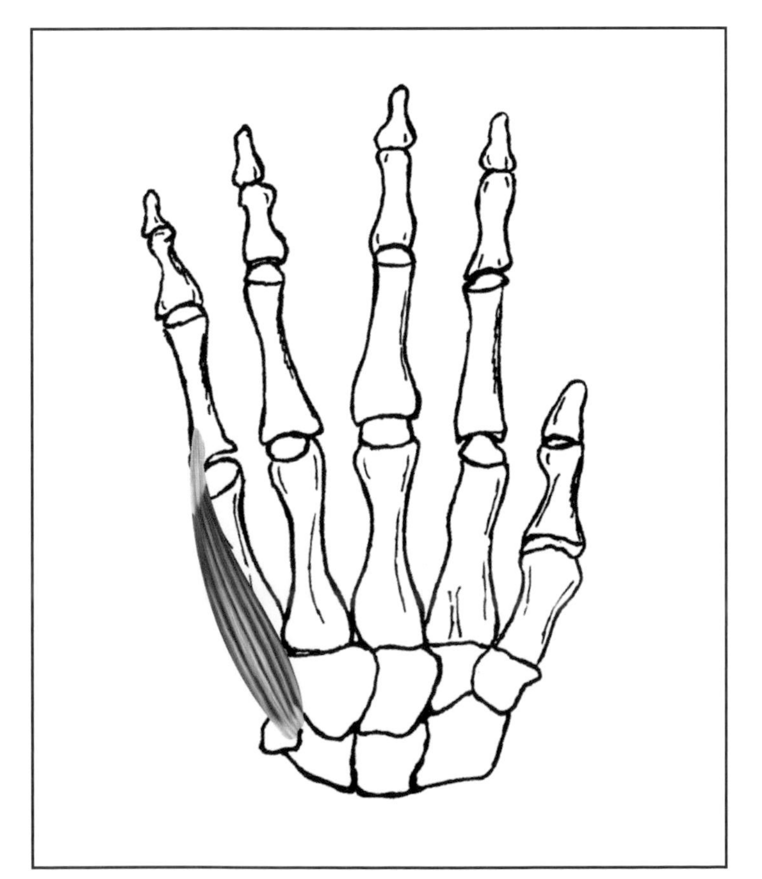

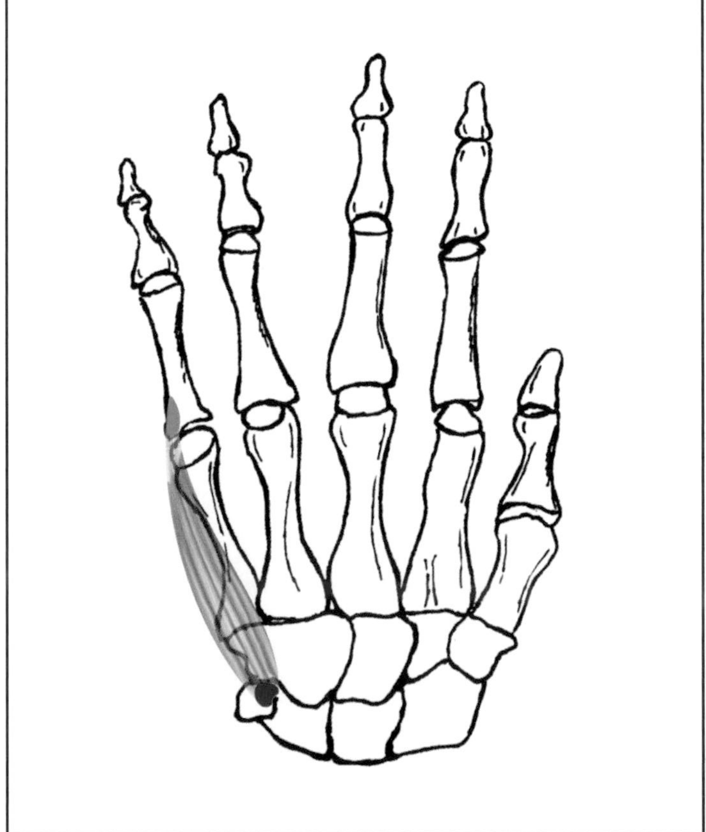

ABDUCTOR POLLICIS BREVIS

Description:
A thin flat muscle of the hand that abducts the thumb at right angles to the plane of the palm.

Joint Crossings:
One

Rank/Body Action:

1 Thumb abduction (carpometacarpal joints)
2 Thumb-finger opposition (slight)
3 Wrist radial flexion (assists)

Blood Supply:
Superficial palmar branches of radial artery

Insertion:
Outside of the base of the first phalange of the thumb

Nerves:
Recurrent branch of median nerve C6, 7

Origin:
- Anterior surface of the transverse carpal ligament
- Trapezium
- Scaphoid

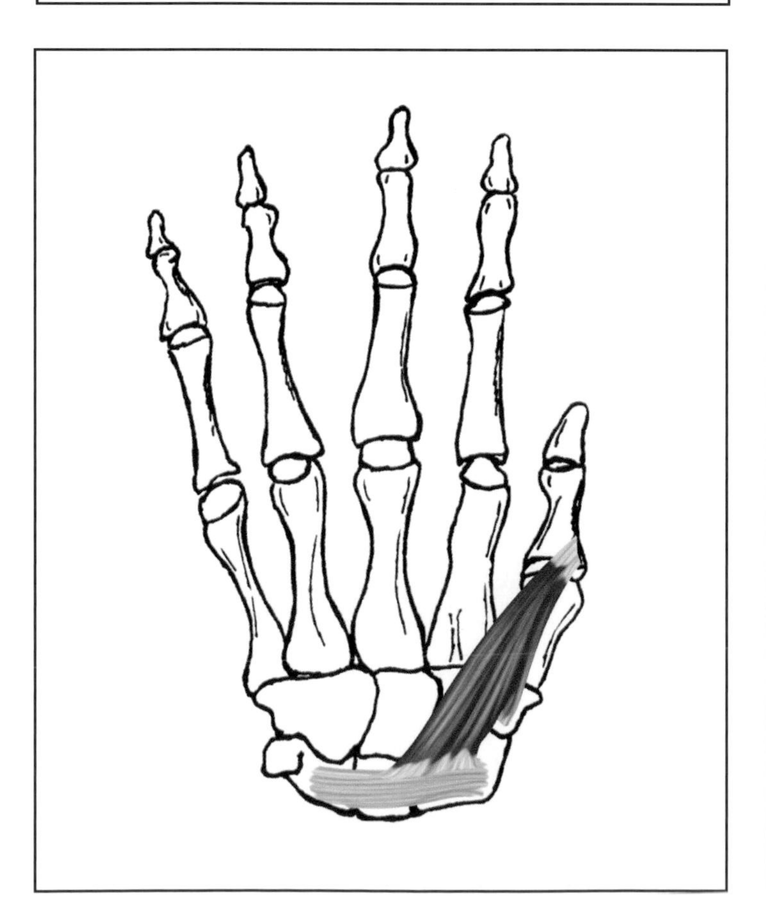

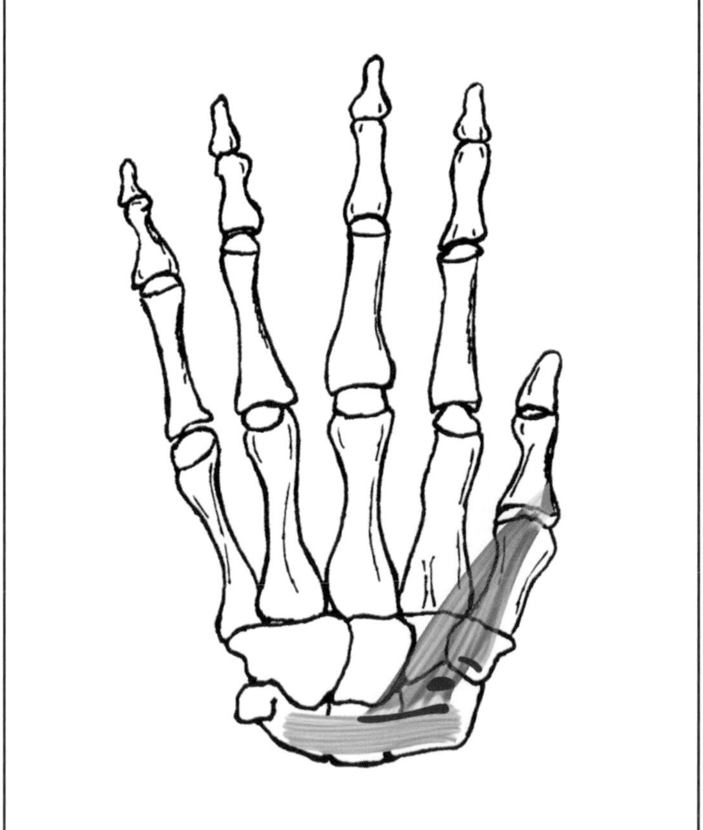

Adductor Pollicis

Description:
A muscle of the hand with two heads that adducts the thumb by bringing it toward the palm. Composed of two parts – caput obliquus and the caput transversus.

Joint Crossings:
One

Body Action:
Thumb adduction (carpometacarpal joint)

Blood Supply:
Superficial palmar branches of radial artery

Insertion:
Ulnar aspect of the base of the first phalange of the first metacarpal (thumb)

Nerves:
Deep branch of ulnar nerve C8, T1

Origin:
- Capitate bone
- Base of second and third metacarpals
- The intercarpal ligaments
- The tendon of the Flexor Carpi Radialis

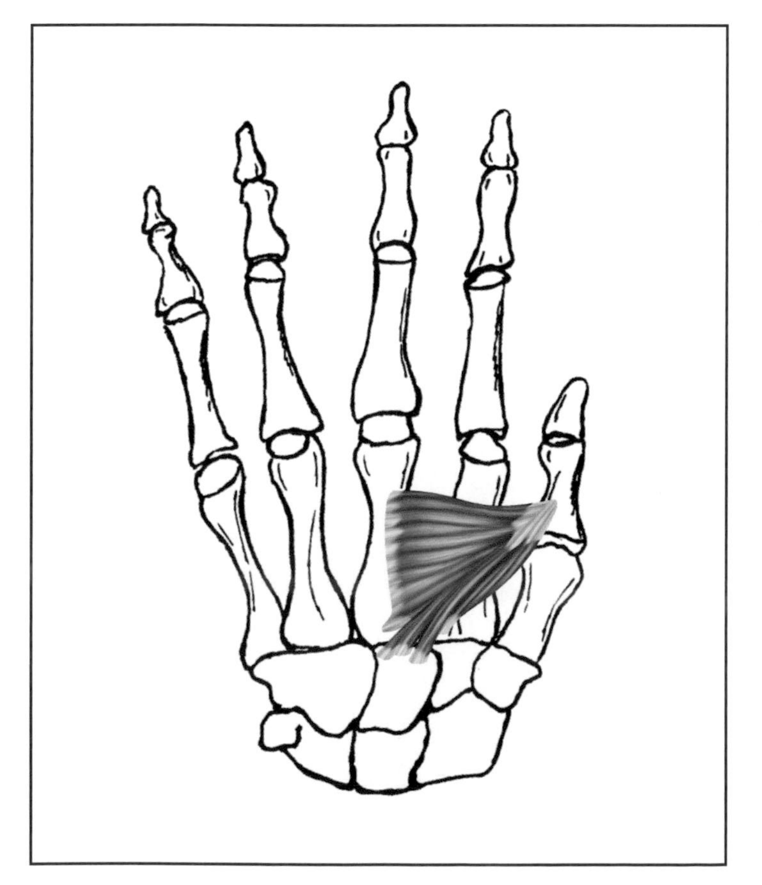

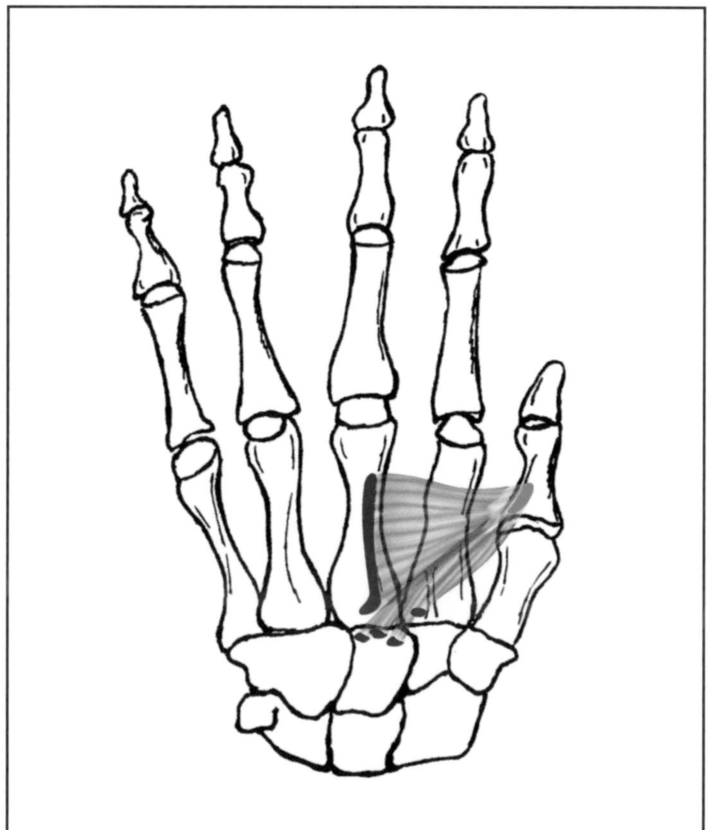

Dorsal Interossei

Description:
A group of four muscles that that both extend and abduct the index, middle, and ring fingers.

Joint Crossings:
One

Rank/Body Action:
1 Finger abduction
2 Finger extension (two distal phalanges)
3 Finger hyperextension (two distal phalanges)

Blood Supply:
Palmar metacarpal artery of deep palmar arch

Insertion:
Bases of 2nd, 3rd, and 4th proximal phalanges onto the tendons of the extensor digitorum.

Nerves:
Palmer branch of ulnar nerve C8, T1

Origin:
Two heads from each side of the metacarpal bones

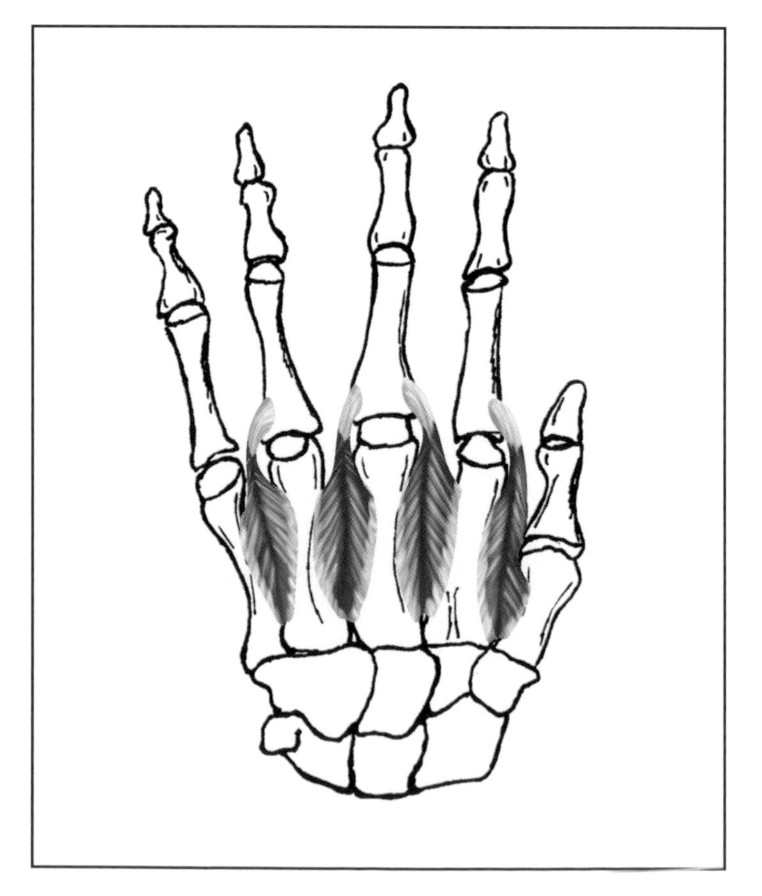

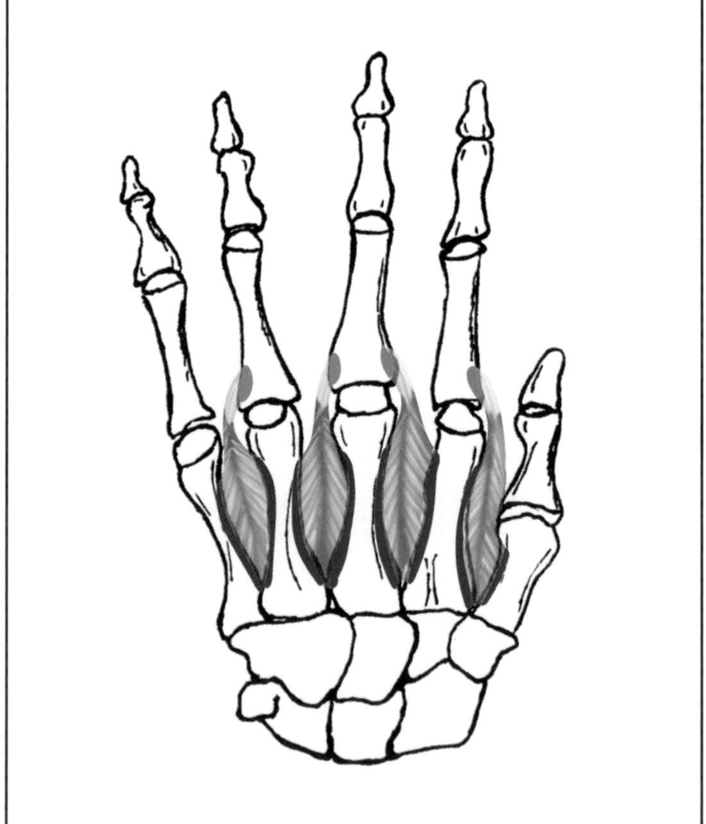

Flexor Digiti Minimi Brevis

Description:
A muscle of the ulnar side of the palm that flexes the little finger. Also known as the flexor digiti quinti brevis.

Joint Crossings:
One

Body Action:
Finger flexion (little finger)

Blood Supply:
Deep palmar branches of ulnar artery

Insertion:
Ulnar side of the first phalange of the little finger

Nerves:
Deep branch of ulnar nerve C8, T1

Origin:
Hamate
Palmar surface of the flexor retinaculum

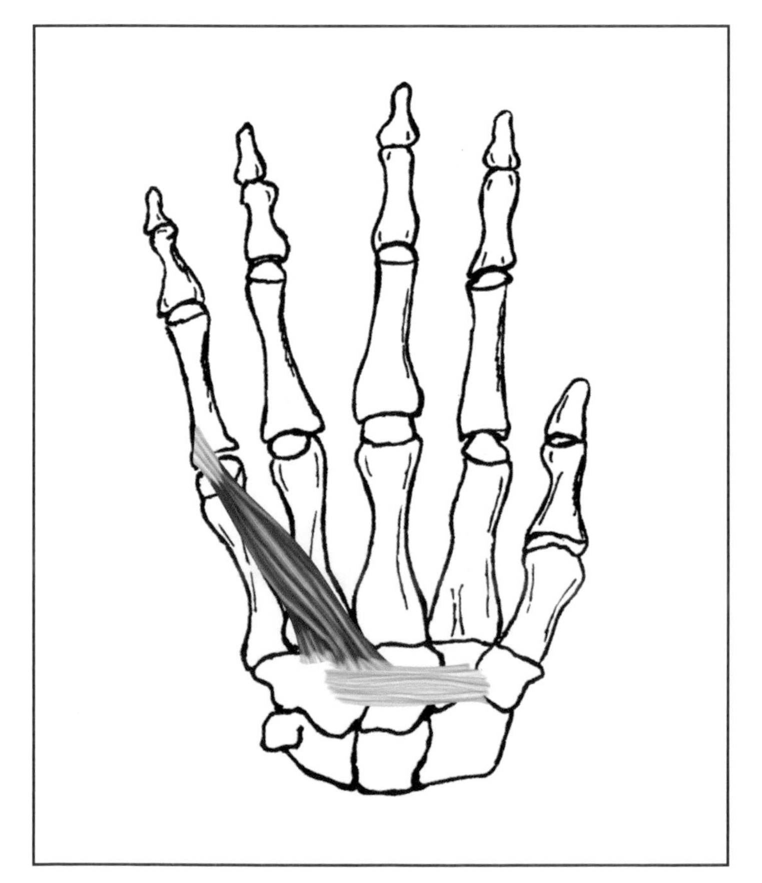

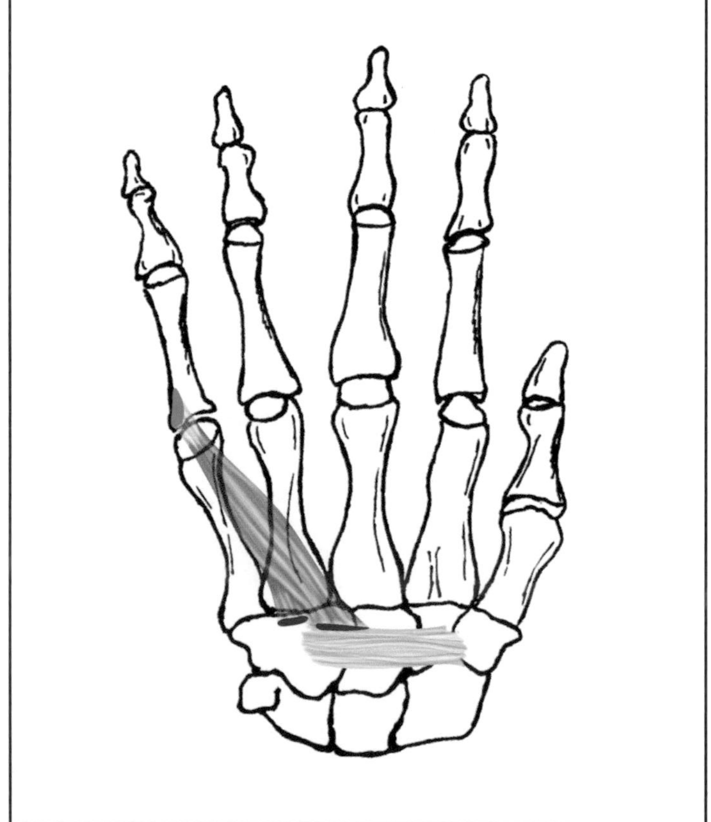

Flexor Pollicis Brevis

Description:
A short muscle of the palm that flexes and adducts the thumb. It is especially active in a firm grip.

Joint Crossings:
One

Rank/Body Action:
1 Thumb flexion (carpometacarpal and metacarpophalangeal joints)
2 Thumb adduction (carpometacarpal joint)

Blood Supply:
Superficial palmar branches of radial artery

Insertion:
Base of the first phalange of the thumb

Nerves:
Superficial head – median nerve C6, 7
Deep head – deep branch of ulnar nerve C8, T1

Origin:
- Superficial (lateral) head – flexor retinaculum
- First metacarpal

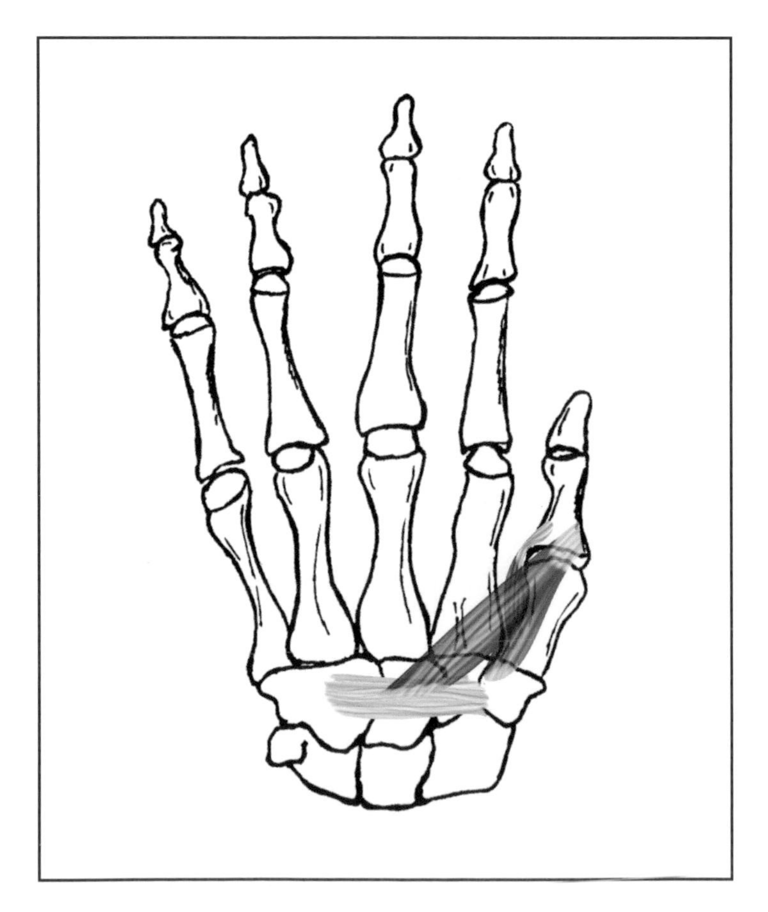

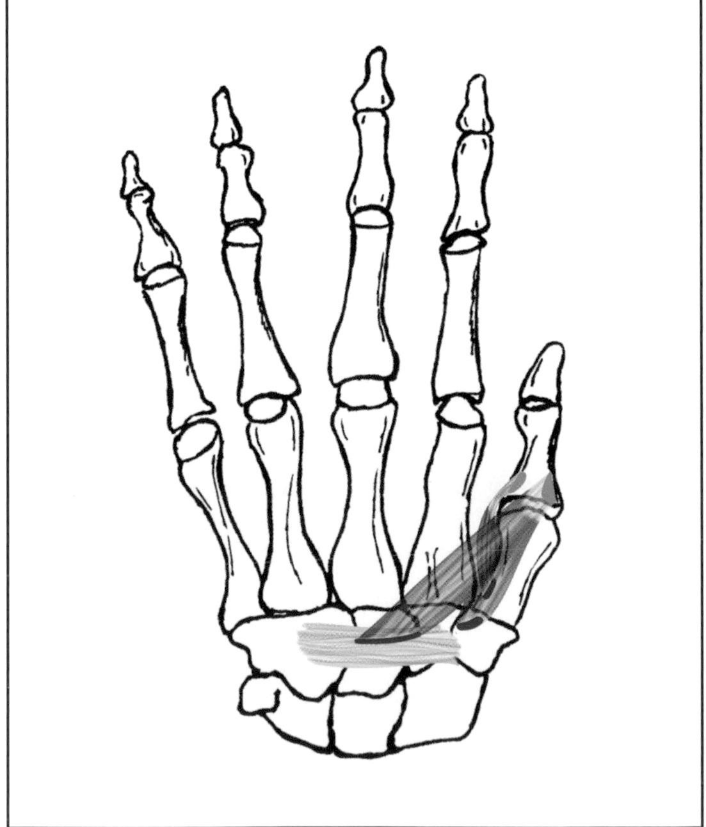

Lumbricals

Description:
Four small muscles of the palm of the hand that flex the proximal phalanges and extend the two terminal phalanges of each finger.

Joint Crossings:
One

Rank/Body Action:

1	Finger flexion (metacarpophalangeal joint)
2	Finger extension (the two distal joints)
3	Finger hyperextension (the two distal joints)

Blood Supply:
Palmar metacarpal artery of deep palmar arch

Insertion:
The tendinous expansion of the extensor digitorum on the dorsal side of the hand

Nerves:
- 1,2 – median nerve C6, 7
- 3,4 – deep branch of ulnar nerve C8, T1

Origin:
- The tendons of the index and middle finger
- The tendons to the little and ring fingers

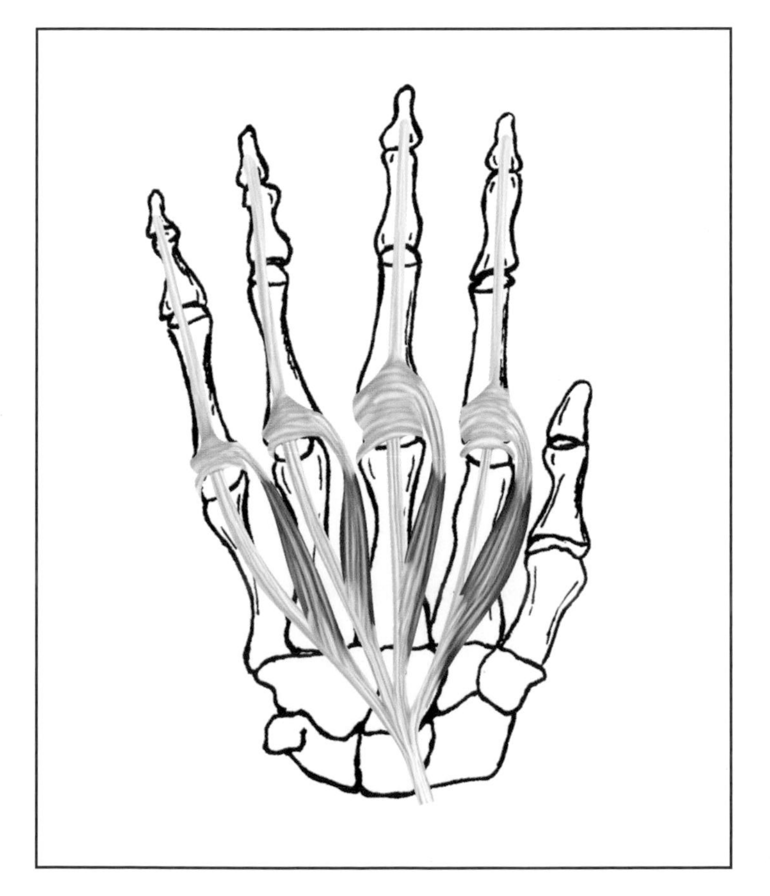

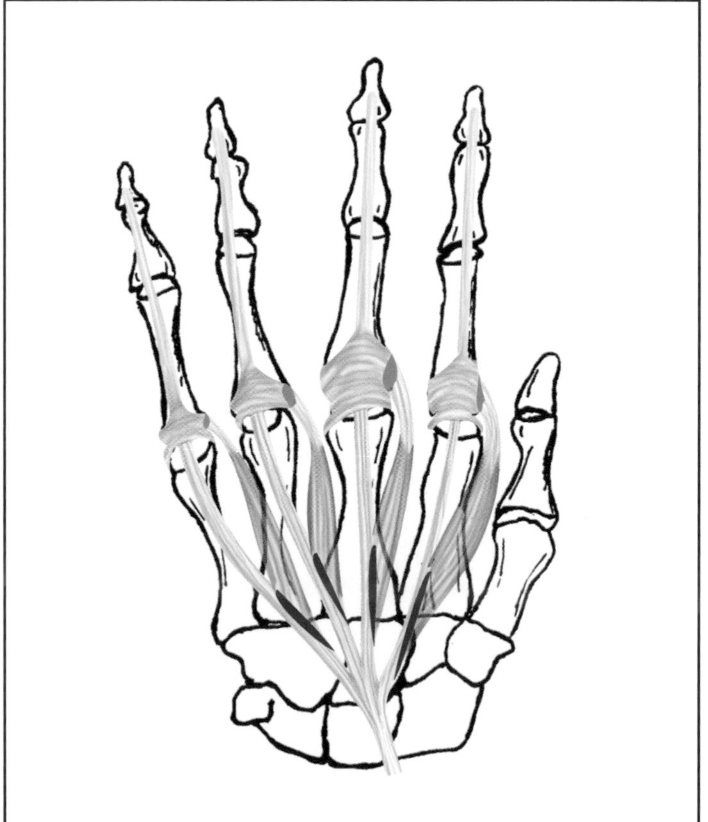

Opponens Digiti Minimi

Description:
A triangular muscle of the hand that functions to abduct, flex, and rotate the fifth metacarpal in opposing the little finger and thumb. Also called the opponens digiti quinti.

Joint Crossings:
One

Rank/Body Action:
1 Thumb-finger opposition (little finger)
2 Finger abduction (little finger)
3 Finger flexion (little finger)

Blood Supply:
Deep palmar branches of ulnar artery

Insertion:
Ulna side of the shaft of the fifth metacarpal

Nerves:
Deep branch of the ulnar nerve C8, T1

Origin:
Hamate and the connecting portion of the flexor retinaculum ligament

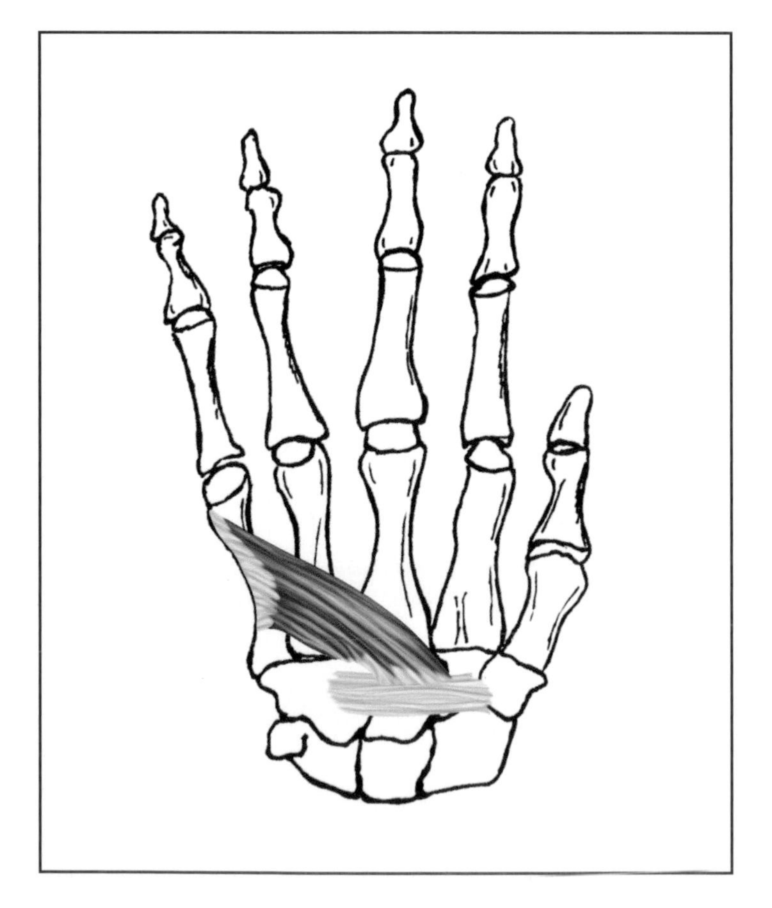

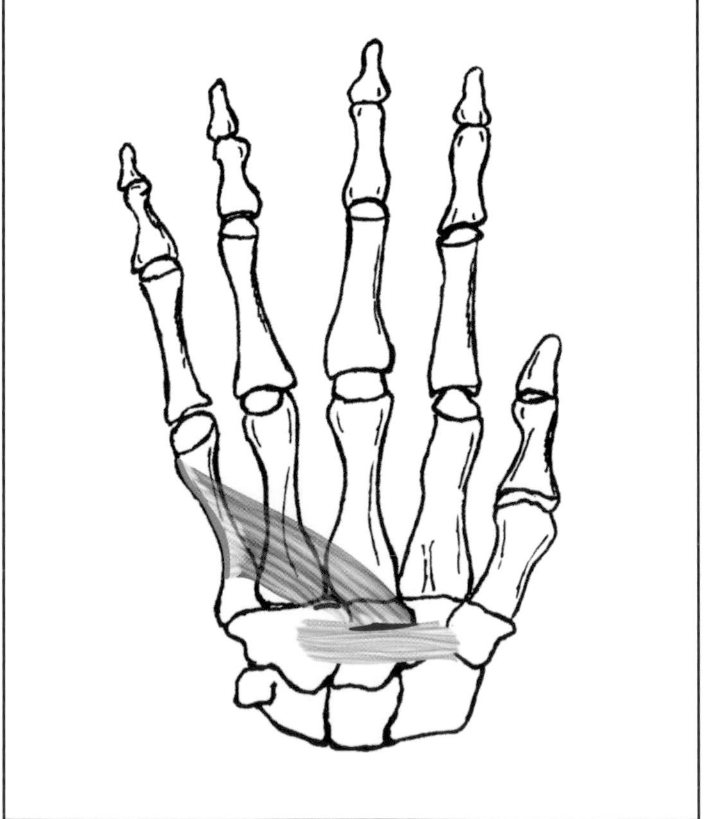

Opponens Pollicis

Description:
A small triangular muscle of the hand, located below the abductor pollicis brevis, that functions to abduct, flex, and rotate the metacarpal of the thumb, while opposing the thumb and fingers.

Joint Crossings:
One

Rank/Body Action:

1	Thumb-finger opposition
2	Thumb abduction
3	Thumb flexion

Blood Supply:
Superficial palmar branches of radial artery

Insertion:
Lateral border of the first metacarpal (thumb)

Nerves:
Recurrent branch of median nerve C6, 7

Origin:
- Trapezium
- The flexor retinaculum

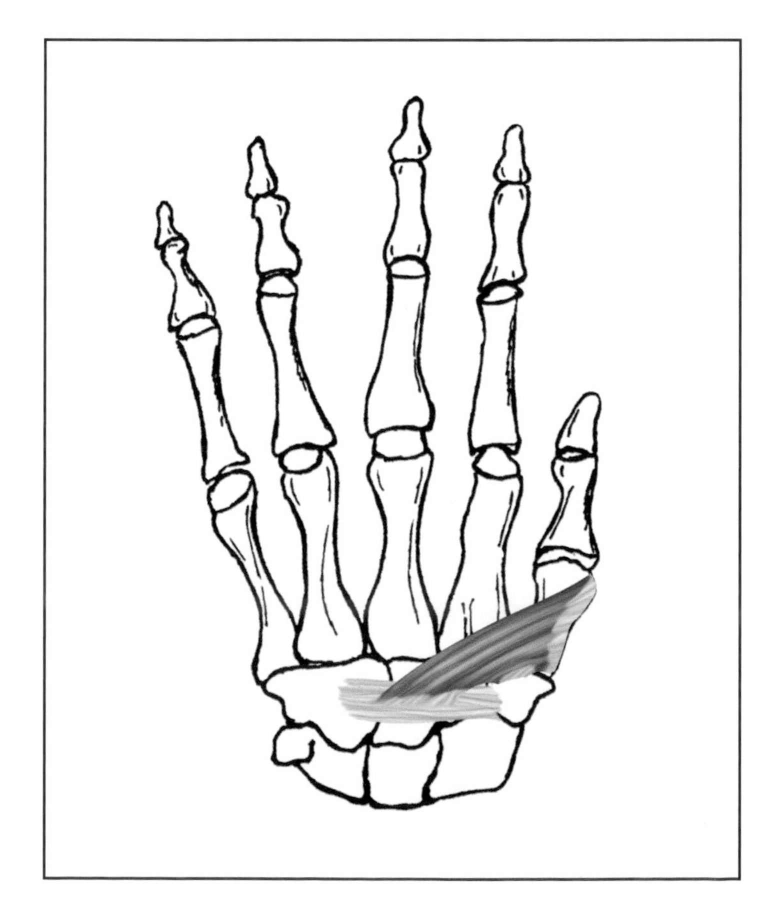

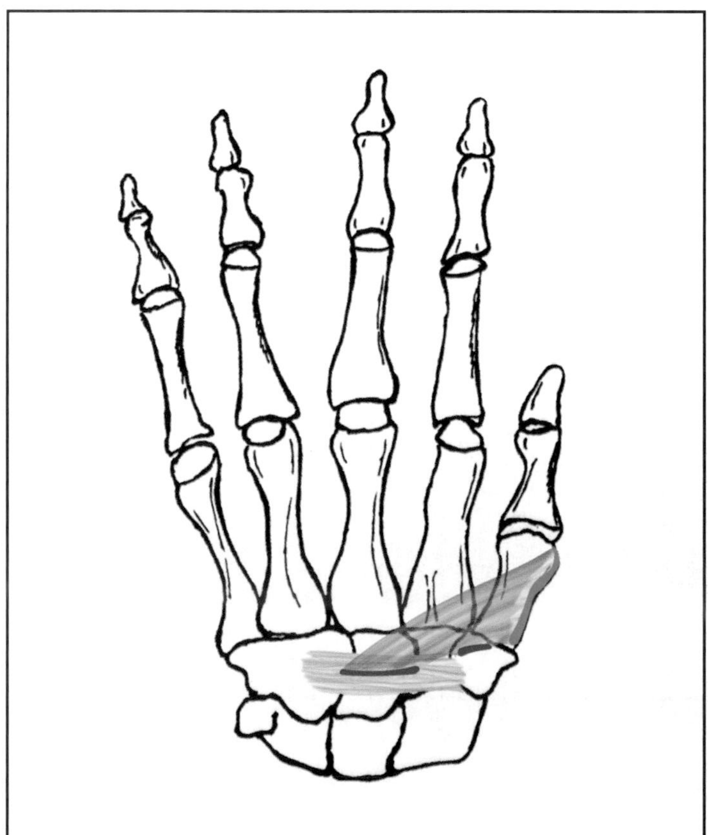

PALMAR INTEROSSEI

Description:
Any of three small muscles of the palmar surface of the hand, each of which acts to adduct its finger toward the middle finger. Also called the interosseus palmaris and volar interossei.

Blood Supply:
Palmar metacarpal artery of the deep palmar arch

Insertion:
The side of the bases of the first phalanges of the 2nd, 4th, and 5th fingers and the extensor digitorum tendon going to the same finger

Joint Crossings:
One

Nerves:
Deep branch of ulnar nerve C8, T1

Rank/Body Action:

1 Finger adduction (2nd, 4th, 5th fingers – metacarpophalangeal joints)
2 Finger flexion (metacarpophalangeal joint)
3 Finger extension (two distal phalanges)
4 Finger hyperextension (two distal phalanges)

Origin:
The length of the 2nd, 4th, and 5th metacarpals

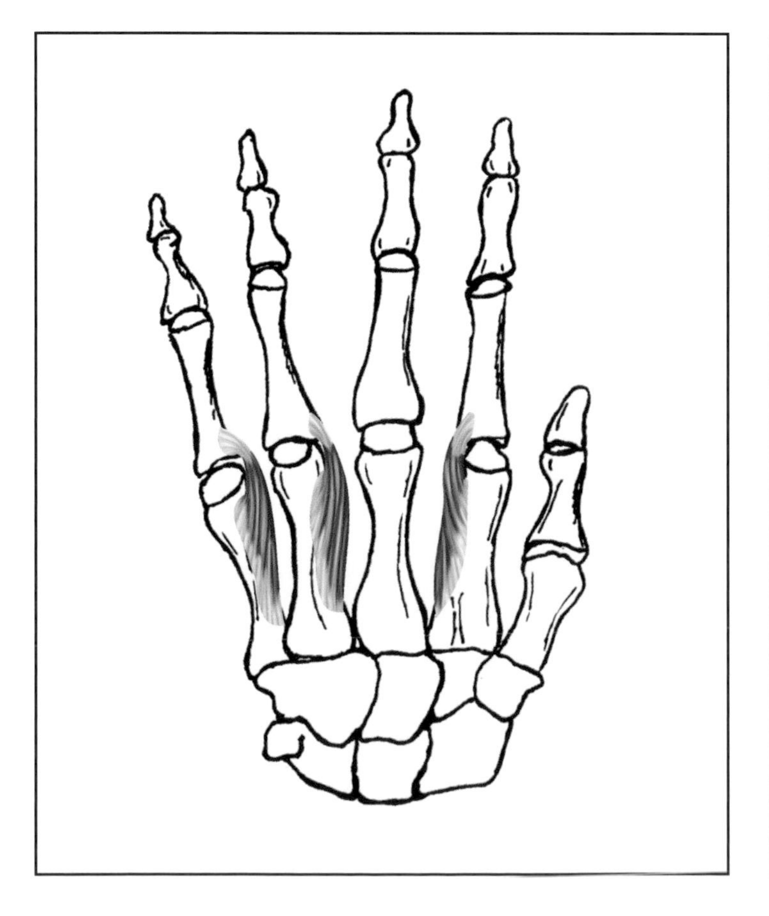

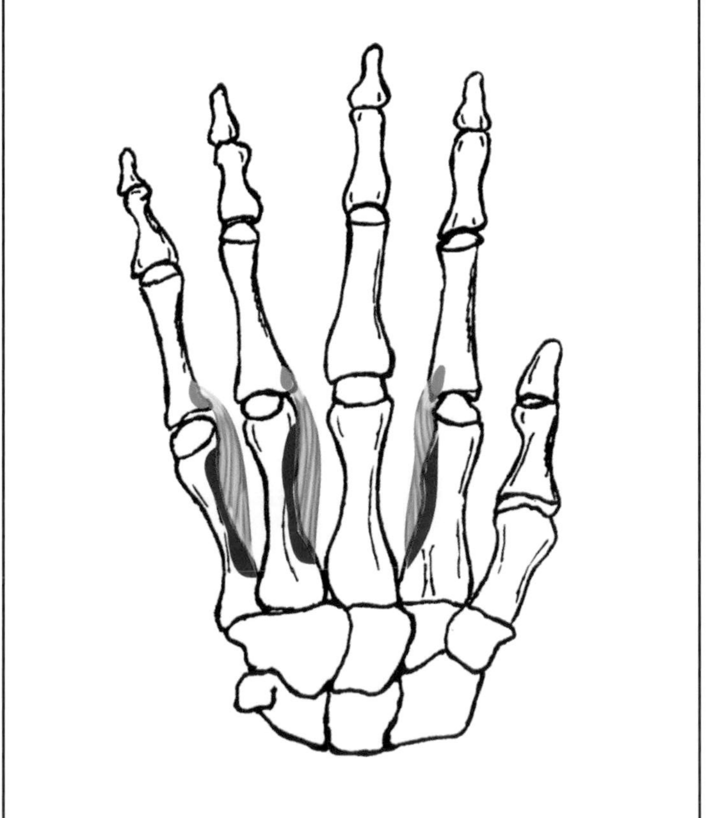

Adductor Brevis

Description:
One of three triangular muscles that contribute to the adduction of the leg.

Joint Crossings:
One

Rank/Body Action:
1 Hip-leg adduction
2 Hip-leg external rotation
3 Hip-leg flexion

Blood Supply:
Muscular branches of femoral artery

Insertion:
Pectineal line of the femur and upper one-third of the linea aspera

Nerves:
Obturator nerve L3, 4

Origin:
Inferior ramus of the pubis

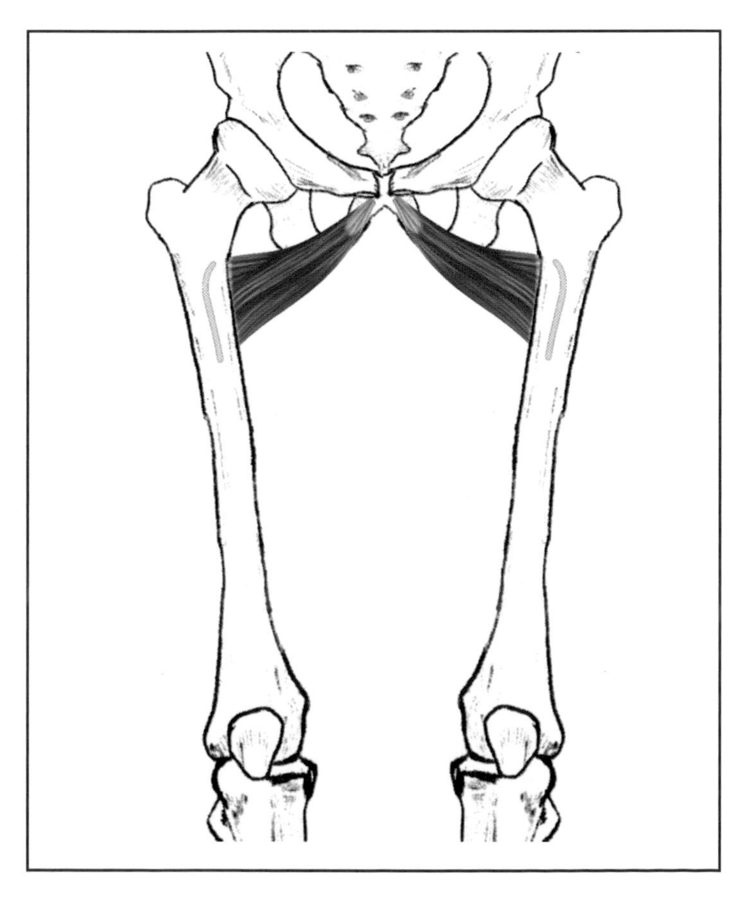

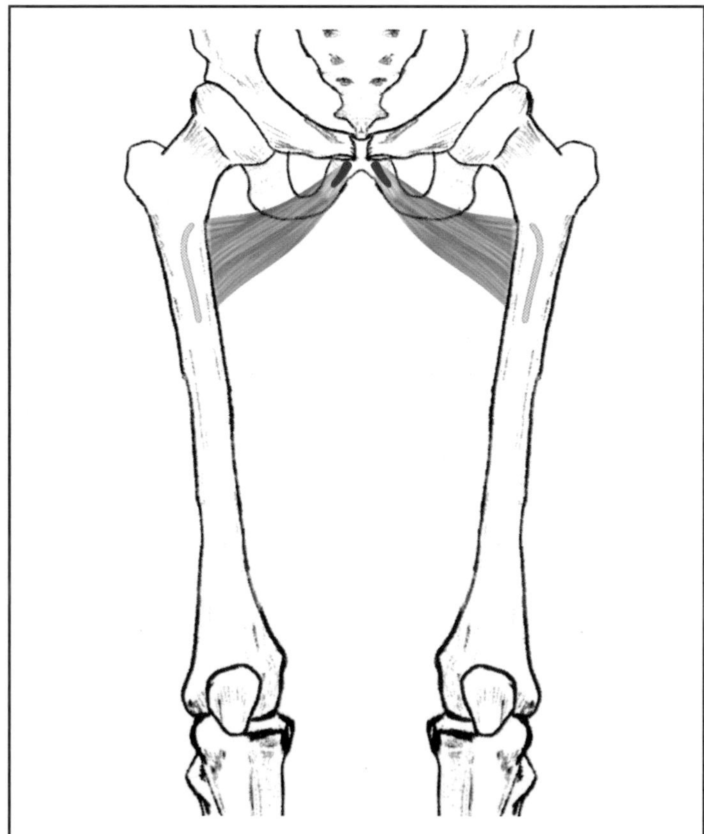

Adductor Longus

Description:
One of three powerful triangular muscles on the inside of the thigh, which adducts the leg.

Joint Crossings:
One

Rank/Body Action:

1	Hip-leg adduction
2	Hip-leg external rotation
3	Hip-leg flexion

Blood Supply:
Muscular branches of femoral artery

Insertion:
Middle one-third of the linea aspera on the posterior femur

Nerves:
Obturator nerve L3, 4

Origin:
Body of the pubis, just inferior to the pubic crest

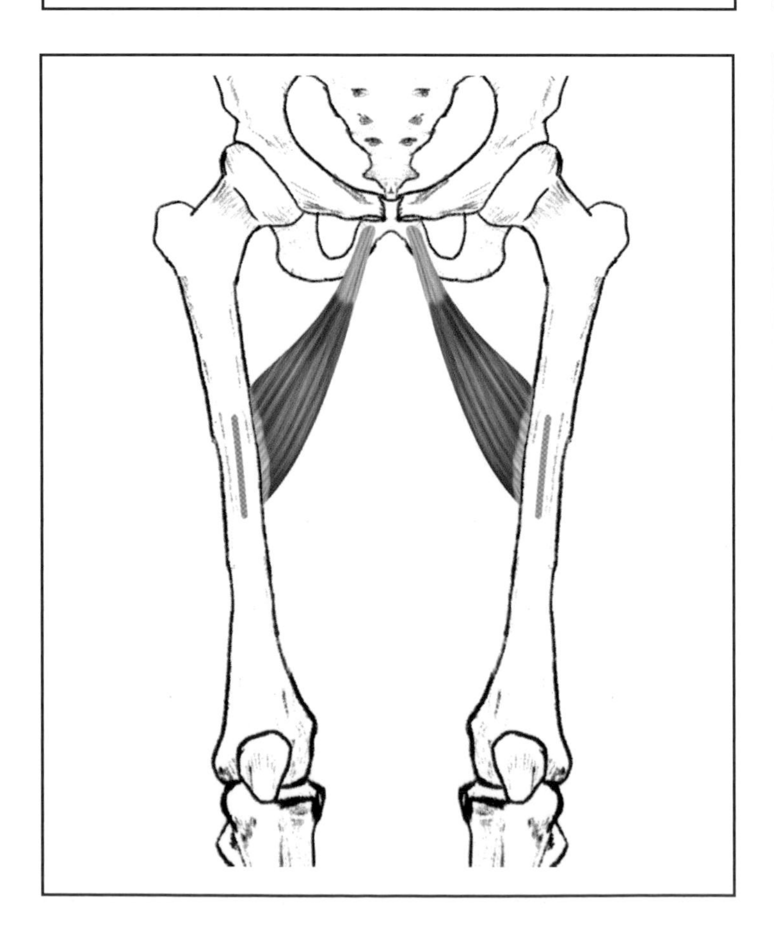

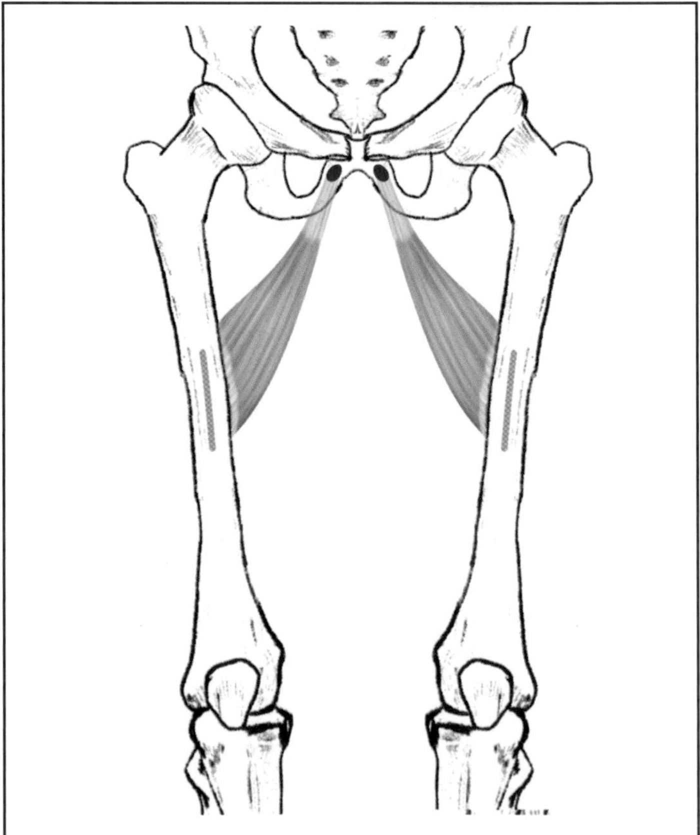

Adductor Magnus

Description:
One of three powerful triangular muscles on the inside of the thigh, which adducts the leg.

Joint Crossings:
One

Rank/Body Action:

1	Hip-leg adduction
2	Hip-leg external rotation (anterior fibers)
3	Hip-leg flexion (anterior fibers)
4	Hip-leg extension (posterior fibers)
5	Hip-leg hyperextension (posterior fibers)
6	Hip-leg internal rotation (posterior fibers)

Blood Supply:
Muscular branches of profunda femoris

Insertion:
- Linea aspera
- Medial supracondylar ridge of femur
- Adductor tubercle of femur

Nerves:
- Anterior fibers: obturator nerve L2, 3, 4
- Posterior fibers: tibial nerve of sciatic bundle L4, 5, S1, 2, 3

Origin:
Inferior ramus of pubis and ischium

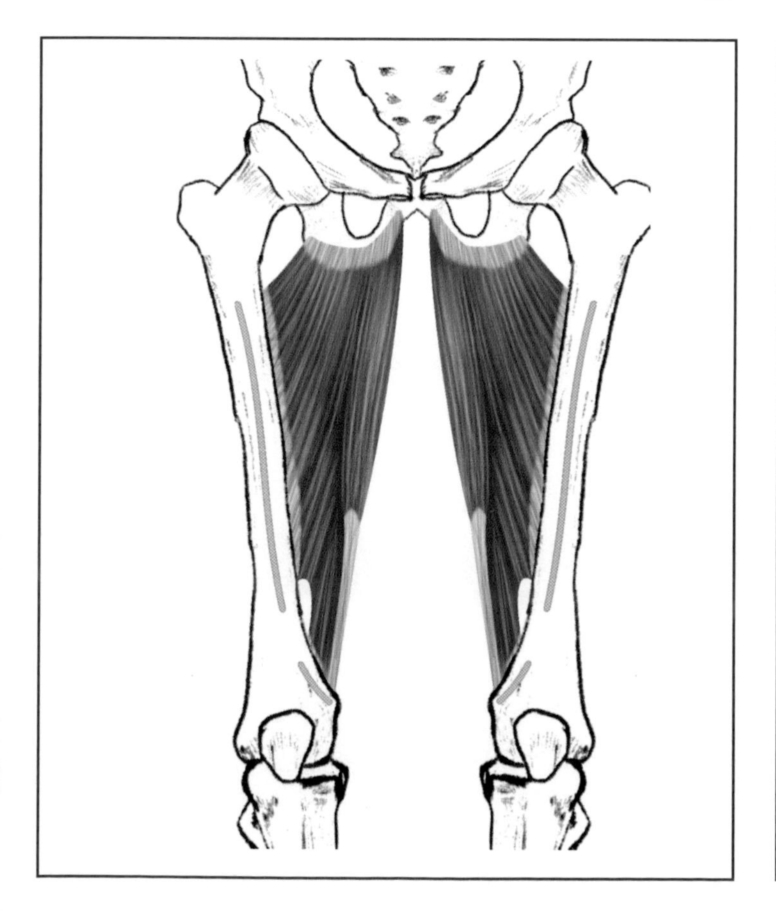

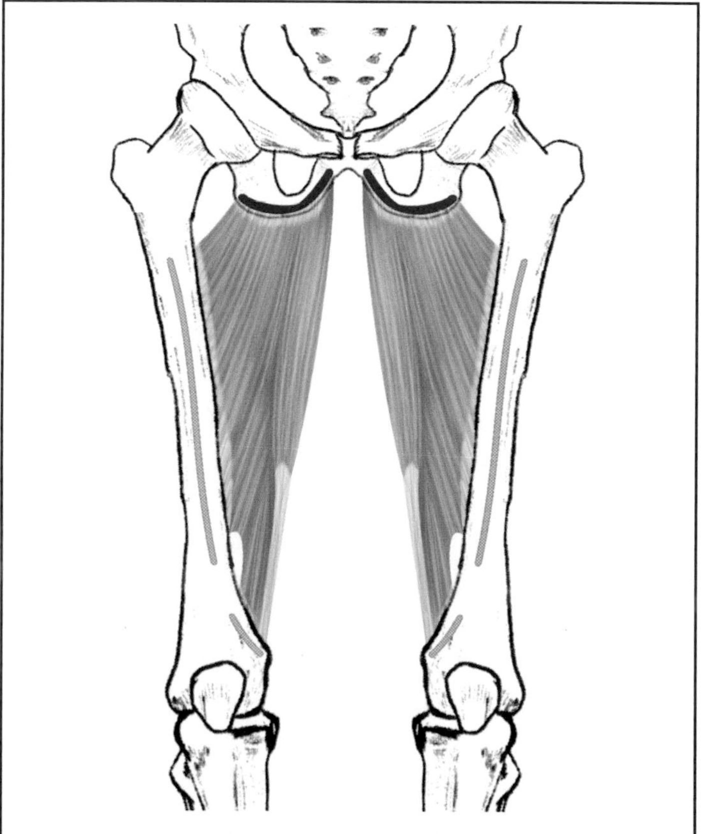

Gluteus Maximus

Description:
A large superficial muscle of the buttocks that is a very strong extensor and external rotator of the leg. The lower fibers of this muscle also adduct the leg.

Joint Crossings:
One

Rank/Body Action:

1	Hip-leg extension
2	Hip-leg hyperextension
3	Hip-leg external rotation
4	Hip-leg adduction

Blood Supply:
- Inferior gluteal artery
- Superior gluteal artery

Insertion:
- Superficial fibers – iliotibial band
- Deep fibers – gluteal line on posterior upper femur

Nerves:
Inferior gluteal nerve L5, S1, 2

Origin:
- Posterior gluteal line of Ilium
- Posterior iliac crest
- Lateral posterior sacrum and coccyx

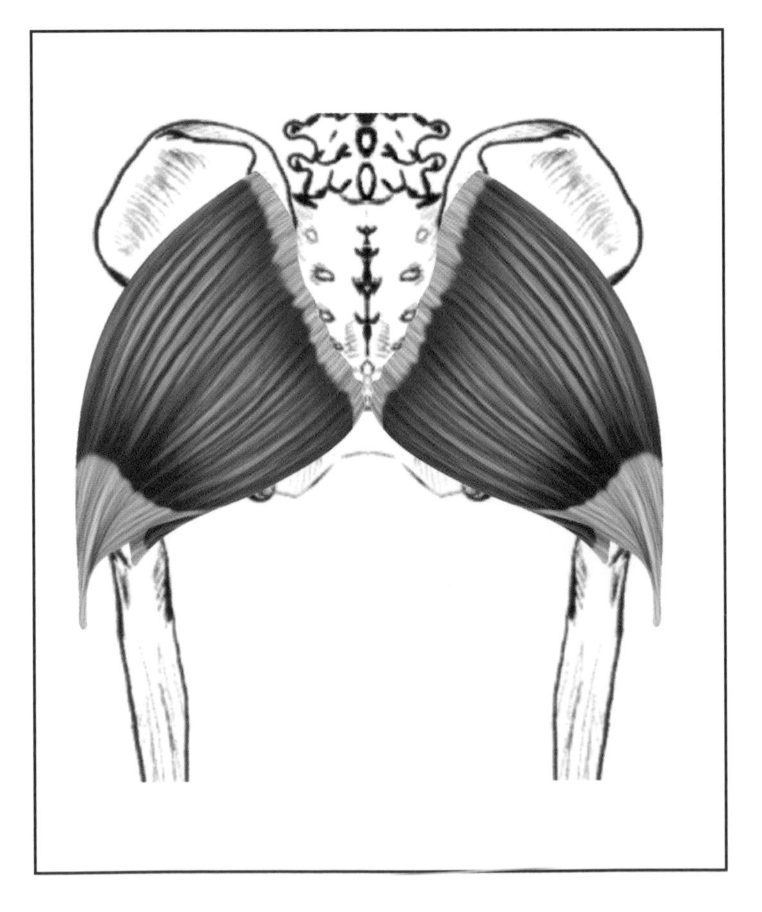

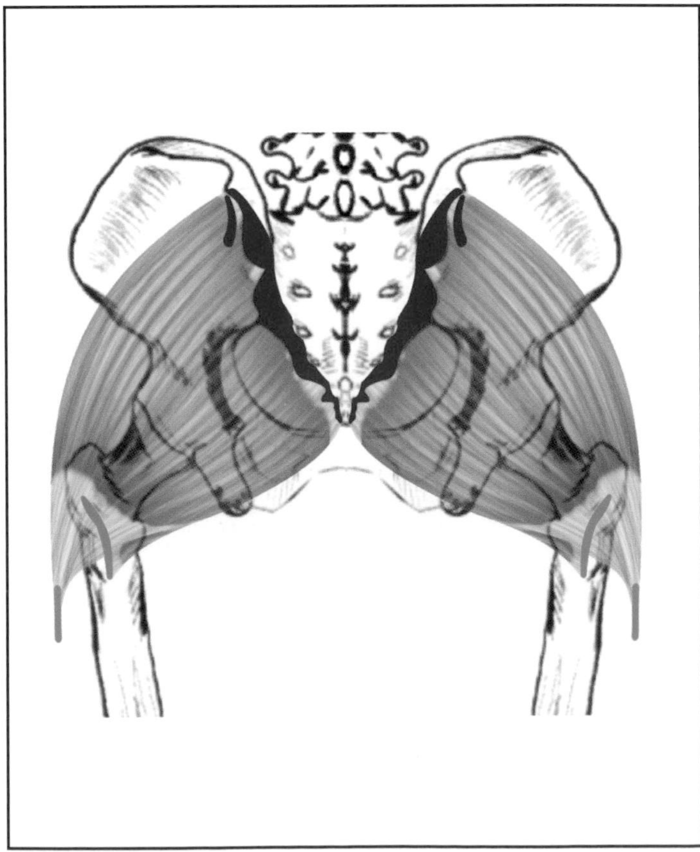

Gluteus Medius

Description:
A fan-shaped muscle that works in conjunction with the gluteus minimus to abduct the leg.

Joint Crossings:
One

Rank/Body Action:

Rank	Body Action
1	Hip-leg abduction
2	Hip-leg external rotation (posterior fibers)
3	Hip-leg internal rotation (anterior fibers)
4	Hip-leg extension (posterior fibers)
5	Hip-leg hyperextension (posterior fibers)
6	Hip-leg flexion (anterior fibers)

Blood Supply:
Superior gluteal artery

Insertion:
Superior and lateral surface of the greater trochanter of the femur

Nerves:
Superior gluteal nerve L4, 5, S1

Origin:
Lateral upper surface of the ilium, above the gluteus minimus

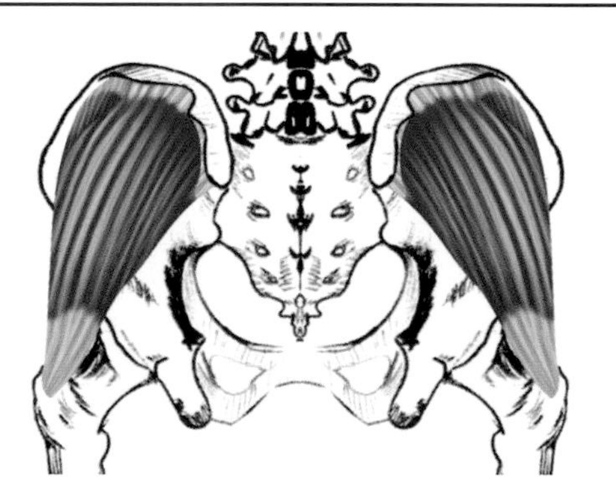

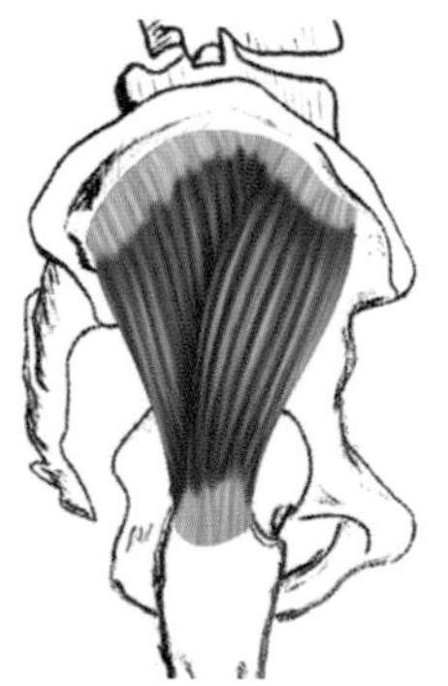

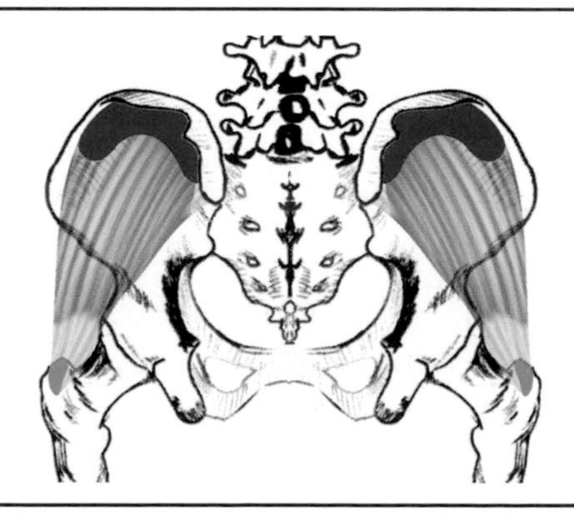

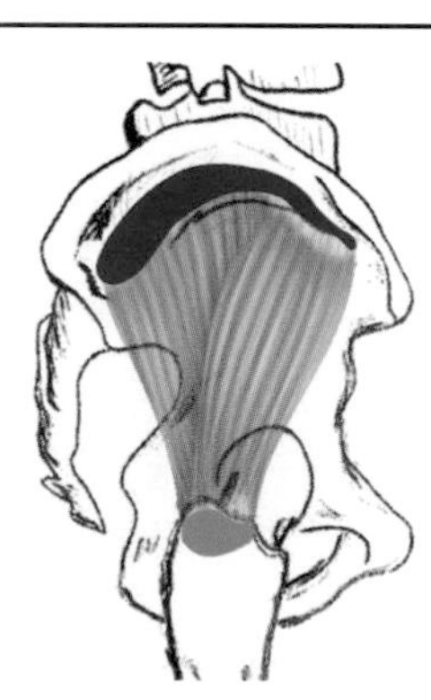

GLUTEUS MINIMUS

Description:
The deepest of the three gluteal muscles. It acts simllarly with the gluteus medius.

Joint Crossings:
One

Rank/Body Action:

1 Hip-leg abduction
2 Hip-leg internal rotation (anterior fibers)
3 Hip-leg flexion (anterior fibers)

Blood Supply:
Superior gluteal artery

Insertion:
Superior and lateral surface of the greater trochanter of the femur

Nerves:
Superior gluteal nerve L4, 5, S1

Origin:
Lateral surface of ilium (between the anterior and inferior gluteal lines)

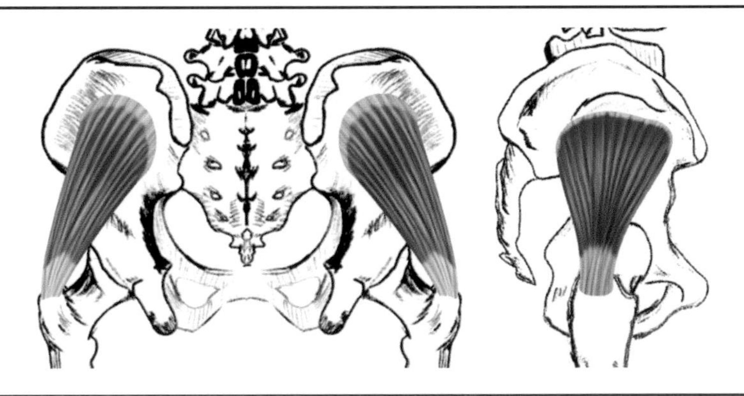

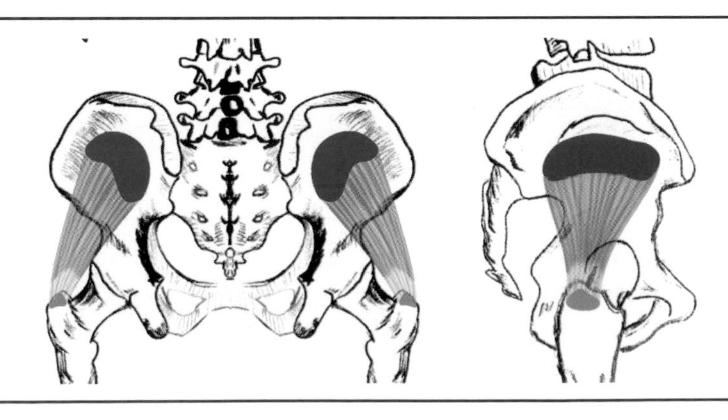

ILIACUS

Description:
A muscle of the iliac region of the abdomen that flexes and externally rotates the leg. Commonly grouped with the psoas major and minor to form a muscle group called the iliopsoas.

Joint Crossings:
One

Rank/Body Action:

1 Hip-leg flexion
2 Hip-leg external rotation
3 Hip-leg adduction

Blood Supply:
- Iliolumbar artery
- Deep circumflex iliac artery

Insertion:
Lesser trochanter of the femur and the shaft just below

Nerves:
- Lumbar nerve
- Femoral nerve L2, 3, 4

Origin:
Iliac fossa

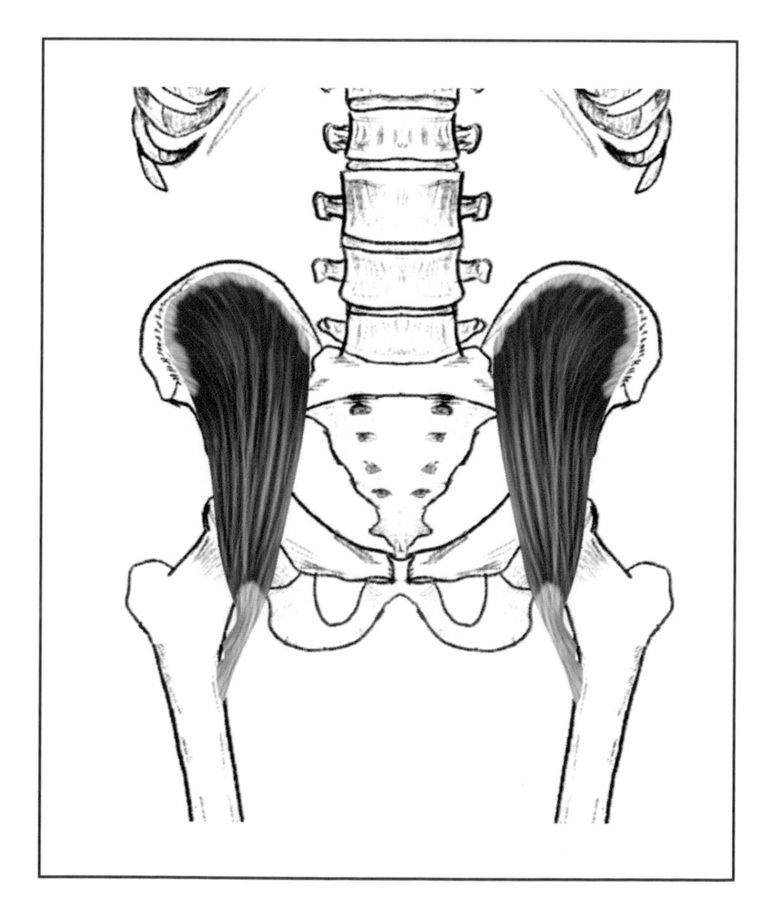

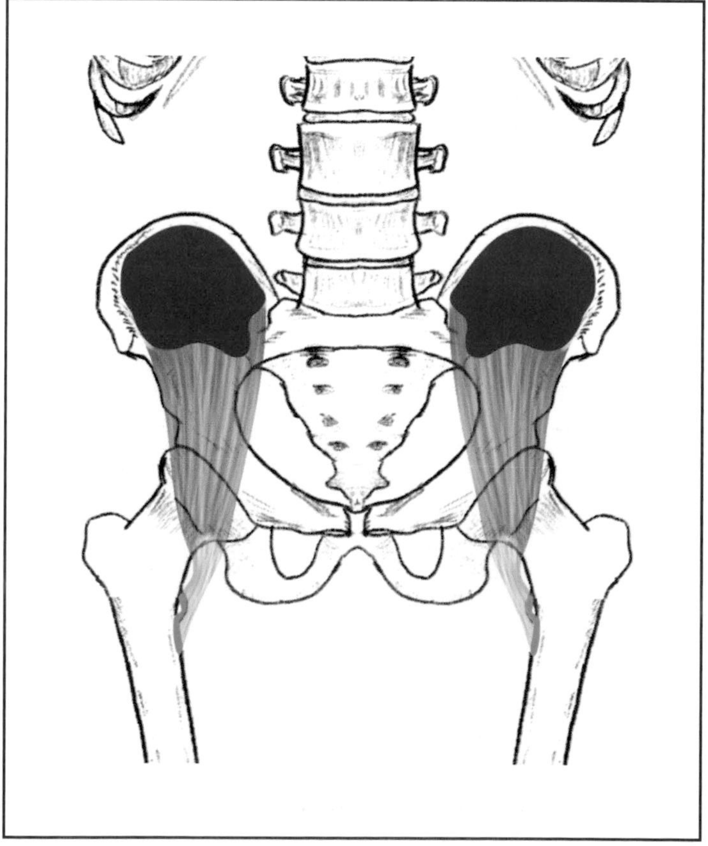

PECTINEUS

Description:
A flat, quadrangular muscle of the upper front and inner aspect of the thigh that flexes, adducts, and externally rotates the leg.

Joint Crossings:
One

Rank/Body Action:

1	Hip-leg flexion
2	Hip-leg adduction
3	Hip-leg external rotation

Blood Supply:
Muscular branches of medial femoral circumflex artery

Insertion:
Pectineal line of the femur

Nerves:
Femoral nerve L2, 3, 4

Origin:
Superior ramus of the pubis

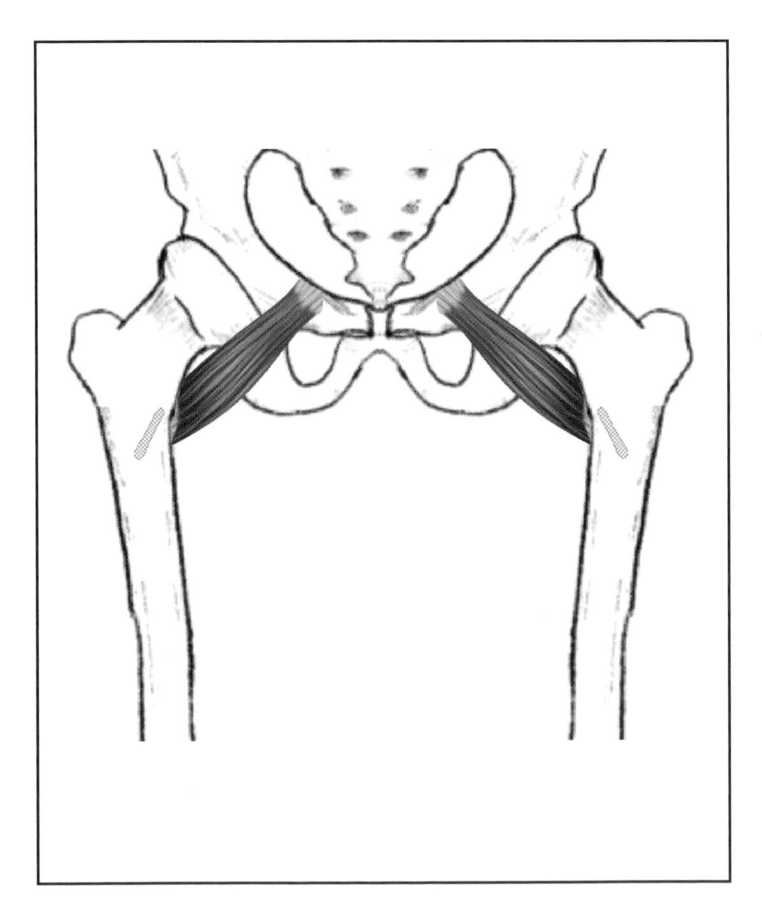

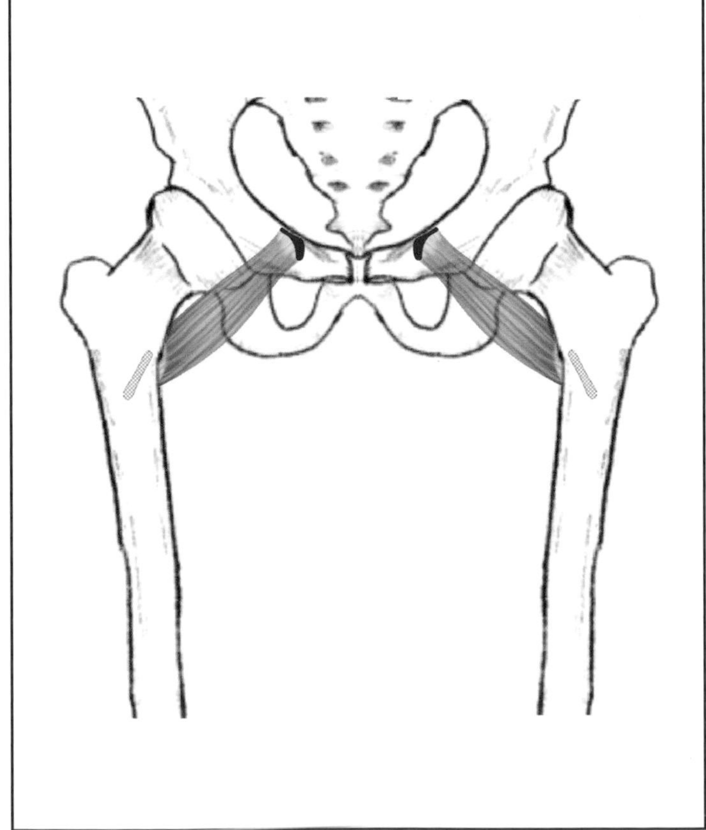

Psoas Major

Description:
The larger of the two psoas muscles that flexes the leg. Commonly grouped with the psoas minor and iliacus to form a muscle group called the iliopsoas.

Joint Crossings:
One

Rank/Body Action:

1	Hip-leg flexion
2	Hip-leg external rotation

Blood Supply:
- Iliolumbar artery
- Lumbar artery
- Subcostal artery

Insertion:
Lesser trochanter of the femur and the shaft just below

Nerves:
- Lumbar nerve
- Femoral nerve L2, 3, 4

Origin:
The transverse processes of the five lumbar vertebrae and the bodies and disks between them

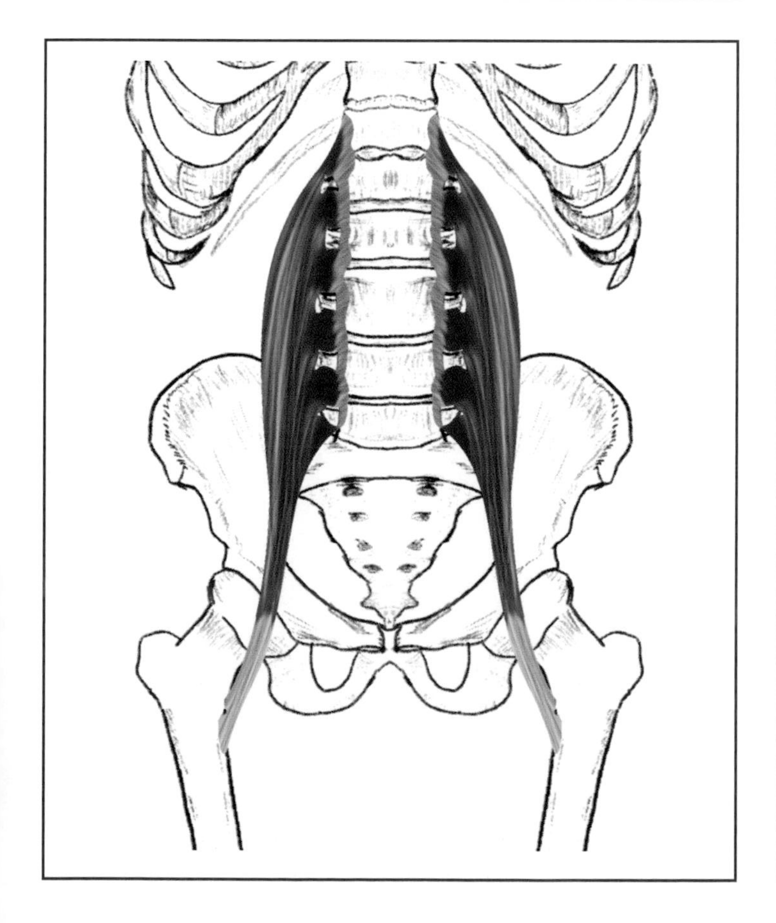

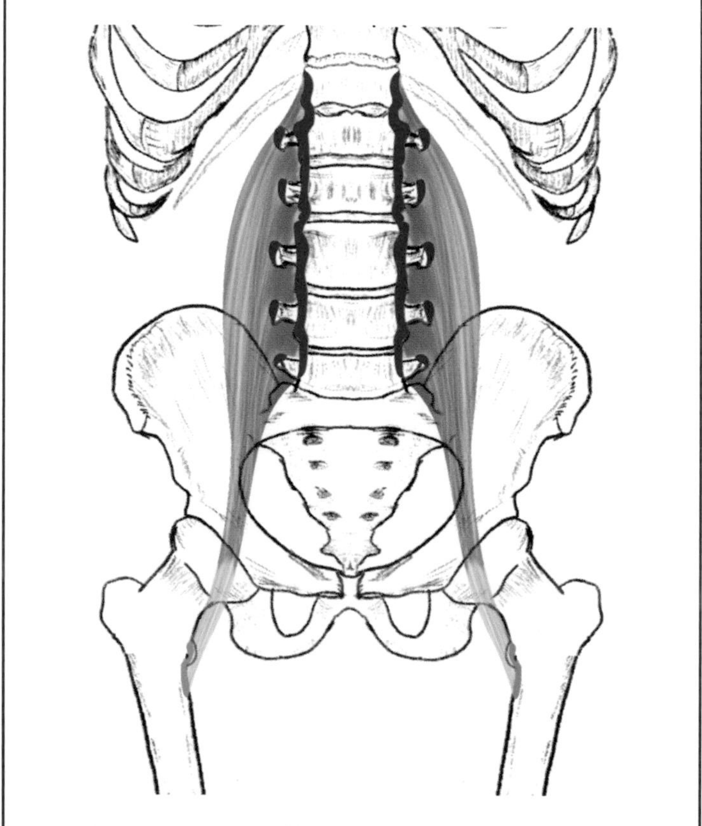

PSOAS MINOR

Description:
The smaller of the two psoas muscles that flexes the hip. It is often absent in some people. Commonly grouped with the psoas major and iliacus to form a muscle group called the iliopsoas.

Joint Crossings:
One

Rank/Body Action:
1 Hip-leg flexion
2 Hip-leg external rotation

Blood Supply:
- Iliolumbar artery
- Lumbar artery
- Subcostal artery

Insertion:
- Pectineal line
- Iliopectineal eminence
- Iliac fascia

Nerves:
- Lumbar nerve
- Femoral nerve L2, 3, 4

Origin:
The bodies of the 12th thoracic and first lumbar vertebrae

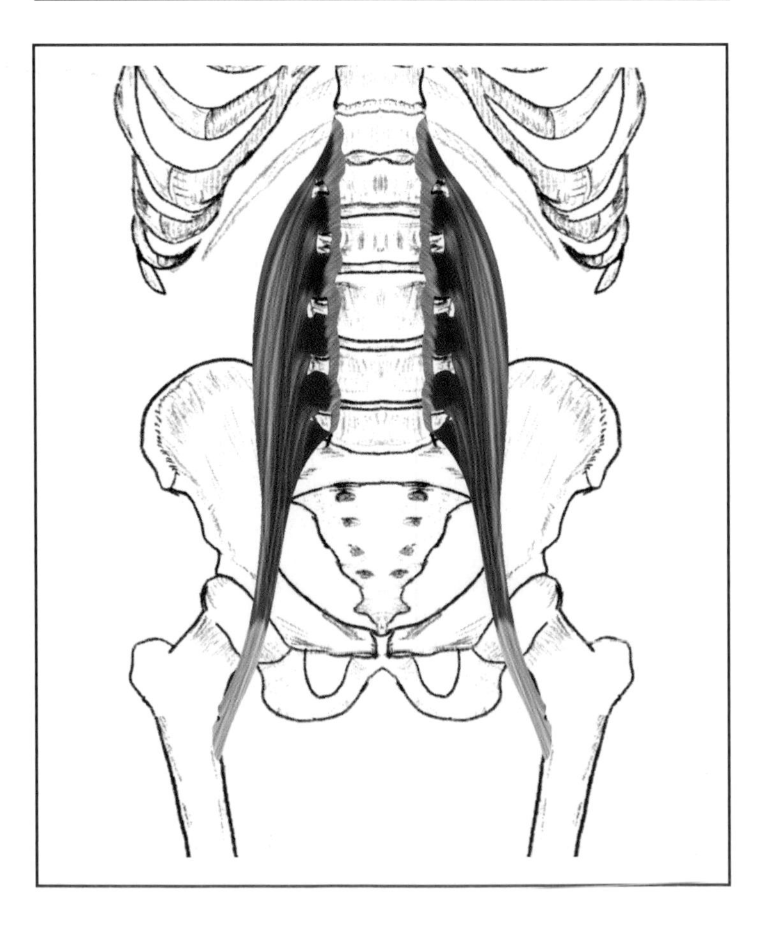

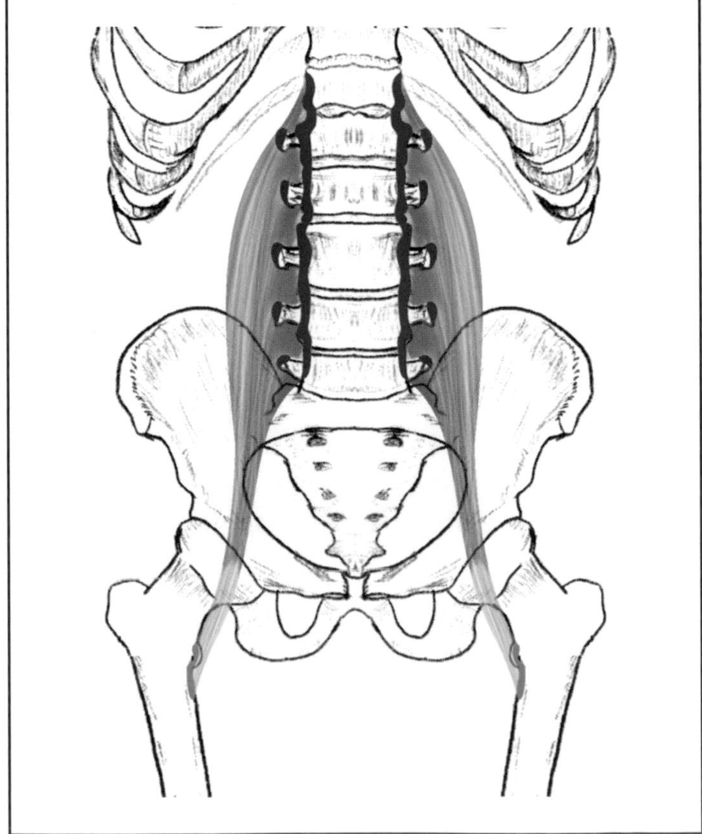

Tensor Fasciae Latae

Description:
A short muscle on the lateral aspect of the thigh that inserts on the ilio-tibial band and continues to the knee joint. It works with the gluteus medius and minimus to abduct the leg.

Joint Crossings:
One

Rank/Body Action:
1 Hip-leg abduction
2 Hip-leg flexion
3 Hip-leg internal rotation

Blood Supply:
- Superior gluteal artery
- Lateral femoral circumflex artery

Insertion:
Ilio-tibial band

Nerves:
Superior gluteal nerve L4, 5, S1

Origin:
Anterior, lateral surface of the ilium below the anterior crest and posterior to the anterior spine of the ilium

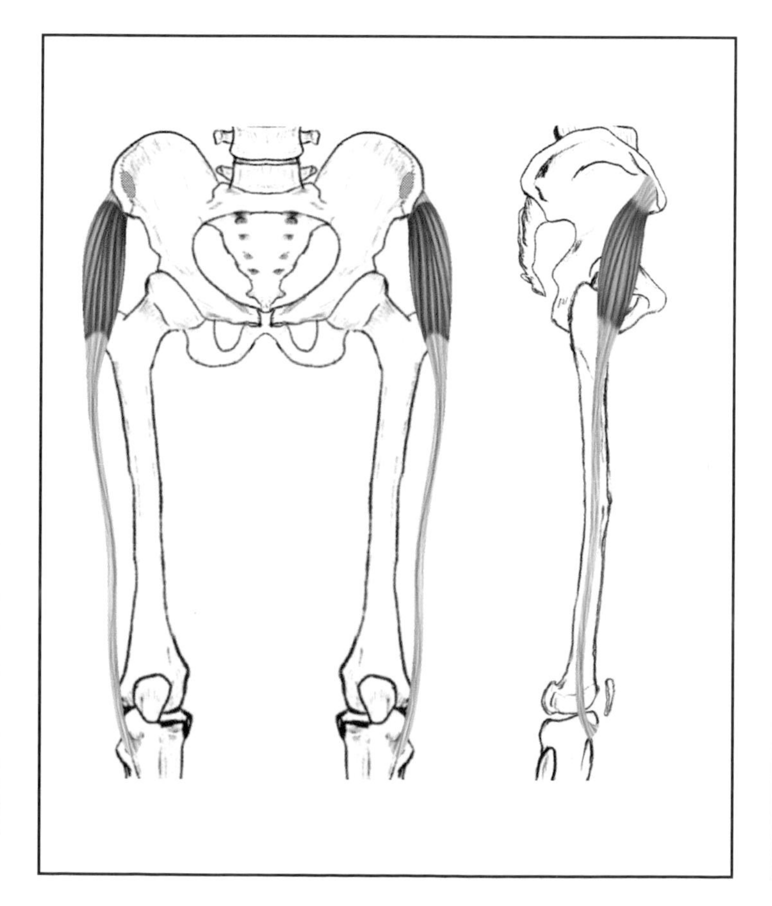

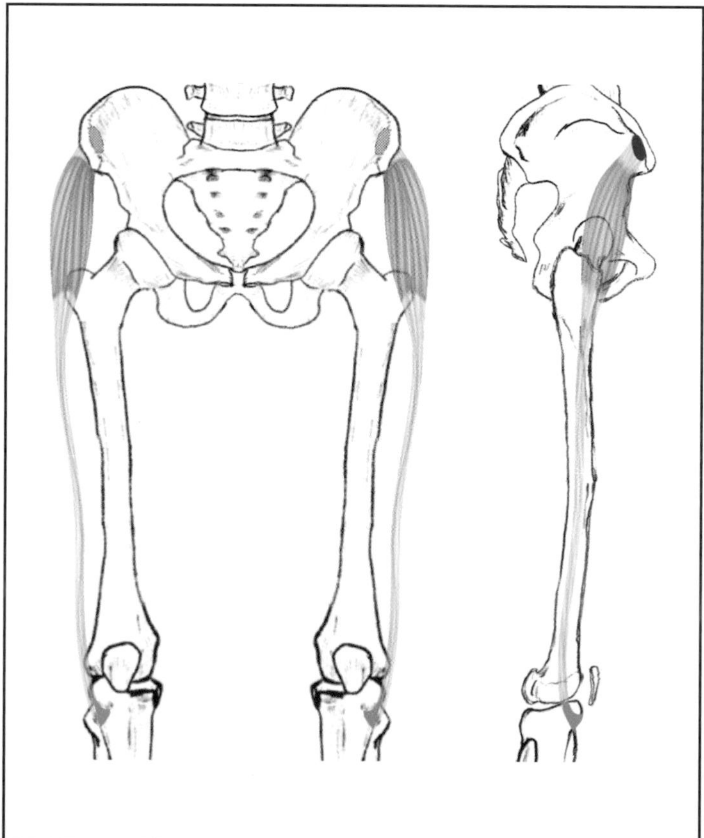

GRACILIS

Description:
The most superficial muscle of the inside thigh that acts to adduct the leg, flex the knee, and assist in rotating the leg internally.

Joint Crossings:
Two

Rank/Body Action:
1 Hip-leg adduction
2 Hip-leg flexion (weak)
3 Knee flexion
4 Hip-leg internal rotation (weak)

Blood Supply:
Obturator artery

Insertion:
Upper, anterior, medial surface of the tibia (with sartorius and semitendinous)

Nerves:
Obturator nerve L2, 3, 4

Origin:
Symphysis pubis and inferior (descending) ramus of the pubis

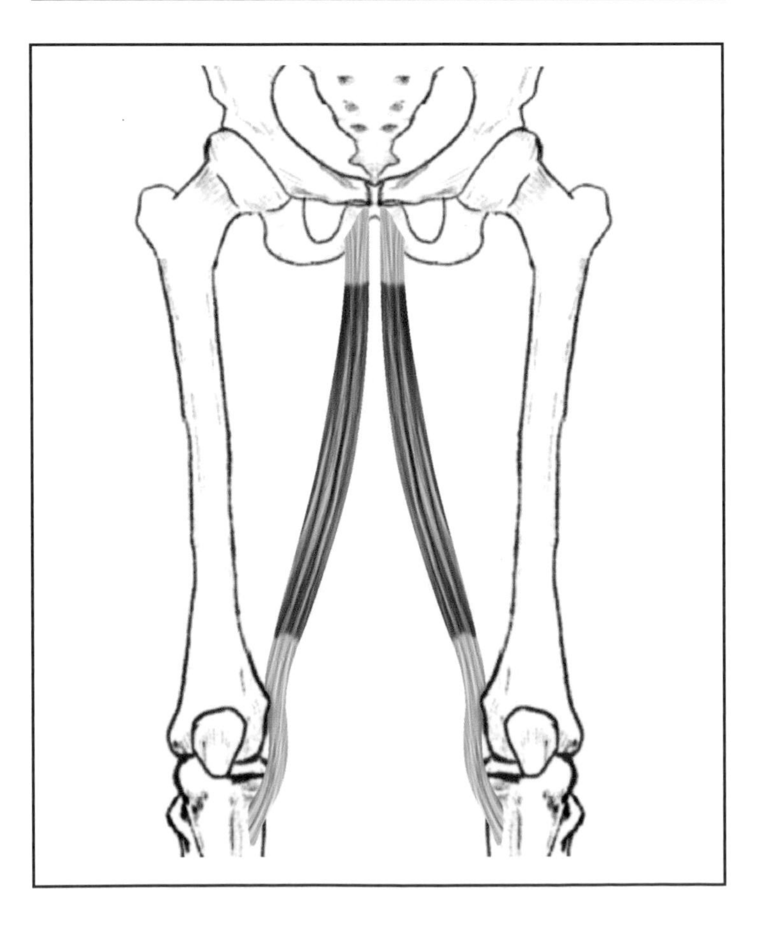

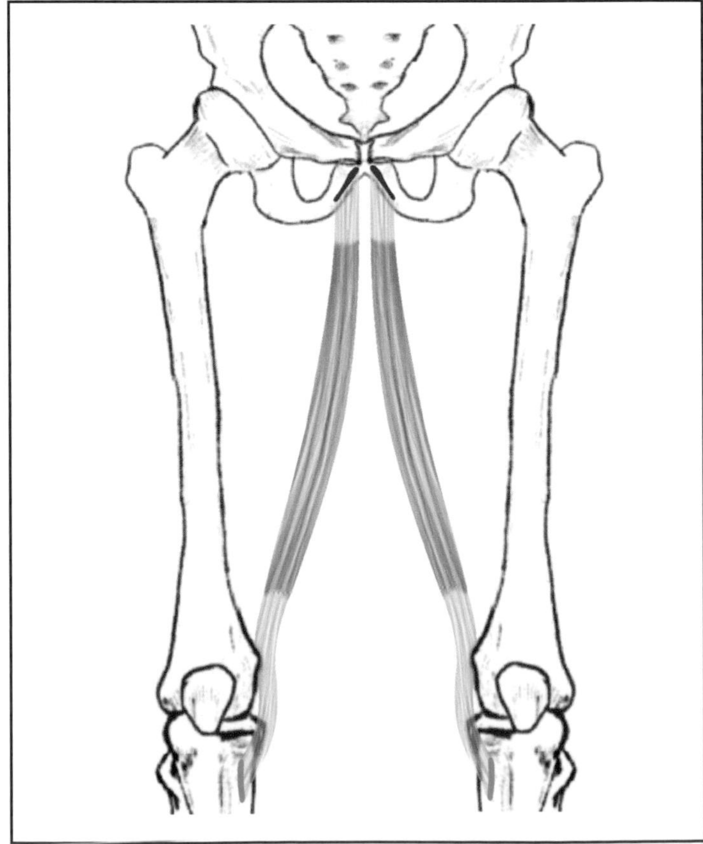

SARTORIUS

Description:
The longest muscle in the body, it acts to flex, abduct, and externally rotate the leg and to flex the knee. Often called the tailor's muscle because early tailors sat cross-legged, which the action of the sartorius enabled them to do.

Joint Crossings:
Two

Rank/Body Action:
1 Hip-leg flexion
2 Knee flexion
3 Hip-leg external rotation
4 Hip-leg abduction

Blood Supply:
- Muscular branches of profunda femoris artery
- Branch of descending genicular artery

Insertion:
Upper, anterior, medial surface of the tibia

Nerves:
Femoral nerve L2, 3

Origin:
Anterior superior iliac spine

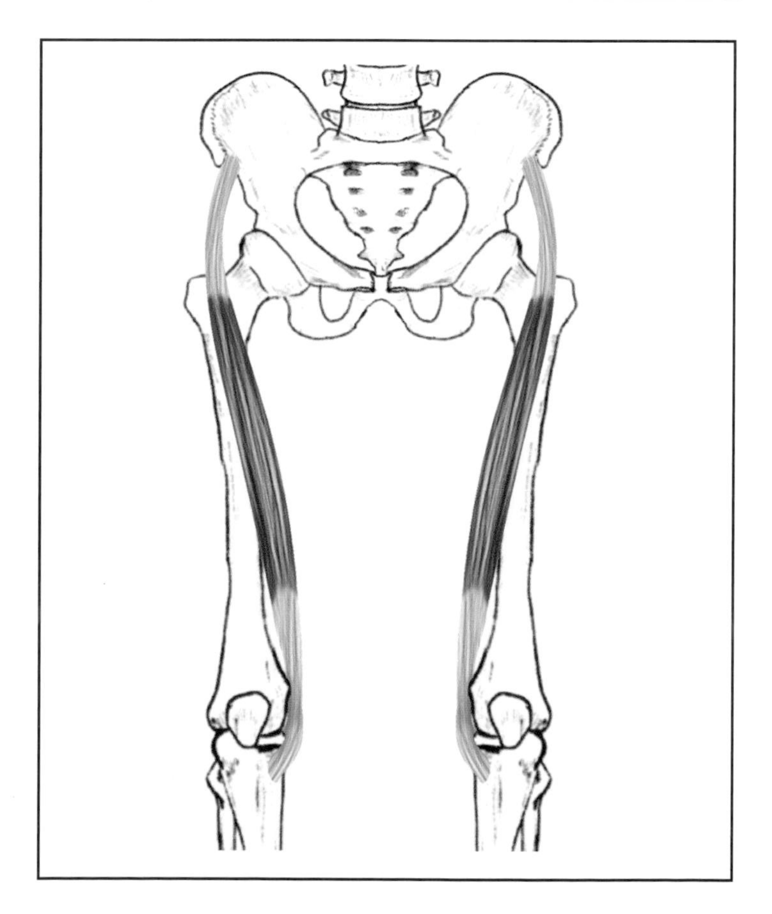

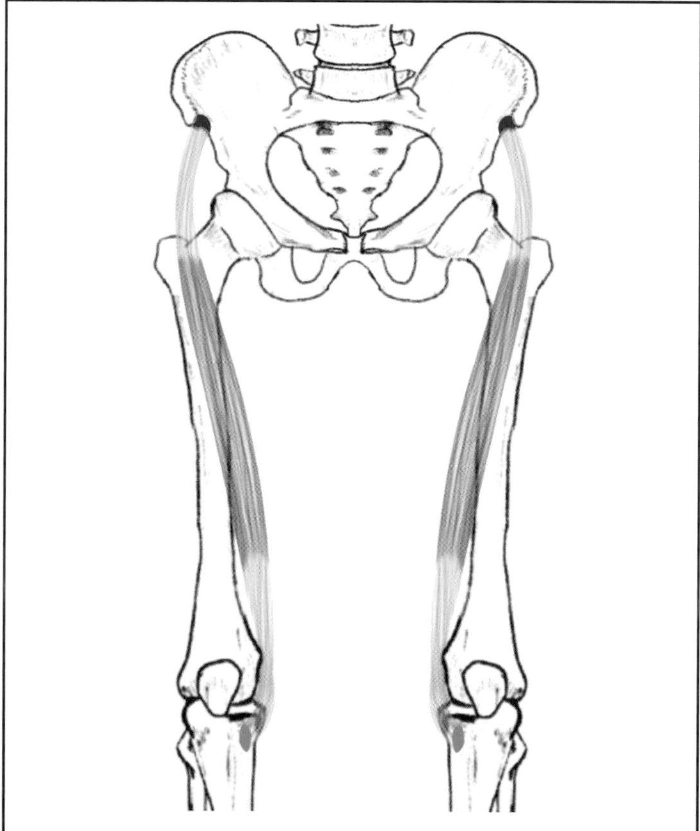

Gemellus Inferior

Description:
One of the small muscles of the hip that holds the femur into the acetabulum and helps to externally rotate the leg.

Joint Crossings:
One

Body Action:
Hip-leg external rotation (holds the femur into the acetabulum)

Blood Supply:
Inferior gluteal artery

Insertion:
Inside surface of the greater trochanter

Nerves:
Branches from sacral plexus L4, 5, S1, 2

Origin:
Tuberosity of the ischium

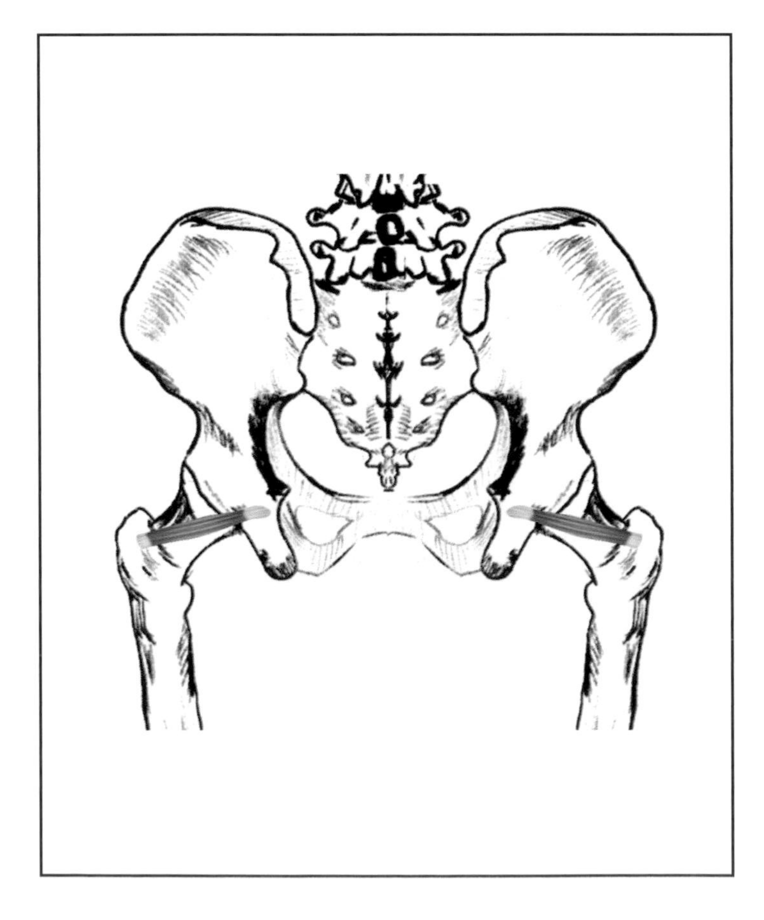

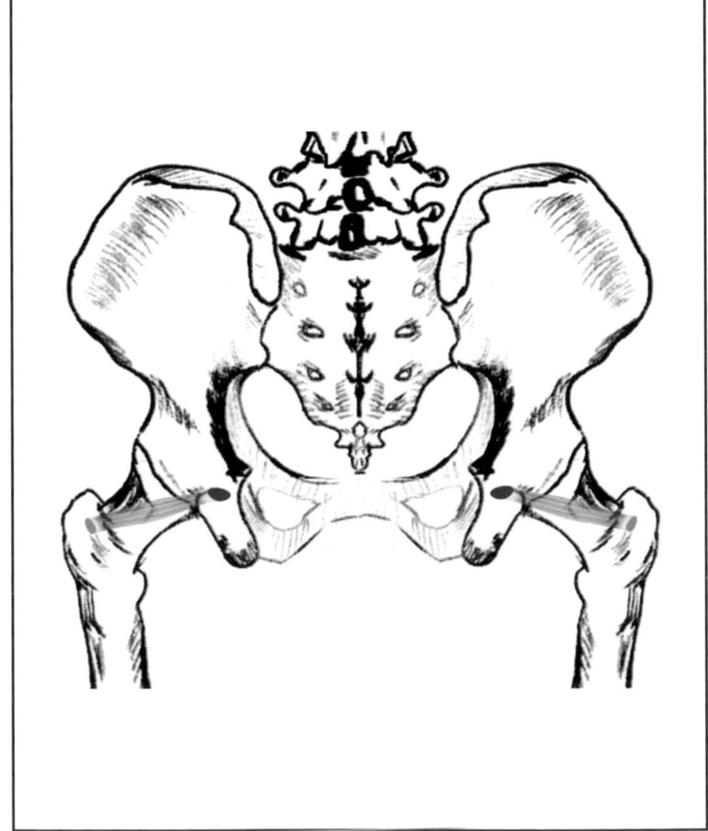

Gemellus Superior

Description:
One of the small muscles of the hip that holds the femur into the acetabulum and helps to externally rotate the leg.

Joint Crossings:
One

Body Action:
Hip-leg external rotation (holds the femur into the acetabulum)

Blood Supply:
Inferior gluteal artery

Insertion:
Generally to the inside surface of the greater trochanter

Nerves:
Sacral nerve L5, S1, 2

Origin:
Outer surface of the spine of the ischium

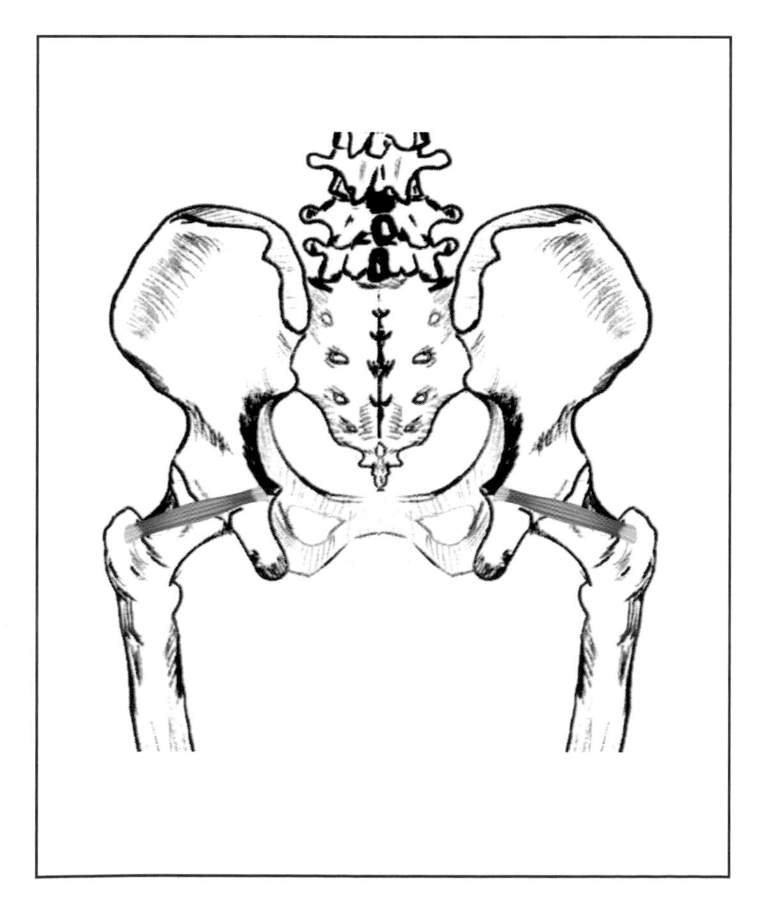

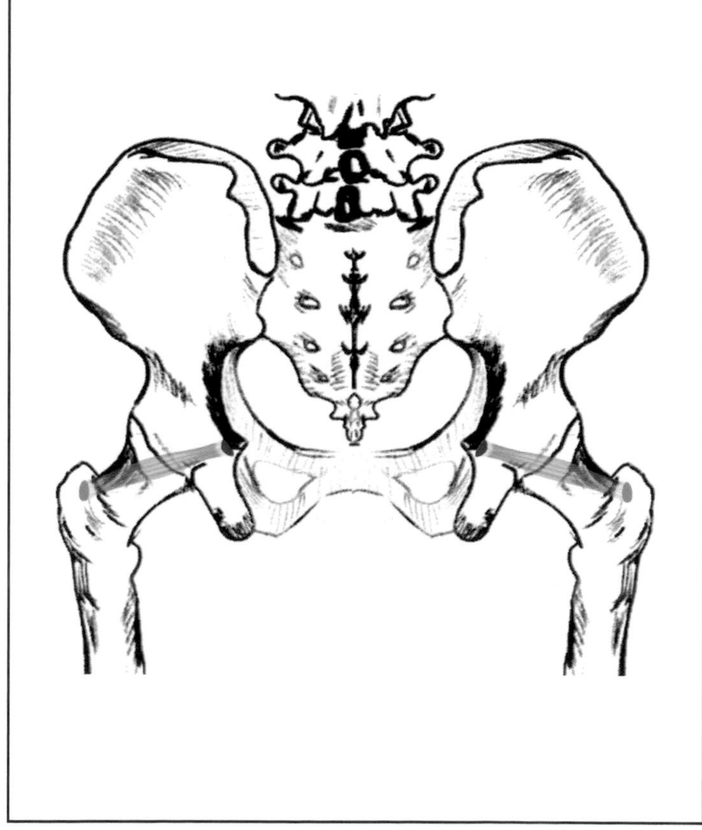

Obturator Externus

Description:
One of the small muscles of the hip that holds the femur into the acetabulum and helps to externally rotate the leg.

Joint Crossings:
One

Body Action:
Hip-leg external rotation (holds the femur into the acetabulum)

Blood Supply:
Obturator artery

Insertion:
Generally to the inside surface of the greater trochanter

Nerves:
Obturator nerve L3, 4

Origin:
Generally from the sacrum and inside surfaces of the pubis and ischium

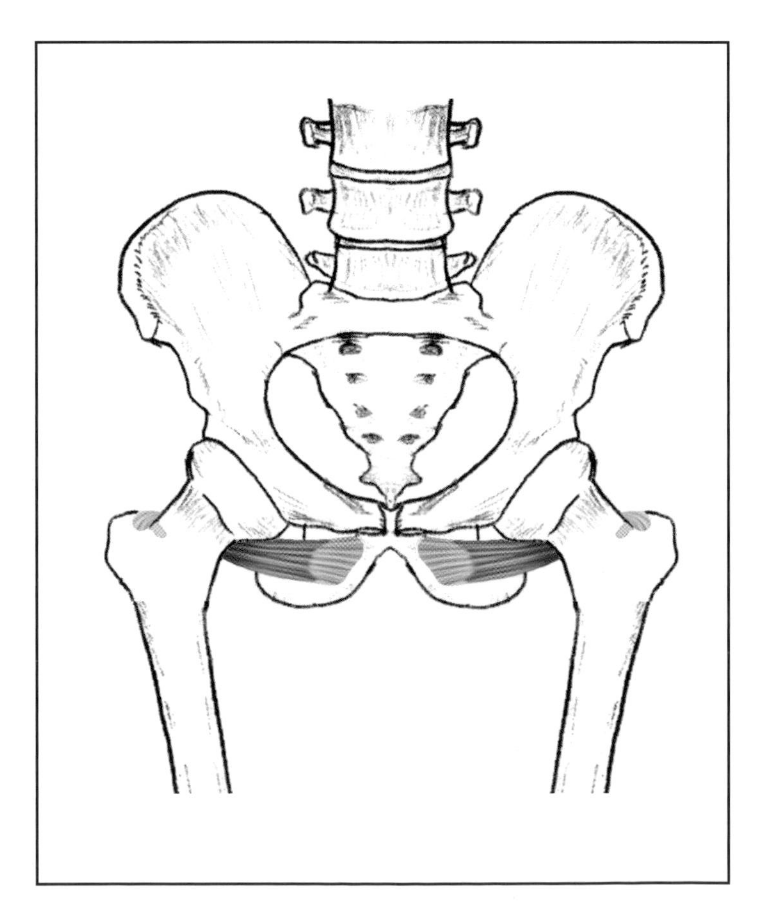

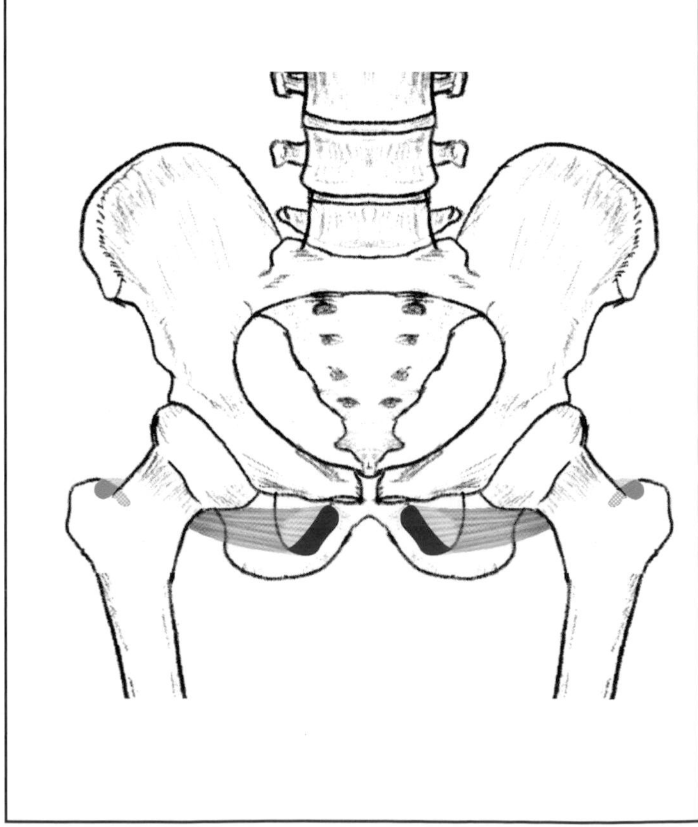

OBTURATOR INTERNUS

Description:
One of the small muscles of the hip that holds the femur into the acetabulum and helps to externally rotate the leg.

Joint Crossings:
One

Body Action:
Hip-leg external rotation (holds the femur into the acetabulum)

Blood Supply:
Inferior gluteal artery

Insertion:
Generally to the inside surface of the greater trochanter

Nerves:
Branches from sacral plexus L4, 5, S1, 2

Origin:
Generally from the sacrum and inside surfaces of the pubis and ischium

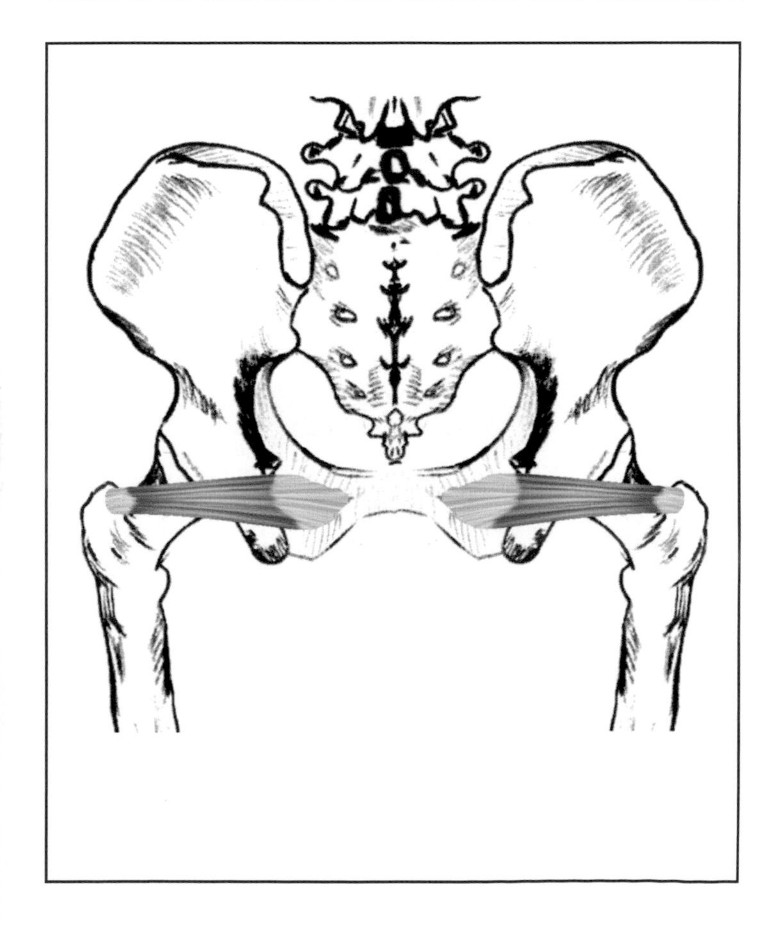

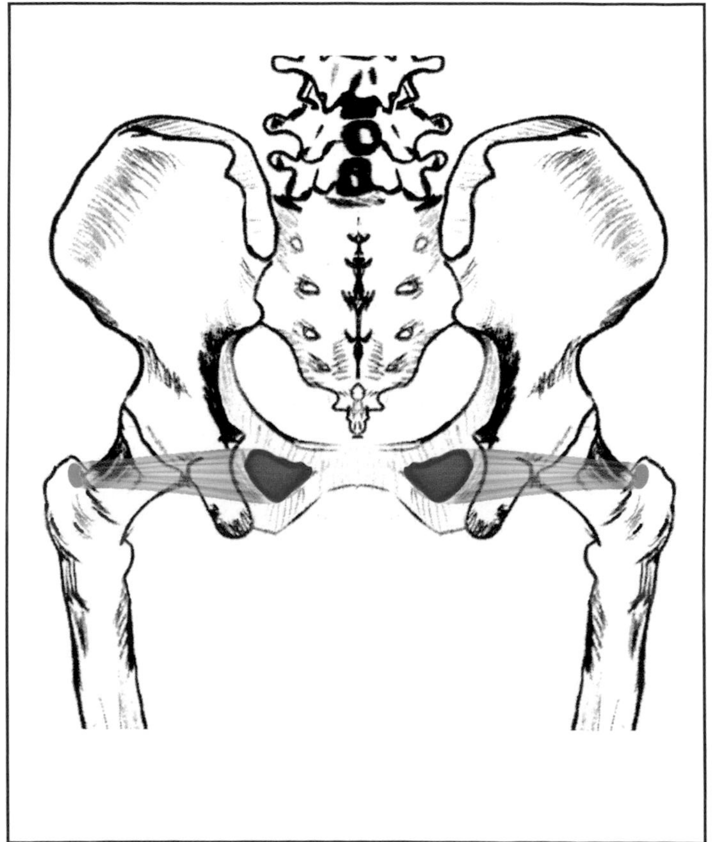

Piriformis

Description:
One of the small muscles of the hip that holds the femur into the acetabulum and helps to externally rotate the leg.

Blood Supply:
- Superior gluteal artery
- Inferior gluteal artery

Insertion:
Generally to the inside surface of the greater trochanter

Joint Crossings:
One

Nerves:
First or second sacral nerve S1, 2

Body Action:
Hip-leg external rotation (holds the femur into the acetabulum)

Origin:
Generally from the sacrum and inside surfaces of the pubis and ischium

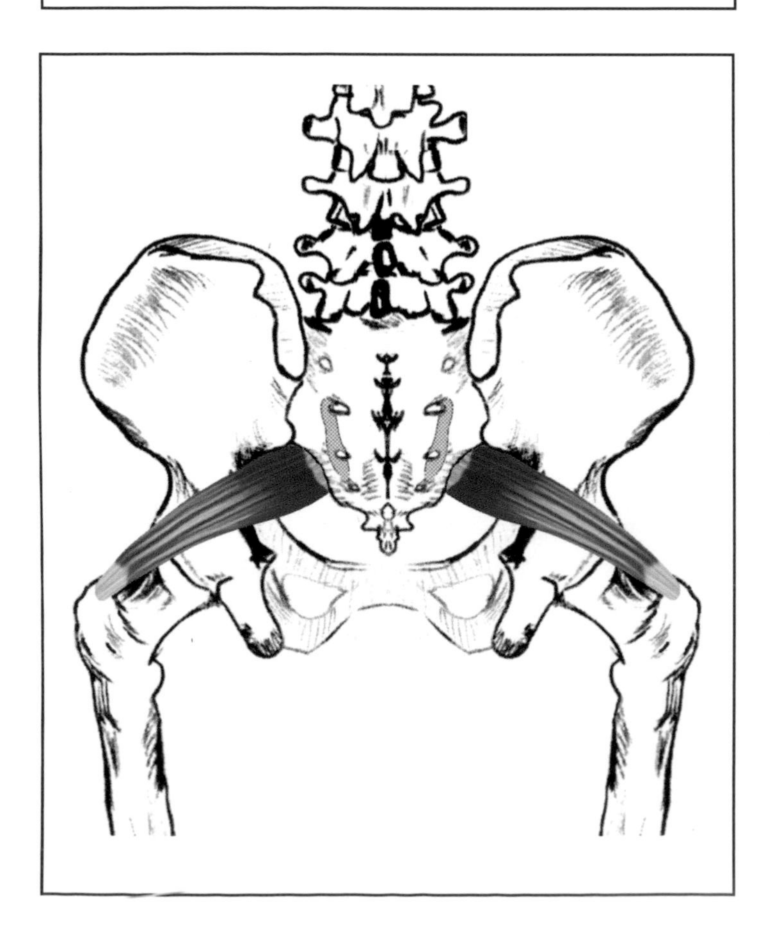

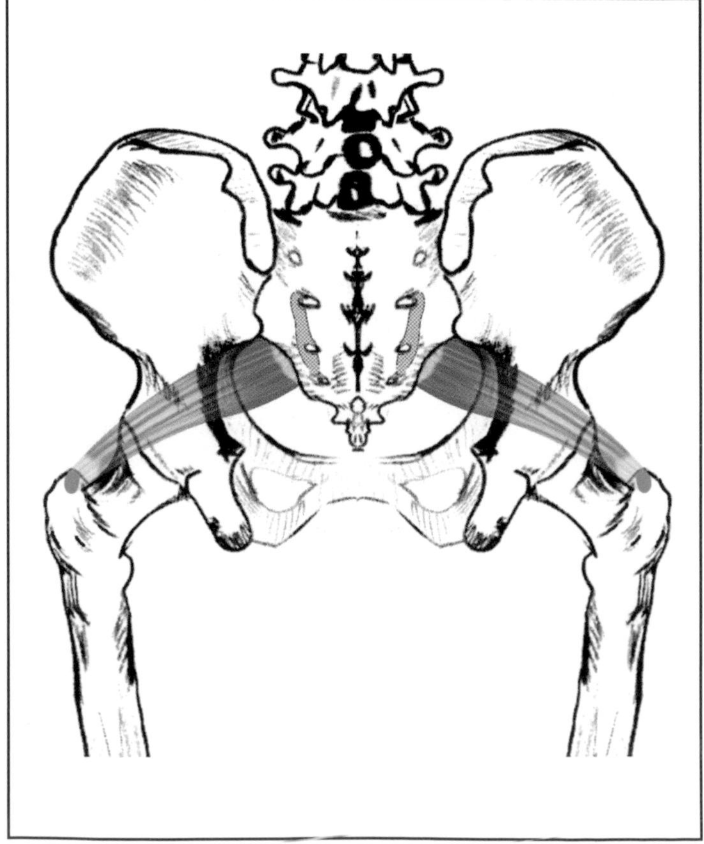

Quadratus Femoris

Description:
One of the small muscles of the hip that holds the femur into the acetabulum and helps to externally rotate the leg.

Joint Crossings:
One

Body Action:
Hip-leg external rotation (holds the femur into the acetabulum)

Blood Supply:
Inferior gluteal artery

Insertion:
Proximal part of the line that extends from the intertrochanteric crest of the femur

Nerves:
Branches from sacral plexus L4, 5, S1

Origin:
Generally from the proximal part of the external border of the tuberosity of the ischium

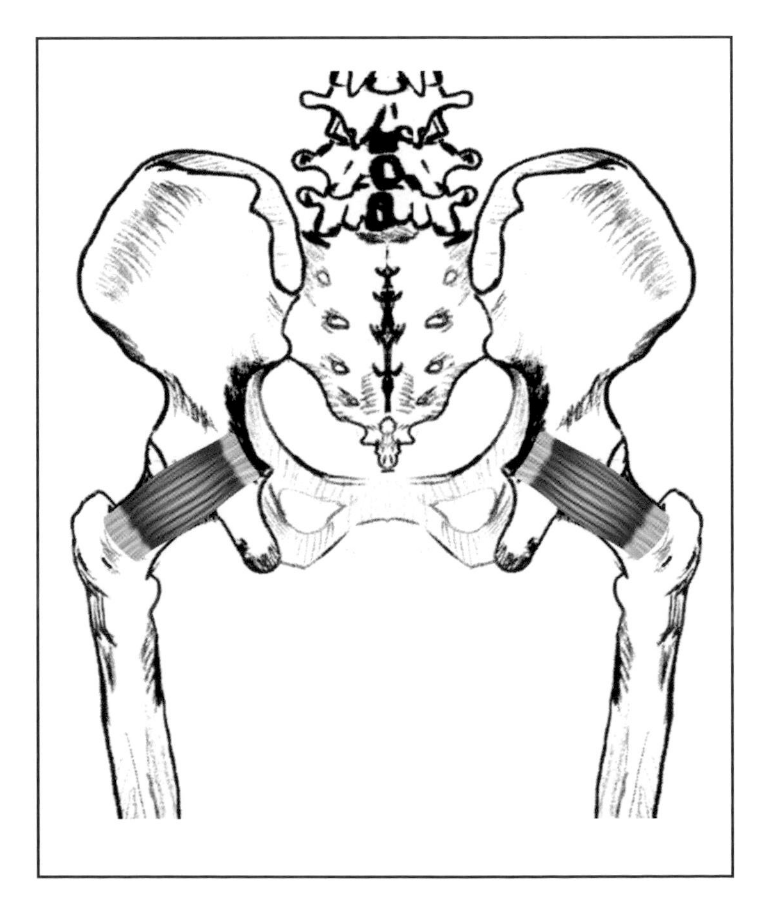

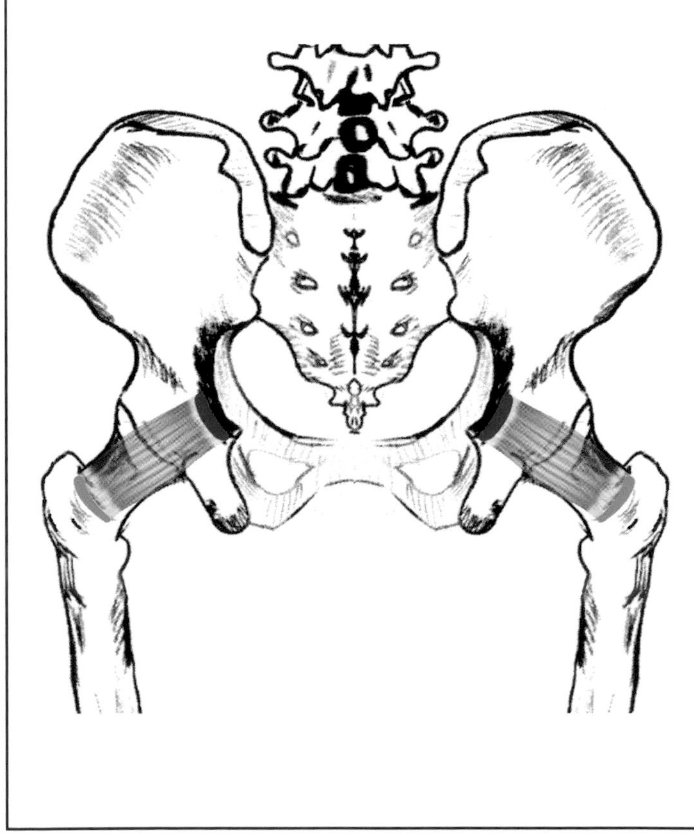

Popliteus

Description:
A flat thin triangular muscle that forms part of the floor of the popliteal space, and functions to flex the knee. It plays a large role in providing postero-lateral stability to the knee.

Joint Crossings:
One

Body Action:
Knee flexion

Blood Supply:
Branches of popliteal artery

Insertion:
Popliteal surface of the tibia

Nerves:
Tibial nerve L5, S1

Origin:
Posterior surface of the lateral epicondyle of the femur

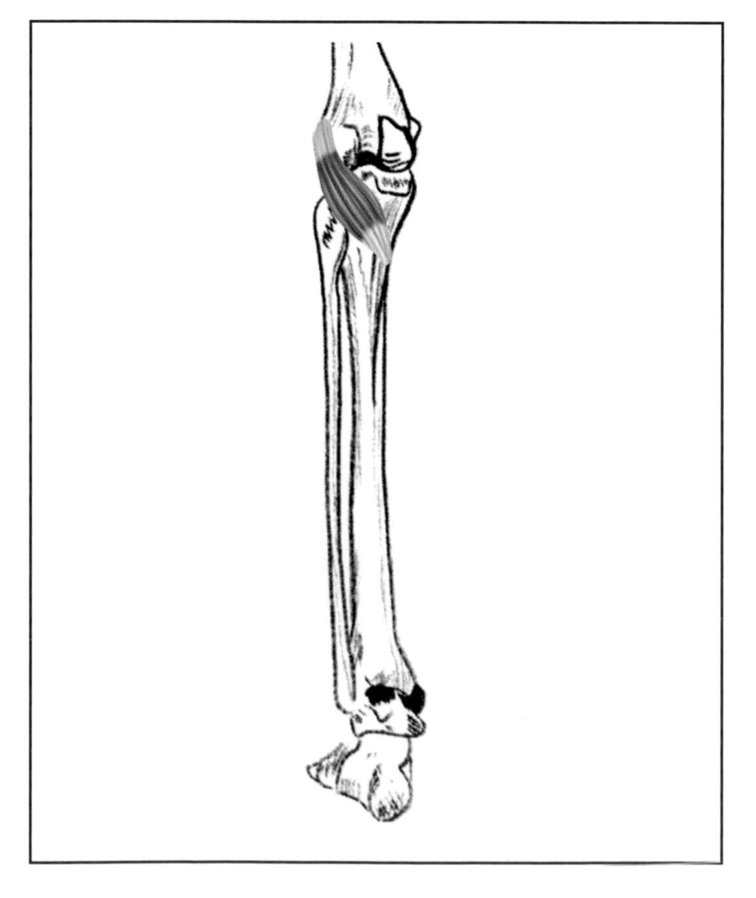

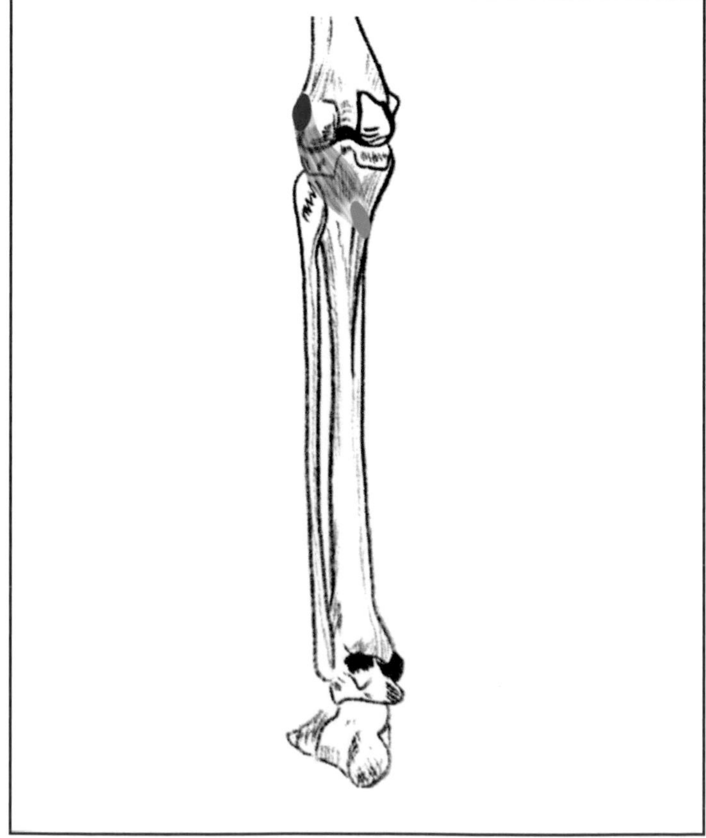

Biceps Femoris (Hamstrings)

Description:
A large superficial muscle of the upper, posterior thigh. One of the hamstring group of muscles, it is a strong hip extensor and knee flexor.

Joint Crossings:
Two

Rank/Body Action:

Rank	Body Action
1	Hip-leg extension
2	Hip-leg hyperextension
3	Knee flexion
4	Hip-leg external rotation
5	Hip-leg adduction (slight)

Blood Supply:
- Perforating branches of profunda femoris artery
- Inferior gluteal artery (upper portion)

Insertion:
- Head of the fibula
- Adjacent (lateral) surface of the tibia

Nerves:
- Long head: sciatic-tibial nerve S1, 2, 3
- Short head: sciatic-peroneal nerve L5, S1, 2

Origin:
- Long head: ischial tuberosity of the innominate bone
- Short head: lower half of lateral lip of the linea aspera on the posterior femur

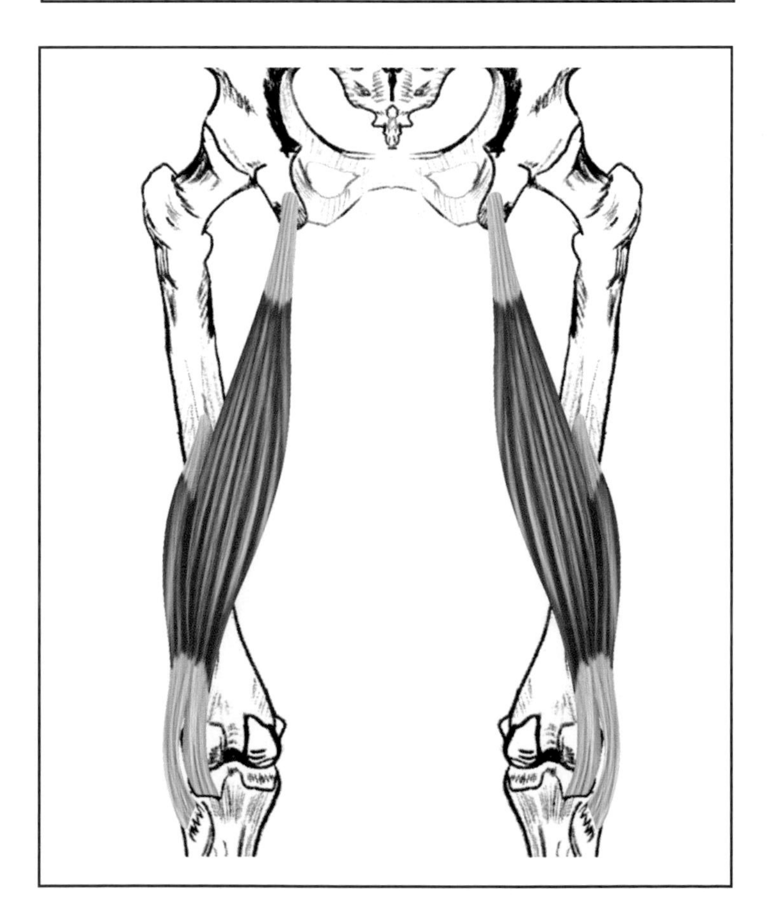

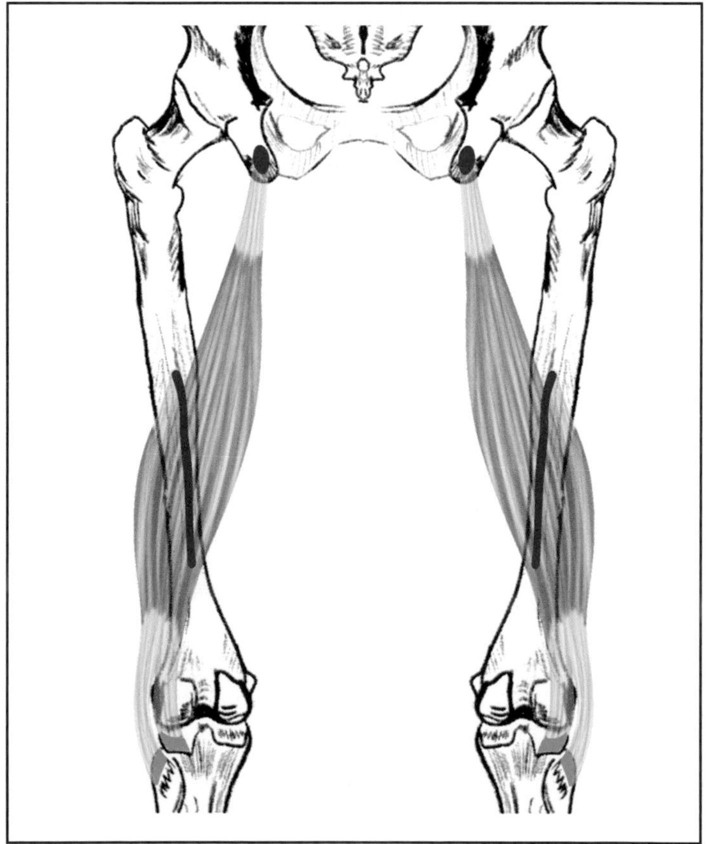

SEMIMEMBRANOSUS (HAMSTRINGS)

Description:
A long, thin muscle on the medial, upper, posterior thigh that's one of the group of hamstring muscles. It crosses the hip and knee joints.

Joint Crossings:
Two

Rank/Body Action:

1	Hip-leg extension
2	Hip-leg hyperextension
3	Knee flexion
4	Hip-leg adduction
5	Hip-leg internal rotation

Blood Supply:
- Perforating branches of profunda femoris artery
- Inferior gluteal artery (upper portion)

Insertion:
Posterior, upper, and medial surface of the condyle of the tibia

Nerves:
Tibial nerve of sciatic bundle L5, S1, 2

Origin:
Ischial tuberosity of the innominate bone

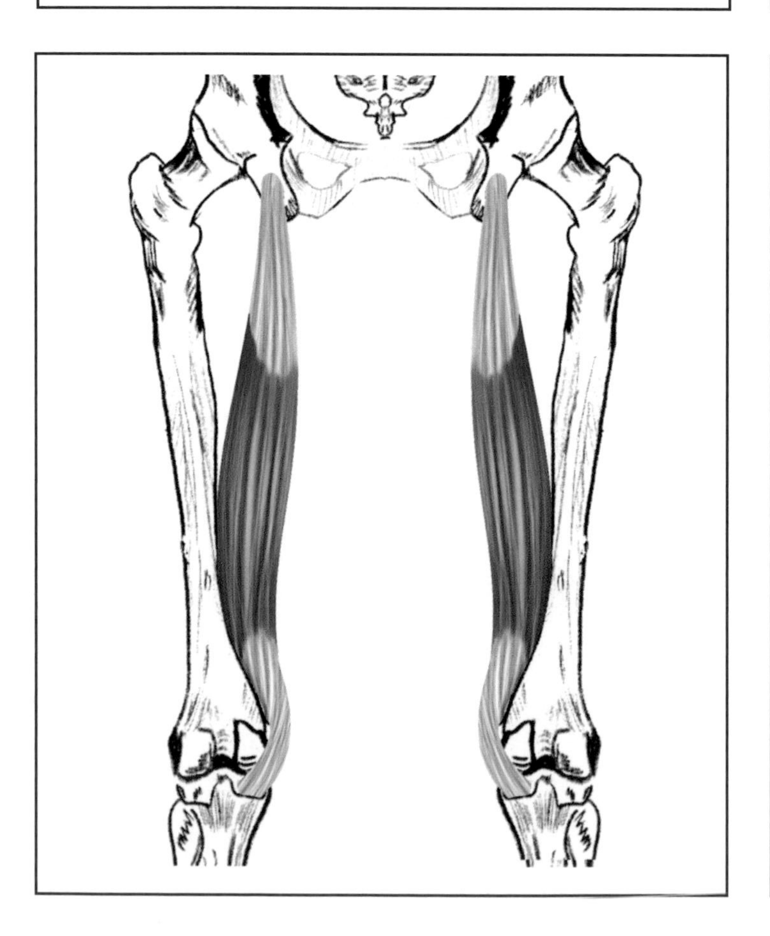

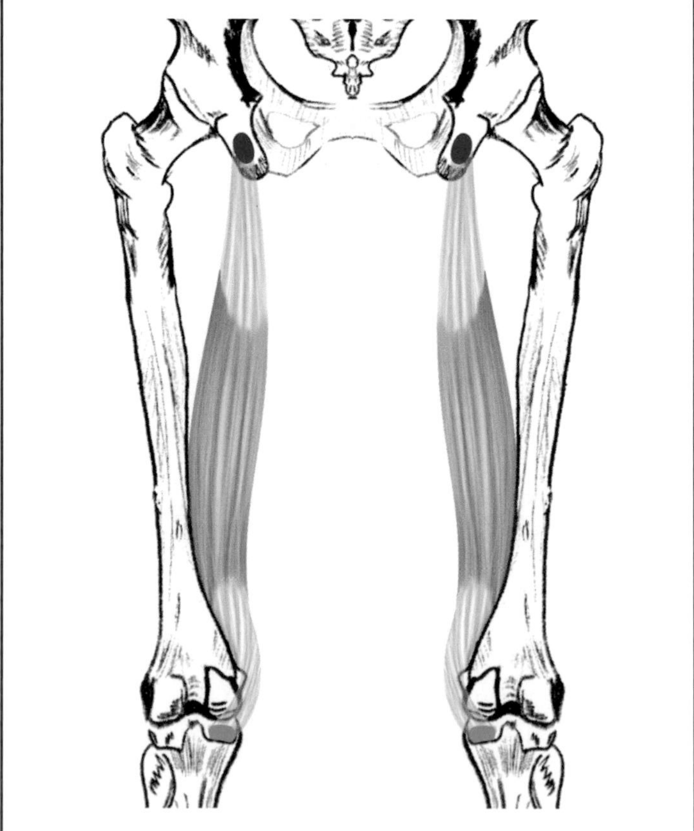

SEMITENDINOSUS (HAMSTRINGS)

Description:
A long band-like muscle which is part of the hamstring muscle group. The distal end of this muscle is held posteriorly by the ligaments around the knee joint, just prior to it being inserted anteriorly on the tibia. This factor allows flexion of the knee.

Joint Crossings:
Two

Rank/Body Action:

1	Hip-leg extension
2	Hip-leg hyperextension
3	Knee flexion
4	Hip-leg adduction
5	Hip-leg internal rotation

Blood Supply:
- Perforating branches of profunda femoris artery
- Inferior gluteal artery (upper portion)

Insertion:
Upper, anterior medial surface of the tibia

Nerves:
Tibial nerve of sciatic bundle L5, S1, 2

Origin:
Ischial tuberosity of the innominate bone

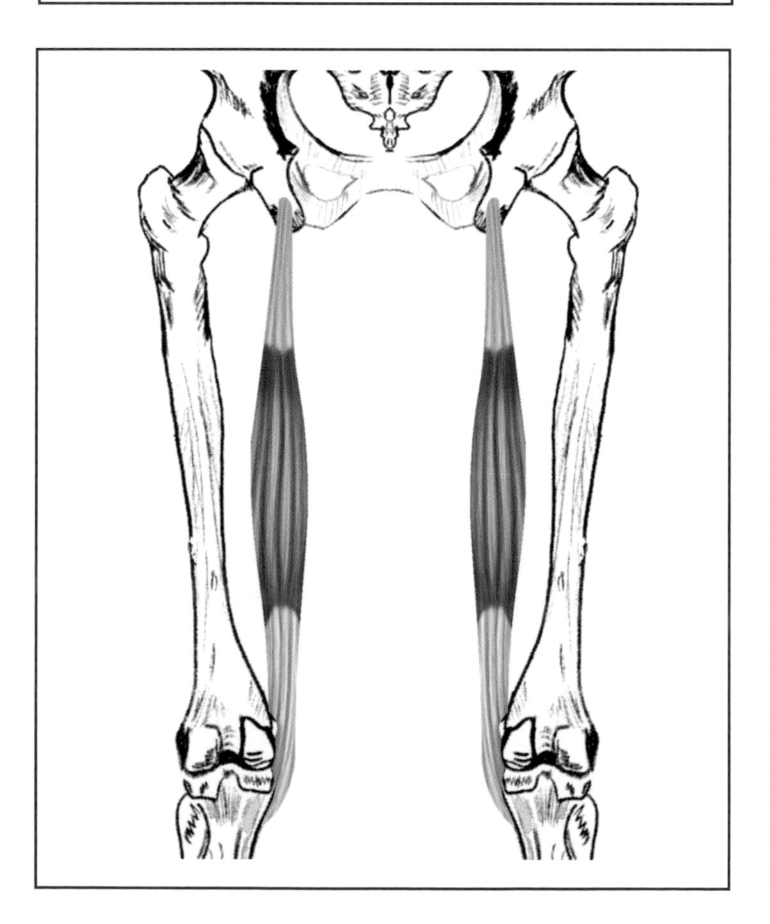

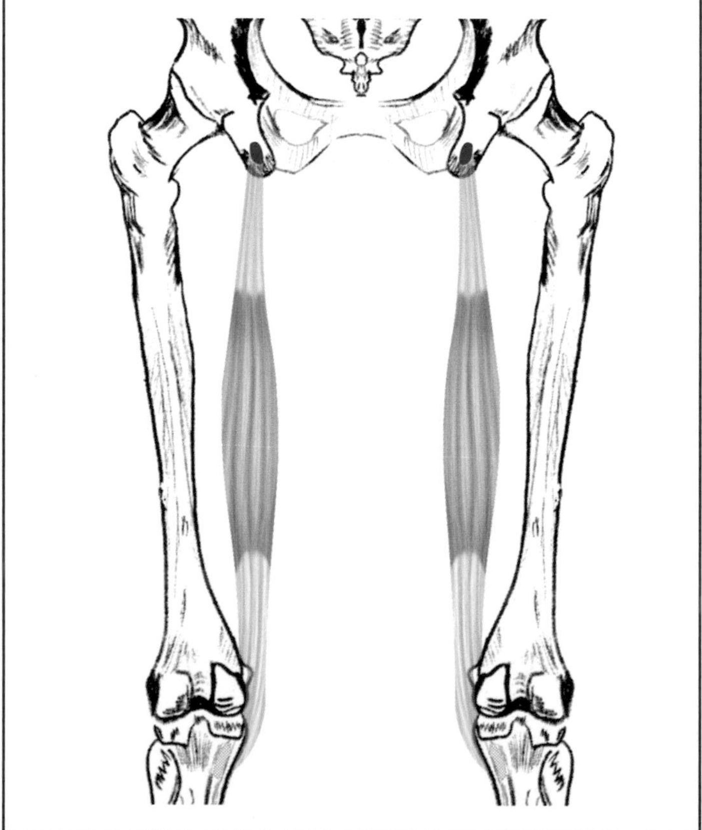

Rectus Femoris (Quads)

Description:
The most superficial muscle of the quadriceps group. Strongly extends the knee with the other three quadriceps muscles. Since it crosses the anterior hip joint, it also flexes the hip.

Joint Crossings:
Two

Rank/Body Action:
1 Knee extension
2 Hip-leg flexion

Blood Supply:
Lateral femoral circumflex artery

Insertion:
Upper borders of the patella by way of the quadriceps tendon. The quadriceps tendon passes over and around the patella and becomes the patellar ligament, which inserts into the tibial tuberosity

Nerves:
Femoral nerve L2, 3, 4

Origin:
- Straight head - anterior inferior iliac spine
- Reflected head - upper lip of the acetabulum

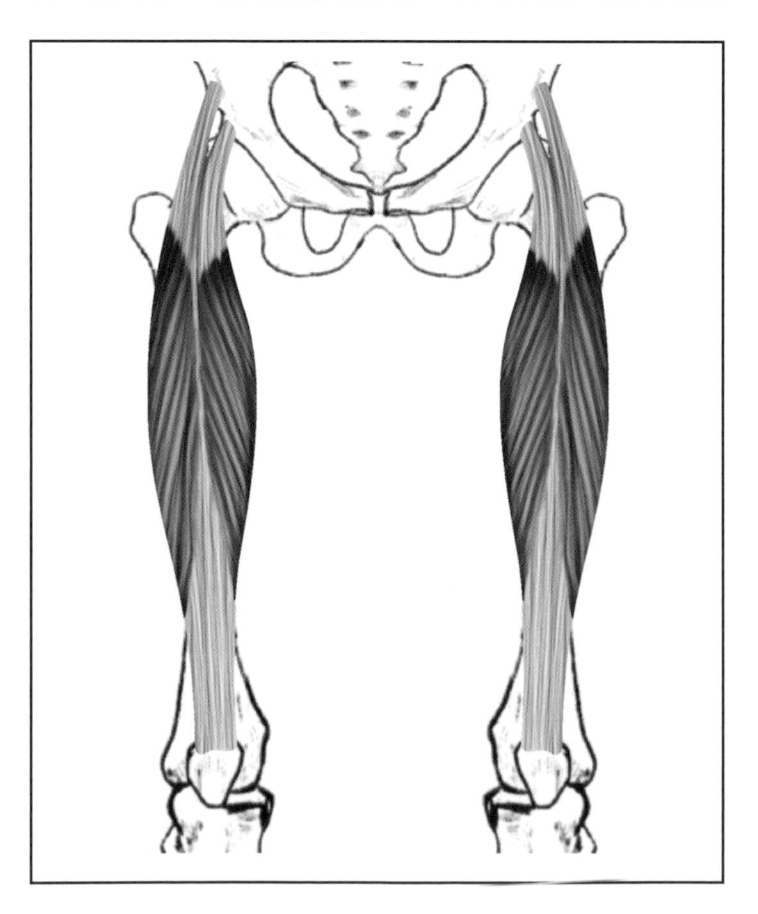

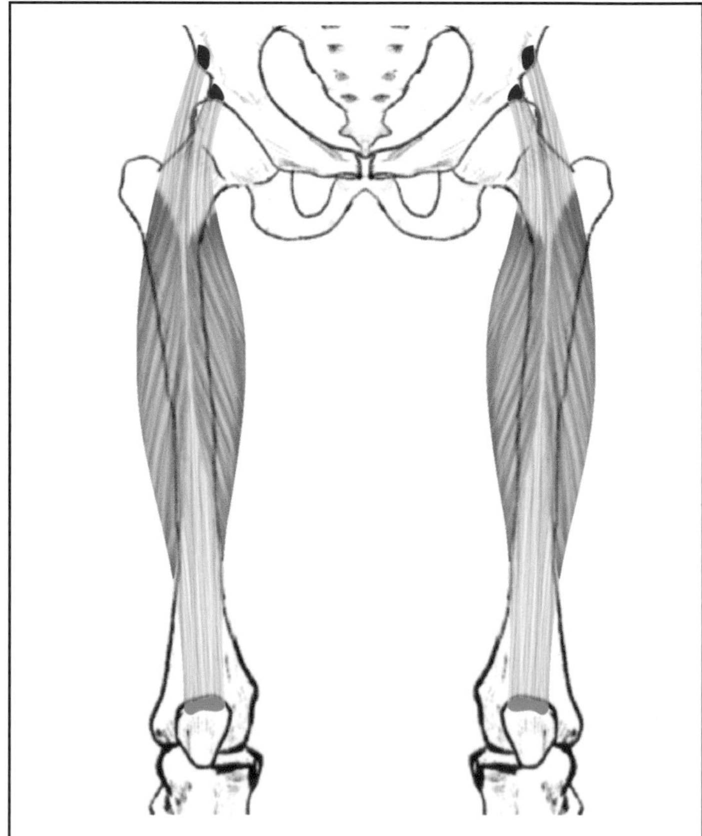

VASTUS INTERMEDIUS (QUADS)

Description:
One of the quadriceps muscle group lying under the rectus femoris. With the other three muscles, it strongly extends the knee. Also known as the crureus.

Joint Crossings:
One

Body Action:
Knee extension

Blood Supply:
Lateral femoral circumflex artery

Insertion:
- Upper borders of the patella by way of the quadriceps tendon
- The quadriceps tendon passes over and around the patella and becomes the patellar ligament, which inserts into the tibial tuberosity

Nerves:
Femoral nerve L2, 3, 4

Origin:
Upper two-thirds of the anterior and lateral surface of the femur

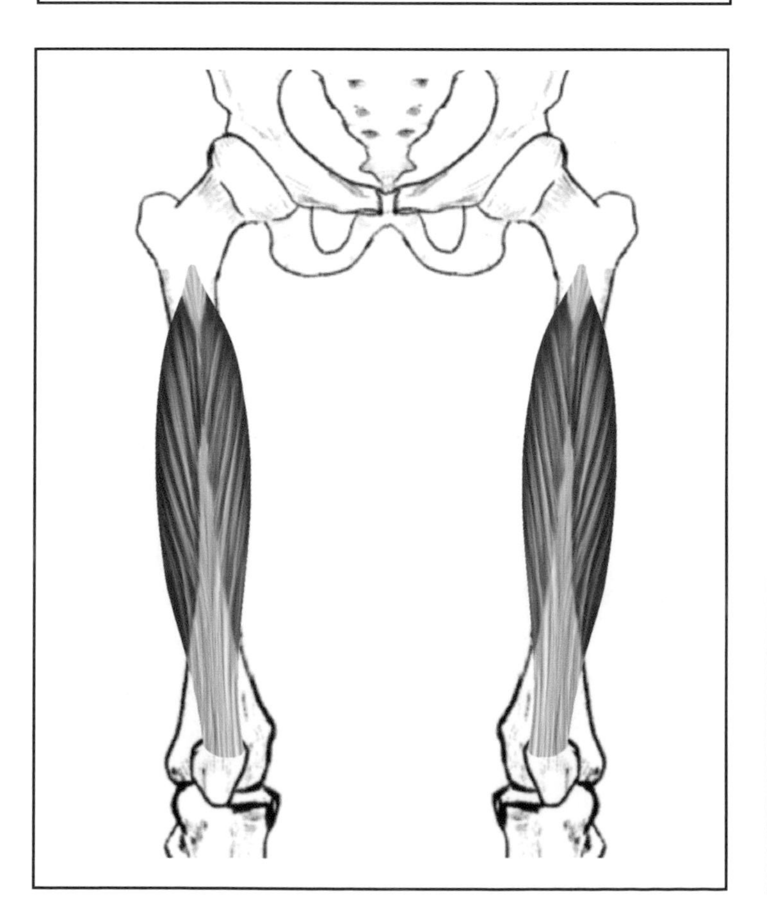

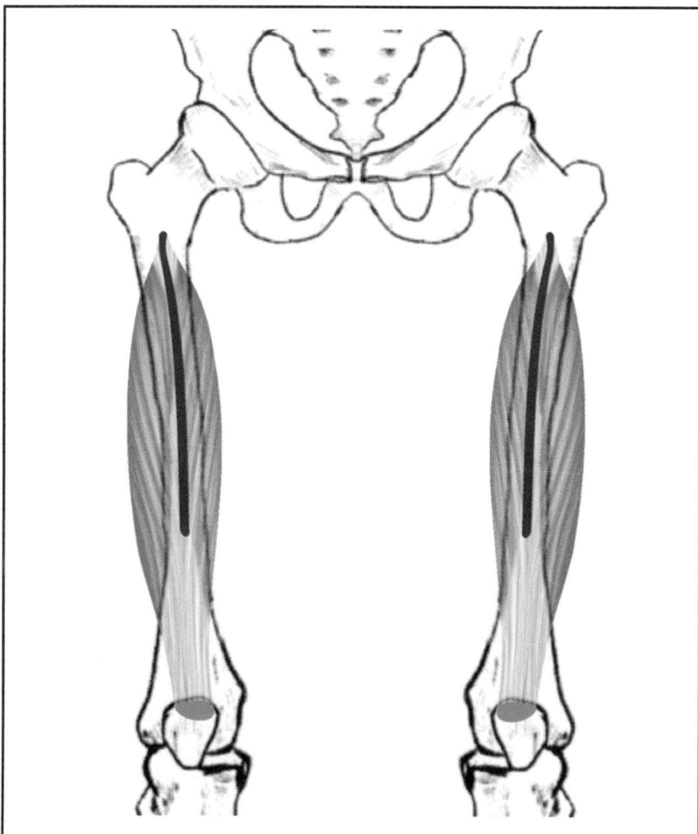

VASTUS LATERALIS (QUADS)

Description:
The largest of the quadriceps muscle group. It forms the lower lateral aspect of the thigh. Also called the vastus externus.

Joint Crossings:
One

Body Action:
Knee extension

Blood Supply:
Lateral femoral circumflex artery

Insertion:
- Upper border of the patella by way of the quadriceps tendon
- The quadriceps tendon passes over and around the patella and becomes the patellar ligament, which inserts into the tibial tuberosity

Nerves:
Femoral nerve L2, 3, 4

Origin:
- Anterior, upper femur, at right angles to the intertrochanteric line
- Lateral lip of the gluteal line (tuberosity)
- The lateral edge of the upper two-thirds of the linea aspera

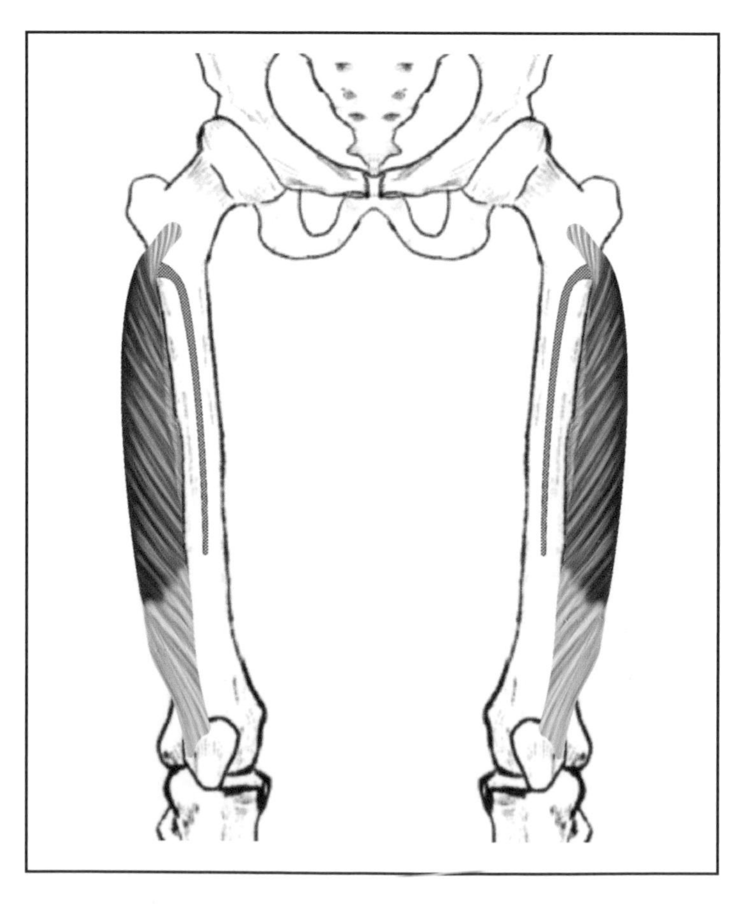

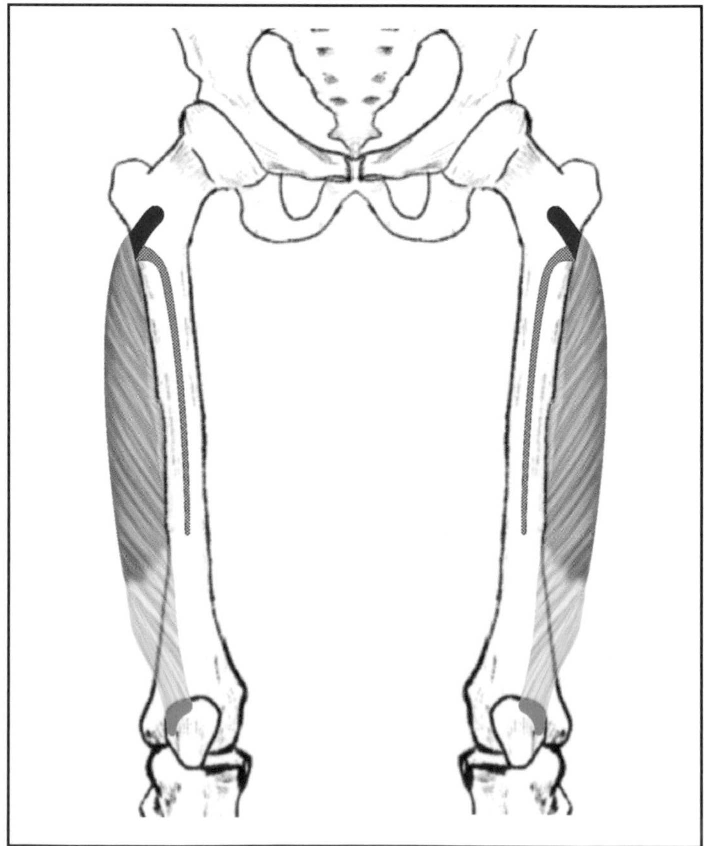

Vastus Medialis (Quads)

Description:
One of the quadriceps muscle group. It makes up the lower medial aspect of the thigh. May be important in maintaining patellofemoral stability, due to the way its fibers are attached to the superior medial patella. Also called the vastus internus.

Joint Crossings:
One

Body Action:
Knee extension

Blood Supply:
- Muscular branches of profunda femoris artery
- Branch of descending genicular artery

Insertion:
- Upper borders of the patella by way of the quadriceps tendon
- The quadriceps tendon passes over and around the patella and becomes the patellar ligament, which inserts into the tibial tuberosity

Nerves:
Femoral nerve L2, 3, 4

Origin:
- Distal half of the intertrochanteric line
- The medial lip of the linea aspera
- The upper two-thirds of the intercondylar line

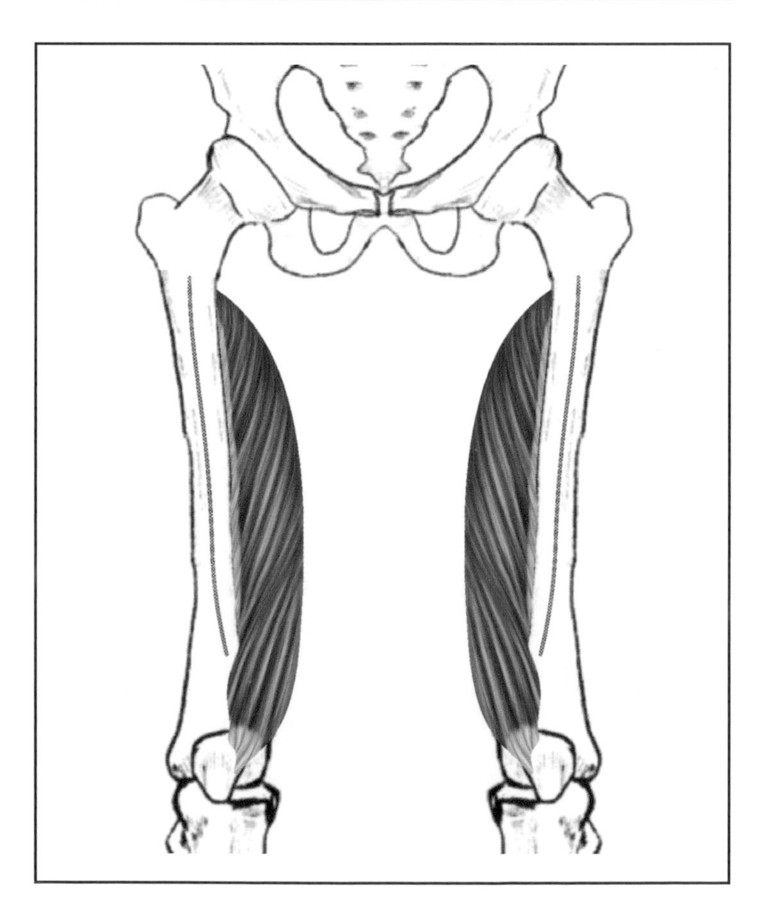

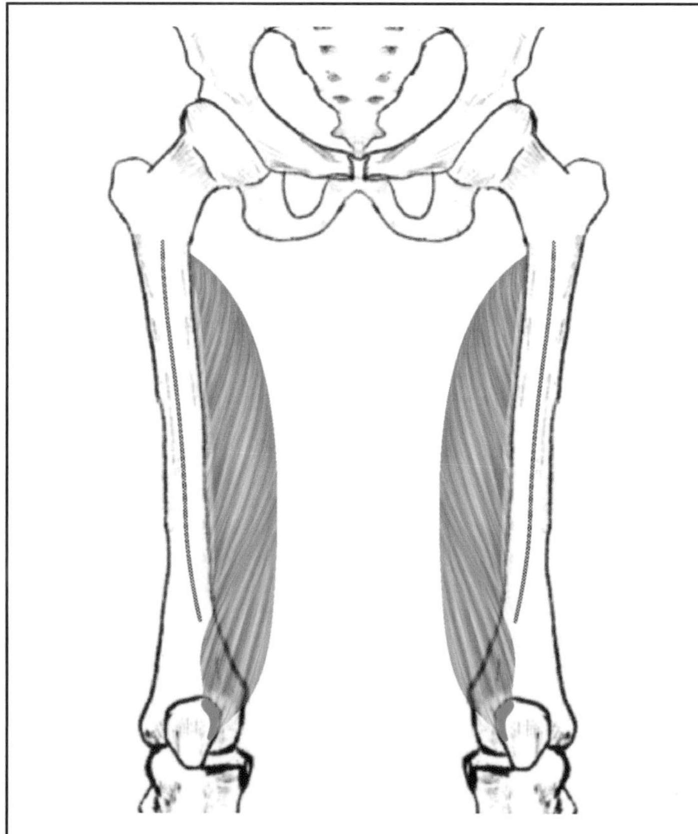

Longus Capitis

Description:
One of the three muscles that make up the anterior vertebral muscles. A muscle on either side of the front and upper portion of the neck that flexes the head. Also known as rectus capitus anticus major.

Joint Crossings:
One

Body Action:
Head flexion

Blood Supply:
Muscular branches of the aorta

Insertion:
Basilar part of occipital bone

Nerves:
Ventral rami C1, 2, 3

Origin:
Transverse processes of C3 - C6

Rectus Capitis Anterior

Description:
One of the three muscles that make up the anterior vertebral muscles. Stabilizes the atlantooccipital joint. Also known as the rectus capitis anticus minor.

Joint Crossings:
One

Body Action:
Head flexion (atlantooccipital joint stabilization)

Blood Supply:
Muscular branches of the aorta

Insertion:
Basilar part of occipital bone anterior to foramen magnum

Nerves:
Ventral rami C1, 2

Origin:
Anterior surface of lateral atlas

Rectus Capitis Lateralis

Description:
One of the three muscles that make up the anterior vertebral muscles. Laterally flexes the head, as well as stabilizing the atlantooccipital joint.

Joint Crossings:
One

Body Action:
Head lateral flexion (atlantooccipital joint stabilization)

Blood Supply:
Muscular branches of the aorta

Insertion:
Jugular process of the occipital bone

Nerves:
Ventral rami C1, 2

Origin:
Superior surface of transverse process of the atlas

STERNOCLEIDOMASTOID

Description:
A thick superficial muscle on each side of the neck that acts to flex and rotate the head.

Joint Crossings:
One

Rank/Body Action:
1 Head flexion (both)
2 Head rotation (to the opposite side - each)
3 Head lateral flexion (to the opposite side - each)

Blood Supply:
- Occipital artery
- Superior thyroid artery

Insertion:
- Mastoid process of temporal bone
- Adjacent portion of occipital bone

Nerves:
Spinal accessory nerve, C2, 3

Origin:
- Anterior, upper surface of the manubrium
- Medial one-third of the clavicle

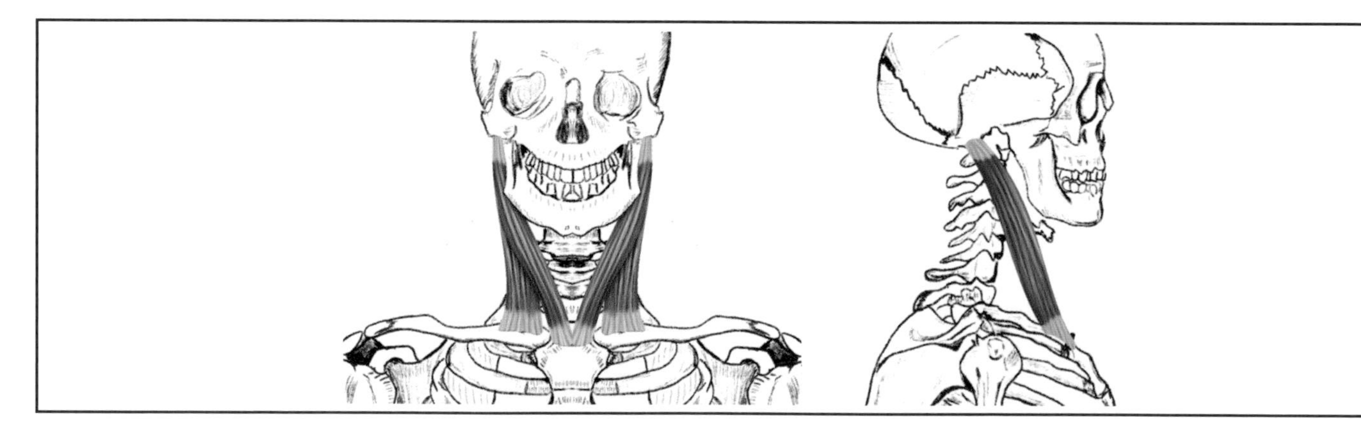

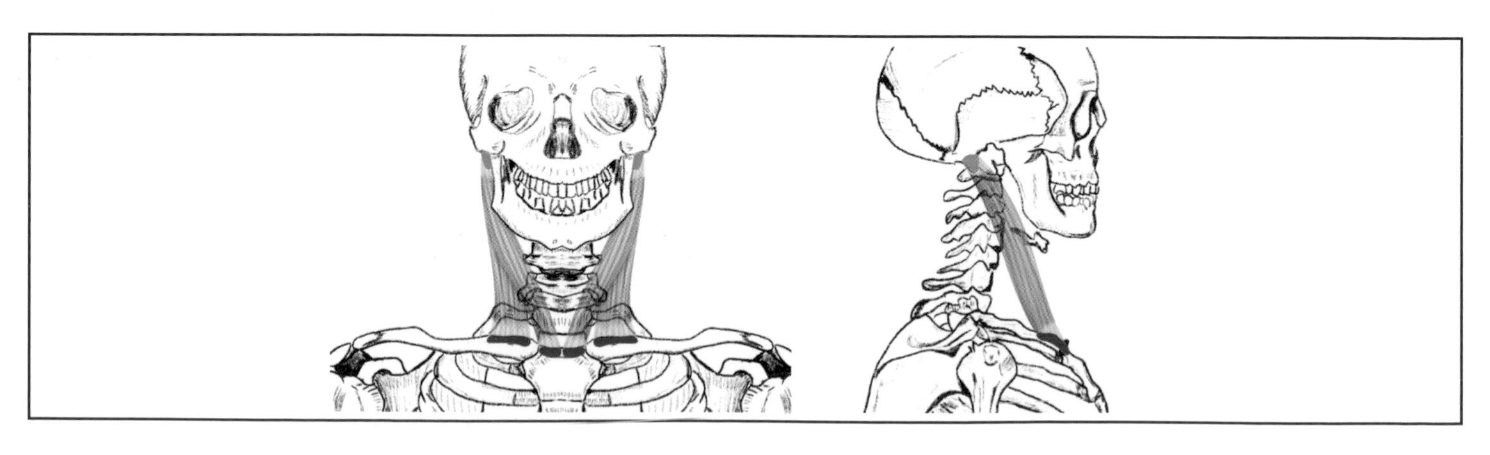

OBLIQUUS CAPITIS INFERIOR

Description:
A small posteriorly positioned muscle that rotates the head.

Joint Crossings:
One

Body Action:
Head rotation

Blood Supply:
Muscular branches of vertebral artery

Insertion:
Transverse process of atlas

Nerves:
Posterior rami of C1

Origin:
Spinous process of axis

OBLIQUUS CAPITIS SUPERIOR

Description:
Located posteriorly, a muscle that extends the head and assists the rectus capitis lateralis in lateral flexion.

Joint Crossings:
One

Rank/Body Action:
1 Head extension
2 Head hyperextension
3 Head lateral flexion

Blood Supply:
Muscular branches of vertebral artery

Insertion:
Between the superior and inferior nuchal line of the occipital bone

Nerves:
Posterior rami of C1

Origin:
Transverse process of atlas

RECTUS CAPITIS POSTERIOR (MAJOR)

Description:
A muscle on each side of the back of the neck that acts to extend and rotate the head.

Blood Supply:
Muscular branches of vertebral artery

Insertion:
Lateral portion of inferior nuchal line of occipital bone

Joint Crossings:
One

Nerves:
Posterior rami of C1

Rank/Body Action:
1 Head extension (both)
2 Head hyperextension
3 Head rotation (to the same side - each)

Origin:
Spinous process of axis

RECTUS CAPITIS POSTERIOR (MINOR)

Description:
A muscle on each side of the back of the neck that extends the head.

Blood Supply:
Muscular branches of vertebral artery

Insertion:
Medial portion of inferior nuchal line of occipital bone

Joint Crossings:
One

Nerves:
Posterior rami of C1

Rank/Body Action:
1 Head extension
2 Head hyperextension

Origin:
Posterior arch of the atlas

Splenius Capitis

Description:
A flat muscle on each side of the back of the neck that rotates and laterally flexes the head to the side on which it is located and, with both muscles, extends the head.

Joint Crossings:
One

Rank/Body Action:
1 Head extension (both)
2 Head hyperextension (both)
3 Head lateral flexion (to the same side - each)
4 Head rotation (to the same side - each)

Blood Supply:
Muscular branches of the aorta

Insertion:
- Mastoid process of the temporal bone
- Adjacent portion of the occipital bone

Nerves:
Posterior branches of cervical nerves C4-8

Origin:
- Lower portion of nuchal ligament
- Spines of the 7th cervical and the first three or four thoracic vertebrae

Splenius Cervicis

Description:
A flat narrow muscle on each side of the back of the neck that acts to rotate and laterally flex the head to the side on which it is located and, with both muscles, extend the head. Also known as the splenius colli.

Joint Crossings:
One

Rank/Body Action:
1 Head extension (both)
2 Head hyperextension (both)
3 Head lateral flexion (to the same side - each)
4 Head rotation (to the same side - each)

Blood Supply:
Deep cervical artery

Insertion:
Transverse process of the first two or three cervical vertebrae

Nerves:
Posterior branches of cervical nerves C4-8

Origin:
Spinous processes of the 3rd - 6th thoracic vertebrae

Coracobrachialis

Description:
A weak muscle whose major function is to move the arm horizontally and across the chest.

Joint Crossings:
One

Rank/Body Action:

1	Arm horizontal adduction
2	Arm flexion
3	Arm adduction
4	Arm external rotation (slight)

Blood Supply:
Muscular branches of the brachial artery

Insertion:
Middle one-third of the medial surface of the humerus

Nerves:
Musculocutaneous nerve C5, 6, 7

Origin:
Coracoid process of the scapula

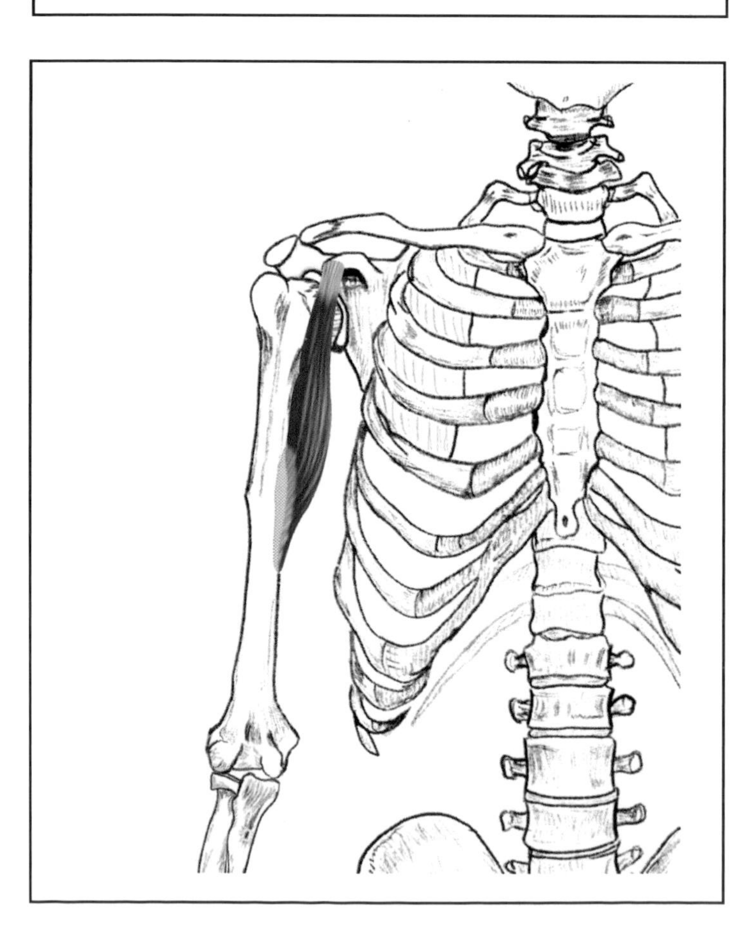

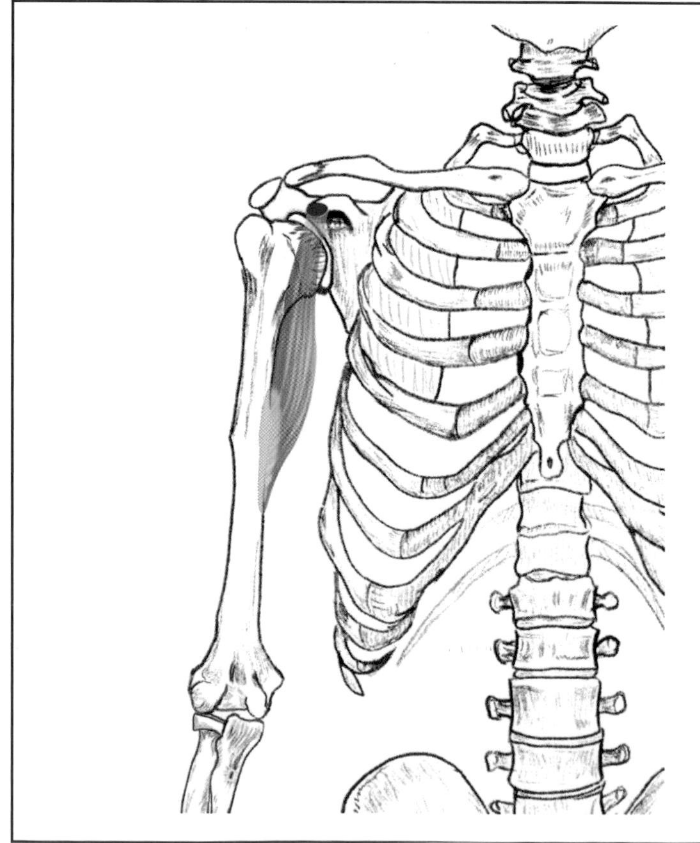

Deltoid Anterior

Description:
The front part of the deltoid muscle, it is a strong abductor of the arm. Its lower fibers weakly adduct and internally rotate the arm.

Joint Crossings:
One

Rank/Body Action:
1 Arm abduction (upper fibers)
2 Arm flexion
3 Arm internal rotation
4 Arm horizontal adduction (lower fibers)
5 Arm adduction (lower fibers)

Blood Supply:
- Posterior humeral circumflex artery
- Deltoid branch of thoracoacromial artery

Insertion:
Deltoid tuberosity on the lateral surface of the humerus

Nerves:
Axillary nerve C5, 6

Origin:
Anterior border of the lateral one-third of the clavicle

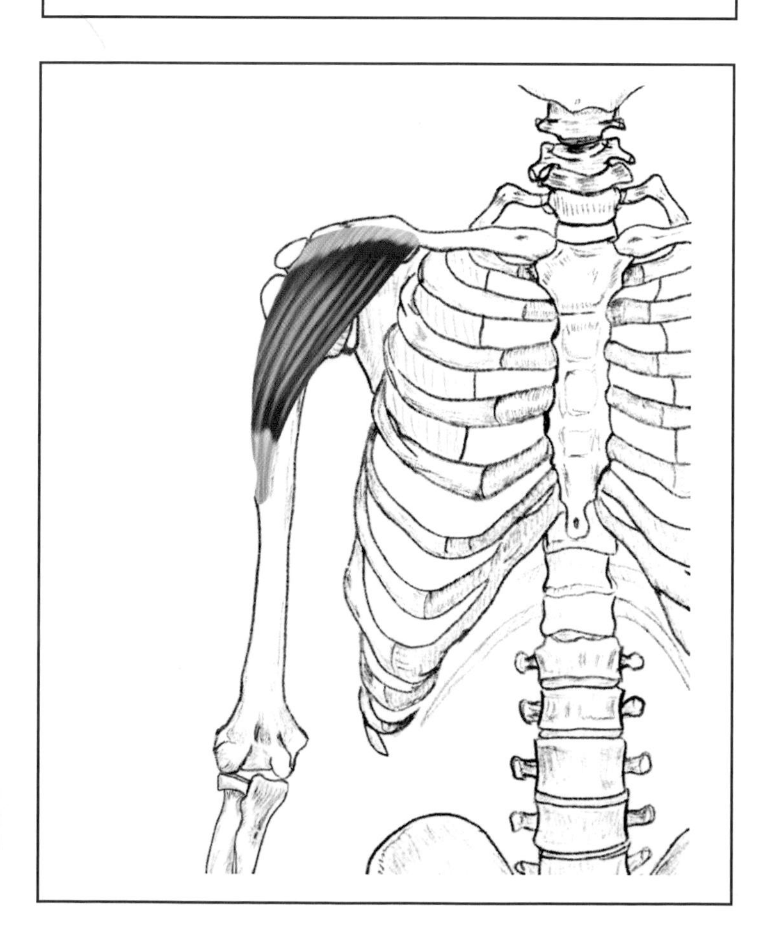

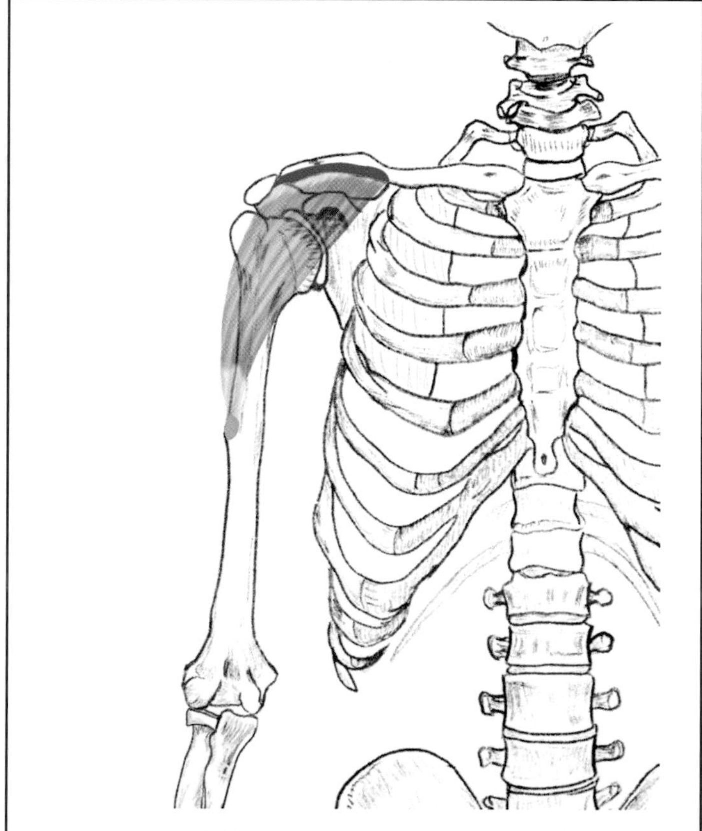

Deltoid Middle

Description:
A strong abductor of the arm. The abduction is assisted by the upper fibers of the posterior and anterior deltoid.

Joint Crossings:
One

Body Action:
Arm abduction

Blood Supply:
- Posterior humeral circumflex artery
- Deltoid branch of thoracoacromial artery

Insertion:
Deltoid tuberosity on the lateral surface of the shaft of the humerus

Nerves:
Axillary nerve C5,6

Origin:
Acromion process of the scapula

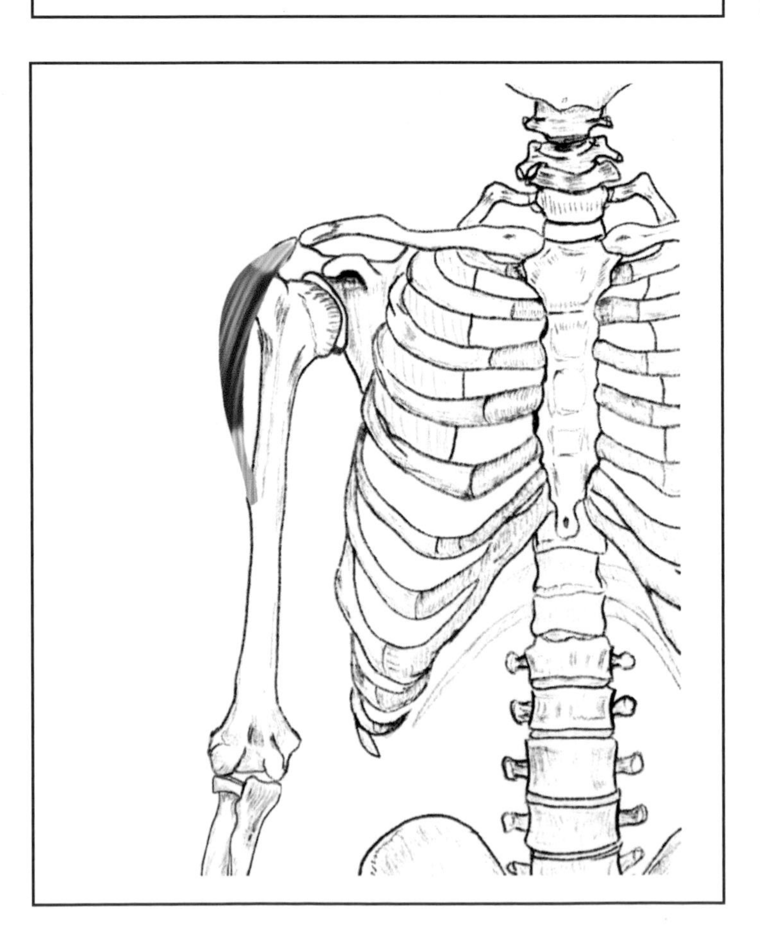

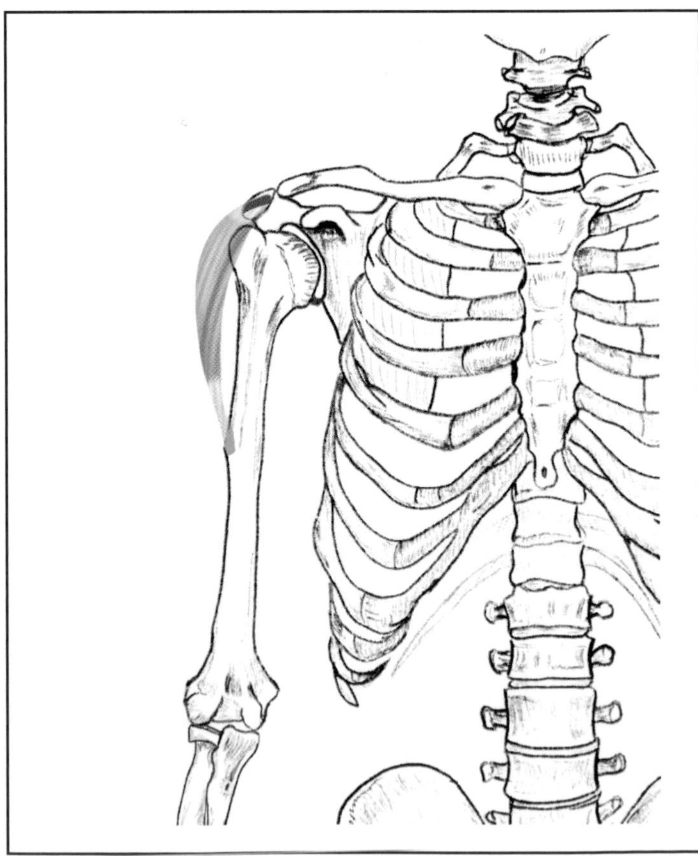

Deltoid Posterior

Description:
The posterior part of the deltoid muscle, it is a strong abductor of the arm. Its lower fibers weakly adduct and externally rotate the arm.

Joint Crossings:
One

Rank/Body Action:

1	Arm abduction (upper fibers)
2	Arm external rotation
3	Arm extension
4	Arm hyperextension
5	Arm horizontal abduction
6	Arm adduction (lower fibers)

Blood Supply:
- Posterior humeral circumflex artery
- Deltoid branch of thoracoacromial artery

Insertion:
Deltoid tuberosity on the lateral surface of the shaft of the humerus

Nerves:
Axillary nerve C5, 6

Origin:
Inferior edge of scapular spine

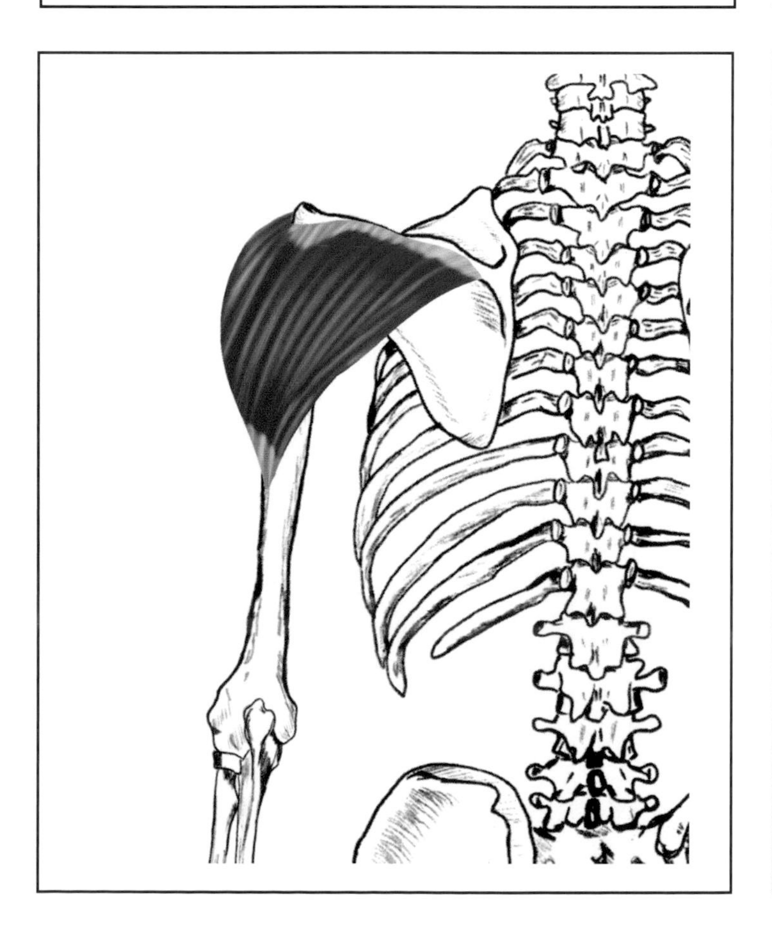

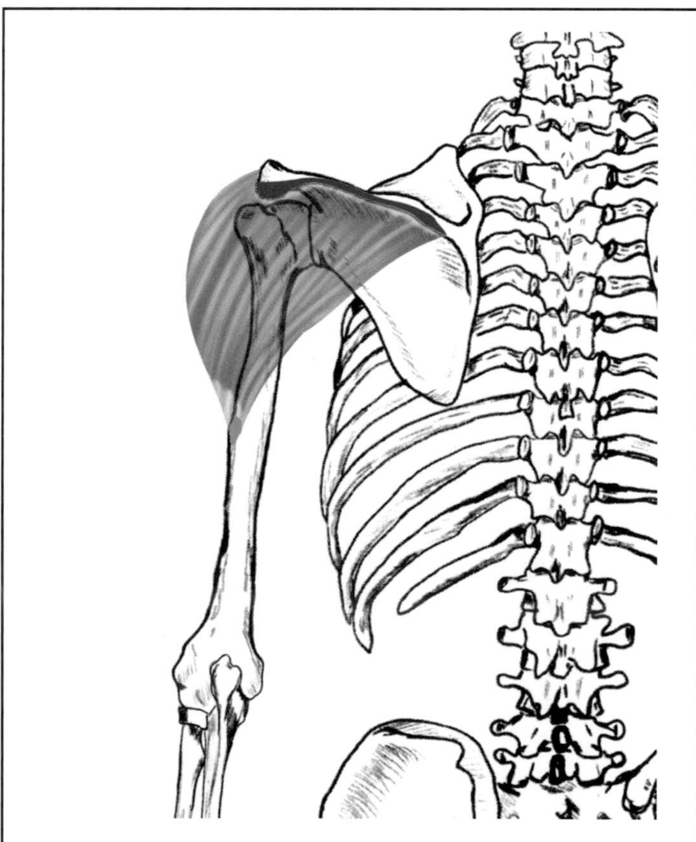

INFRASPINATUS

Description:
A muscle that occupies most of the infraspinous fossa of the scapula and rotates the arm externally. Because of its wide origin, its upper fibers abduct and its lower fibers adduct the arm. It is a strong external rotator, and is one of the four rotator cuff muscles.

Joint Crossings:
One

Rank/Body Action:

Rank	Body Action
1	Arm external rotation
2	Arm horizontal abduction
3	Arm adduction (lower fibers)
4	Arm abduction (upper fibers)

Blood Supply:
- Suprascapular artery
- Scapular circumflex artery

Insertion:
Superior/lateral surface of the greater tuberosity of the humerus

Nerves:
Suprascapular nerve C5, 6

Origin:
Infraspinous fossa of the scapula

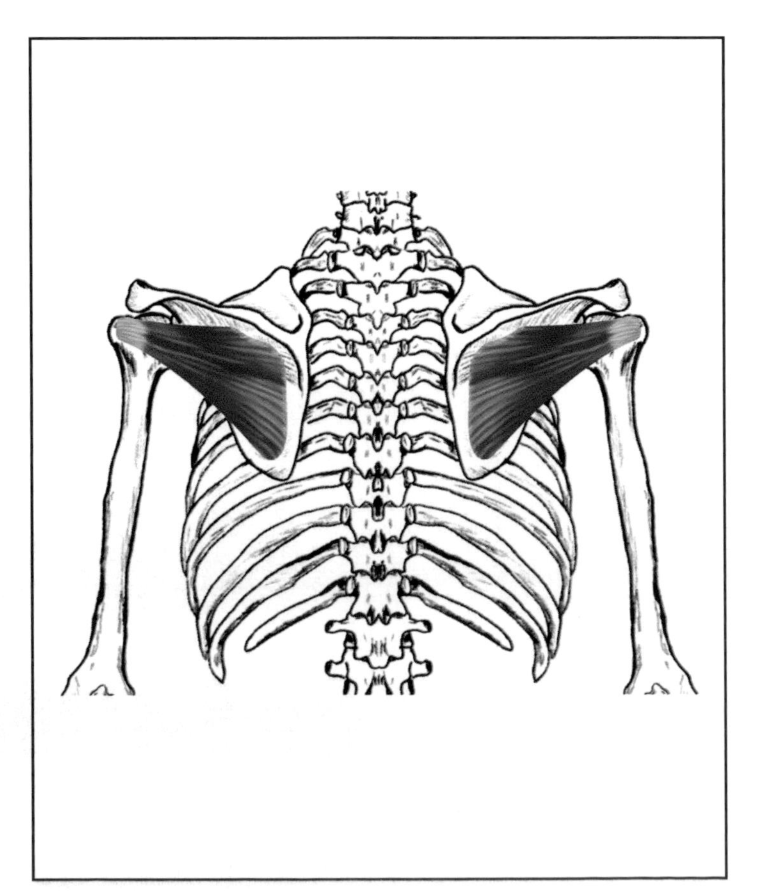

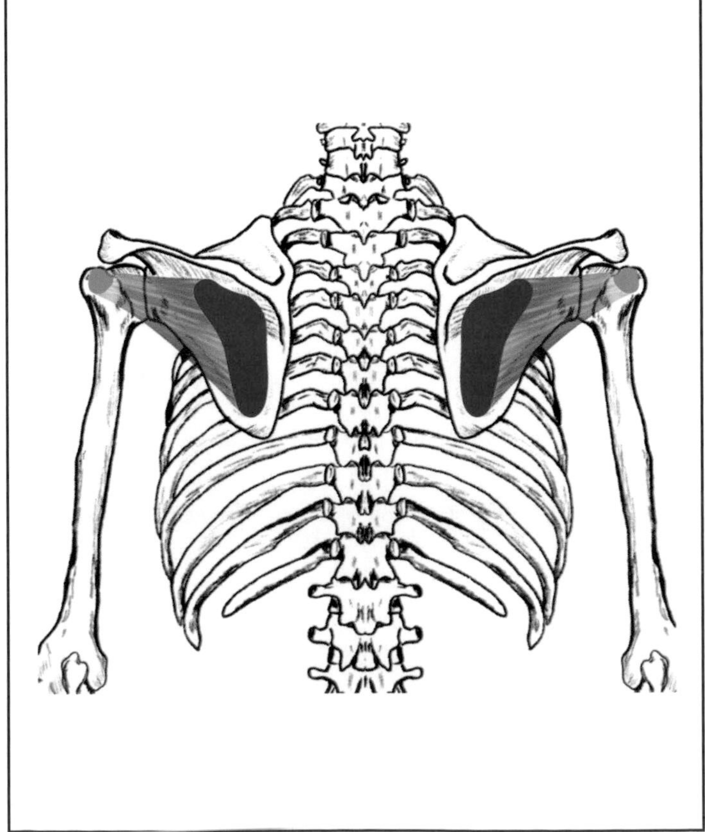

Latissimus Dorsi

Description:
A broad, flat superficial muscle of the lower part of the back that extends, adducts, and internally rotates the arm. Sometimes known as the swimmer's muscle.

Joint Crossings:
One

Rank/Body Action:

1	Arm extension (strong)
2	Arm hyperextension (strong)
3	Arm adduction (strong)
4	Arm internal rotation
5	Arm horizontal adduction

Blood Supply:
Thoracodorsal artery

Insertion:
Intertubular groove on the anterior humerus

Nerves:
Thoracodorsal nerve C6, 7, 8

Origin:
- Spines of the lower six thoracic vertebrae
- Spines of the lumbar vertebrae
- Posterior surface of the sacrum
- Posterior portion of the crest of the ilium
- Lower three or four ribs
- Inferior angle of the scapula

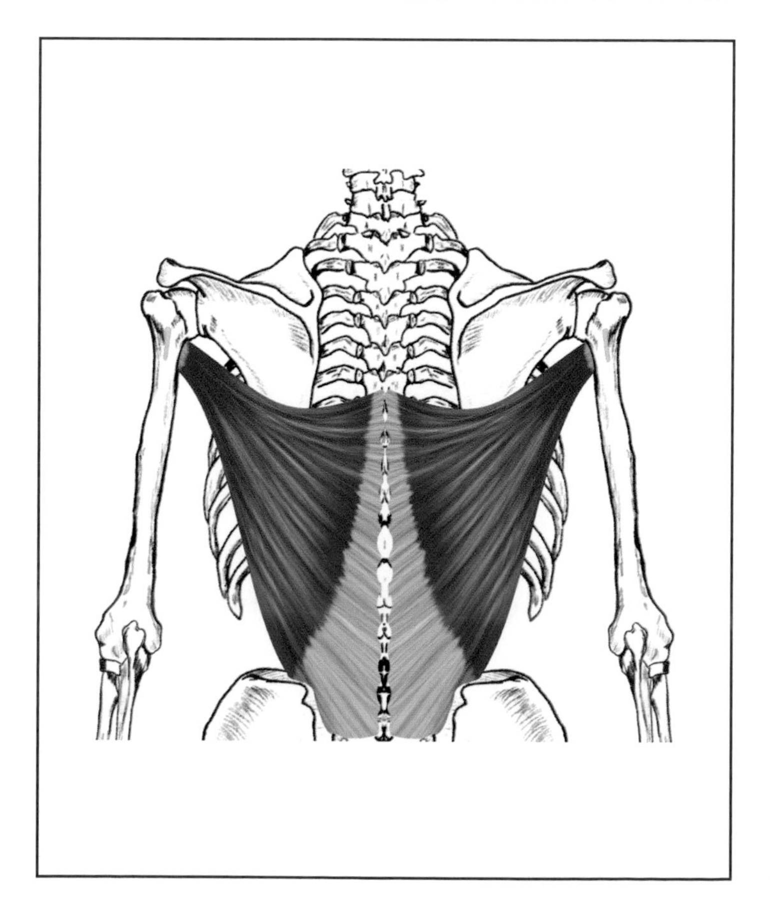

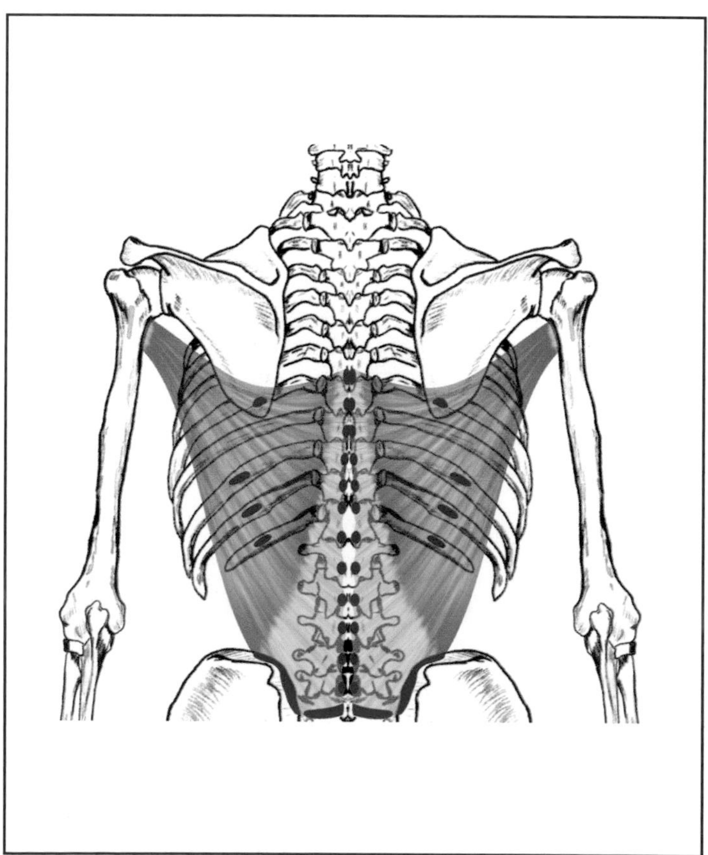

PECTORALIS MAJOR

Description:
A large, superficial fan-shaped muscle of the chest. It is a strong horizontal adductor, flexor, and internal rotator of the arm. It lies superior to the pectoralis minor.

Joint Crossings:
One

Rank/Body Action:

Rank	Body Action
1	Arm horizontal adduction
2	Arm flexion
3	Arm internal rotation
4	Arm adduction

Blood Supply:
- Pectoralis branch of thoracoacromial artery
- Lateral thoracic artery (lesser supply)

Insertion:
Crest of the greater tuberosity of the humerus

Nerves:
- Lateral pectoral nerve C5, 6, 7 (upper)
- Medial pectoral nerve C8, T1 (lower)

Origin:
- Medial two-thirds of the anterior border of the clavicle
- Anterior/lateral surface of the sternum and the costal cartilage's of the first six ribs
- Aponeurosis of the external abdominal oblique muscle

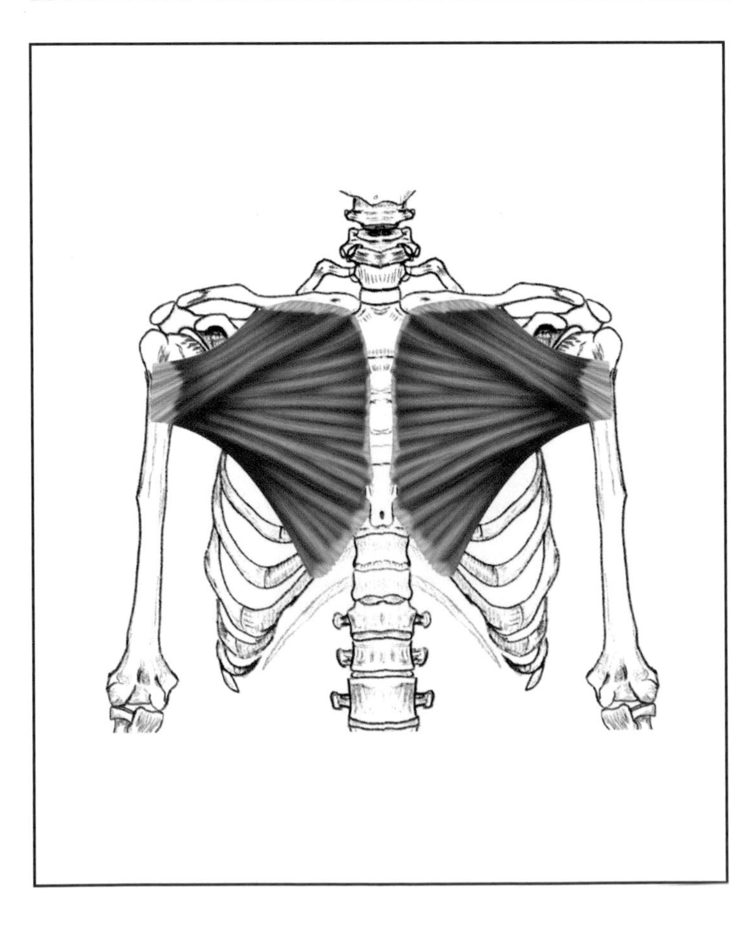

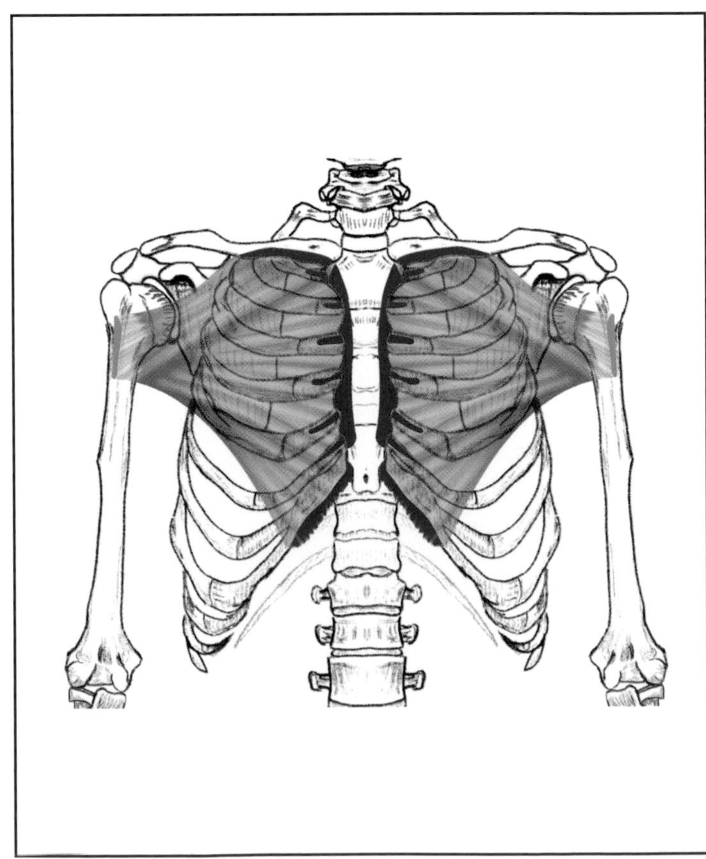

SUBSCAPULARIS

Description:
A triangular-shaped muscle that fills up the subscapular fossa, stabilizes the shoulder, and rotates the arm internally, when the arm is held by the side of the body. It is one of the four rotator cuff muscles.

Joint Crossings:
One

Rank/Body Action:

Rank	Body Action
1	Arm internal rotation
2	Arm extension
3	Arm hyperextension
4	Arm adduction (weak)

Blood Supply:
Branches of subscapular artery

Insertion:
Lesser tuberosity of the humerus

Nerves:
Upper and lower subscapular nerves C5, 6

Origin:
Subscapular fossa of the scapula

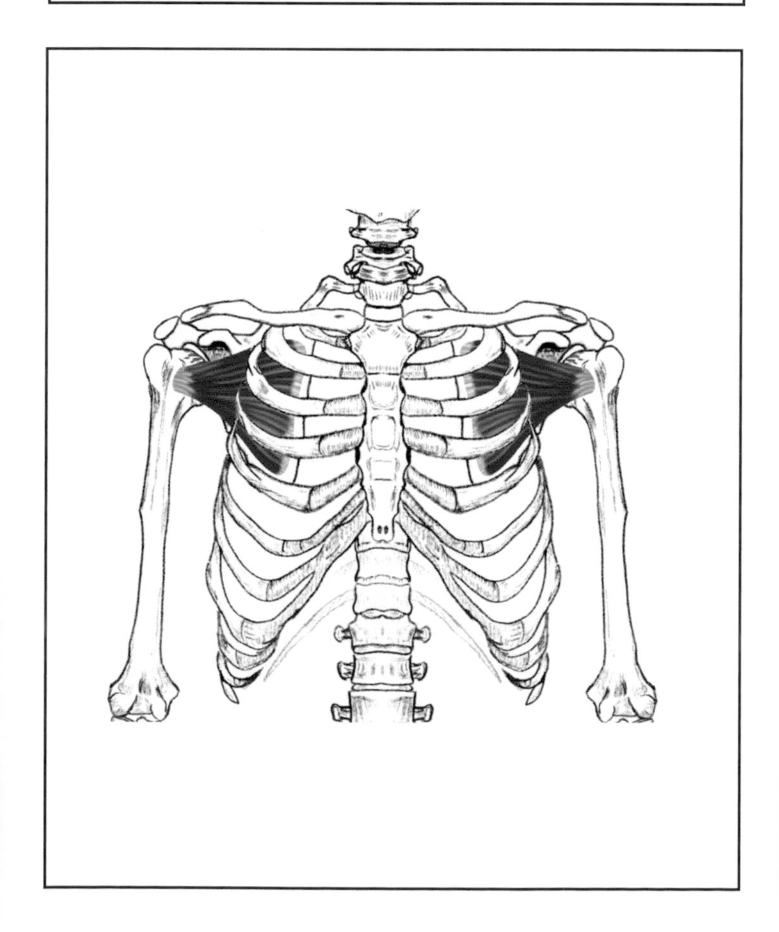

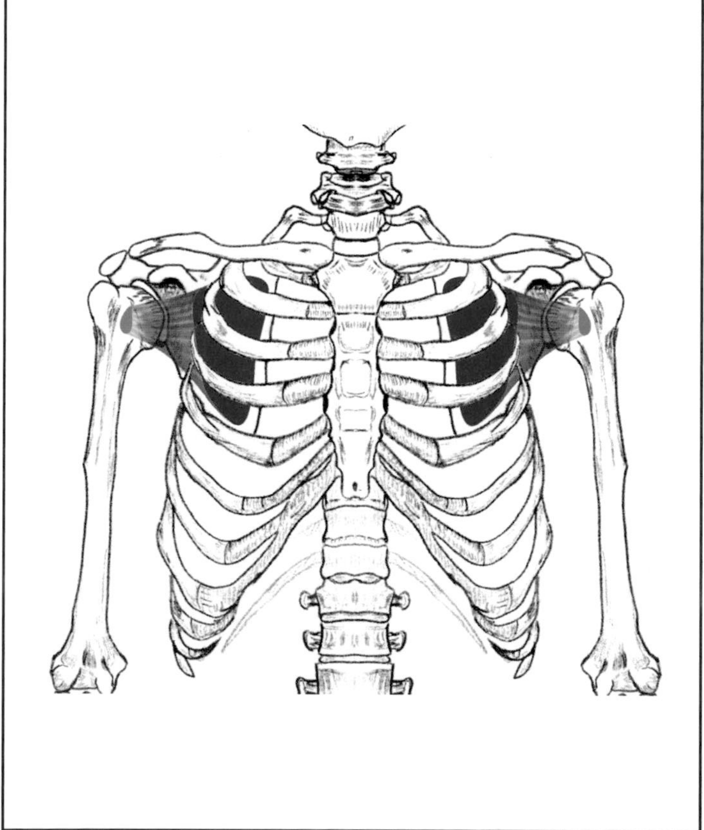

Supraspinatus

Description:
A muscle of the shoulder, that is one of the four rotator cuff muscles. It abducts the arm and is the most often injured of the rotator cuff muscles, especially with activities that raise the arm above the head.

Joint Crossings:
One

Body Action:
Arm abduction

Blood Supply:
Suprascapular artery (poorly supplied)

Insertion:
Superior/lateral surface of the greater tuberosity of the humerus

Nerves:
Suprascapular nerve C5

Origin:
Supraspinous fossa of the scapula

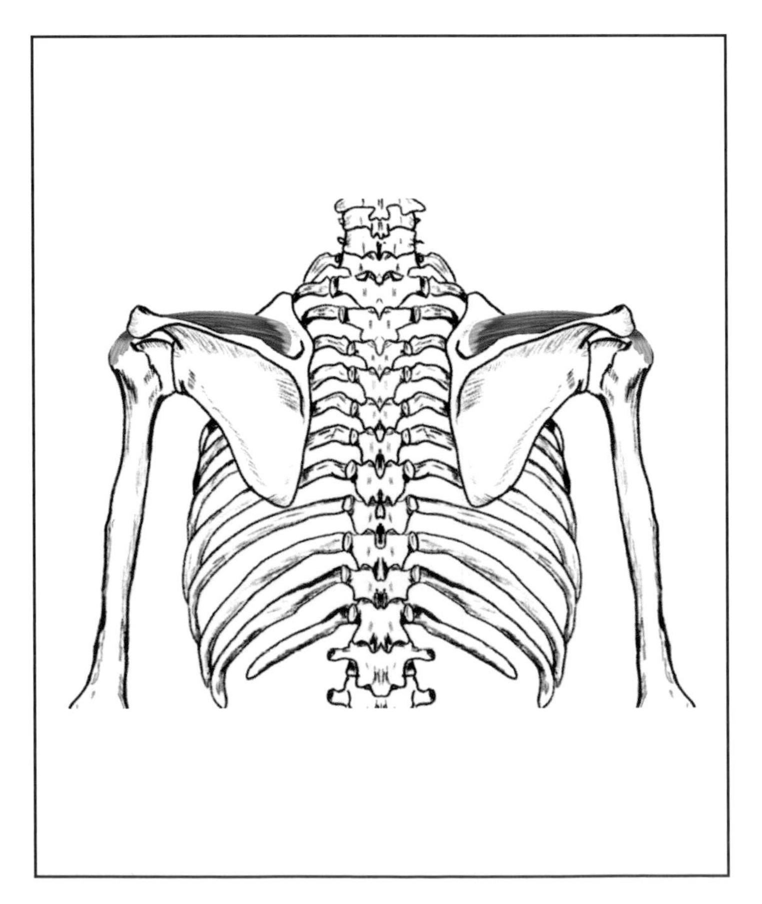

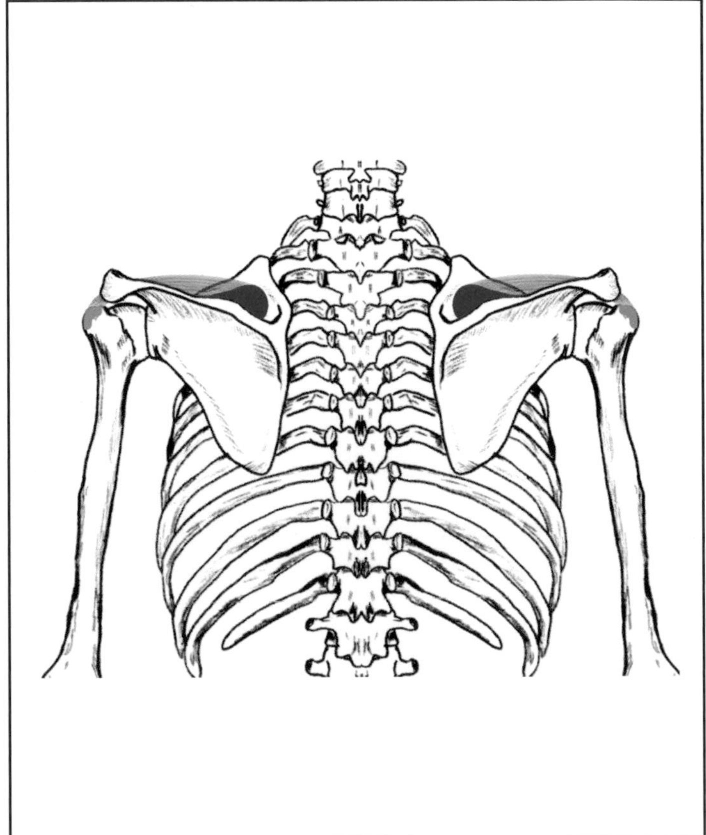

TERES MAJOR

Description:
Similar to a small latissimus dorsi, having the same fiber direction and, hence, the same actions.

Joint Crossings:
One

Rank/Body Action:

Rank	Body Action
1	Arm extension
2	Arm hyperextension
3	Arm adduction
4	Arm internal rotation

Blood Supply:
Thoracodorsal artery

Insertion:
Crest of the lesser tuberosity of the humerus

Nerves:
Lower subscapular nerve C5, 6

Origin:
- Lower one-third of the axillary border of the scapula
- Inferior angle of the scapula

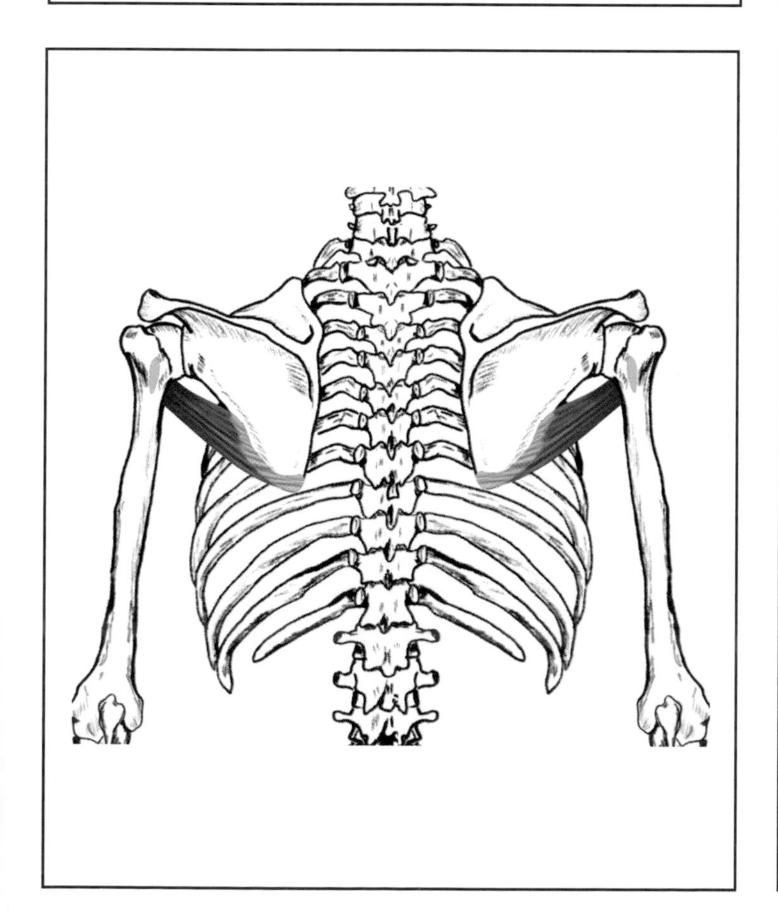

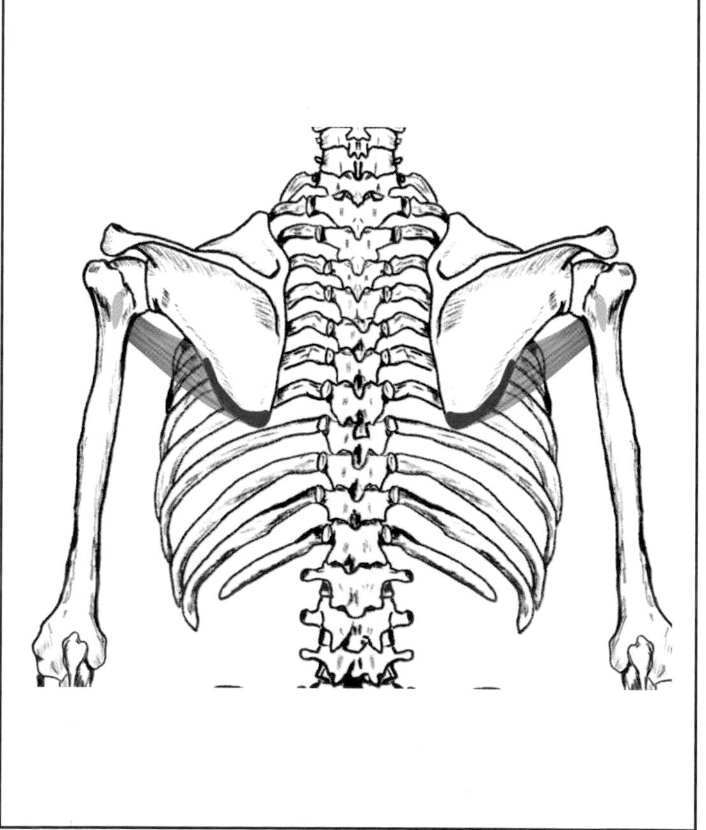

TERES MINOR

Description:
A long cylindrical muscle that contributes to the formation of the four rotator cuff muscles of the shoulder, and acts to extend and rotate the arm externally.

Joint Crossings:
One

Rank/Body Action:
1 Arm external rotation
2 Arm extension
3 Arm hyperextension
4 Arm adduction
5 Arm horizontal abduction

Blood Supply:
Scapular circumflex artery

Insertion:
Posterior surface of the greater tuberosity of the humerus

Nerves:
Axillary nerve C5,6

Origin:
Upper two-thirds of the axillary border of the scapula

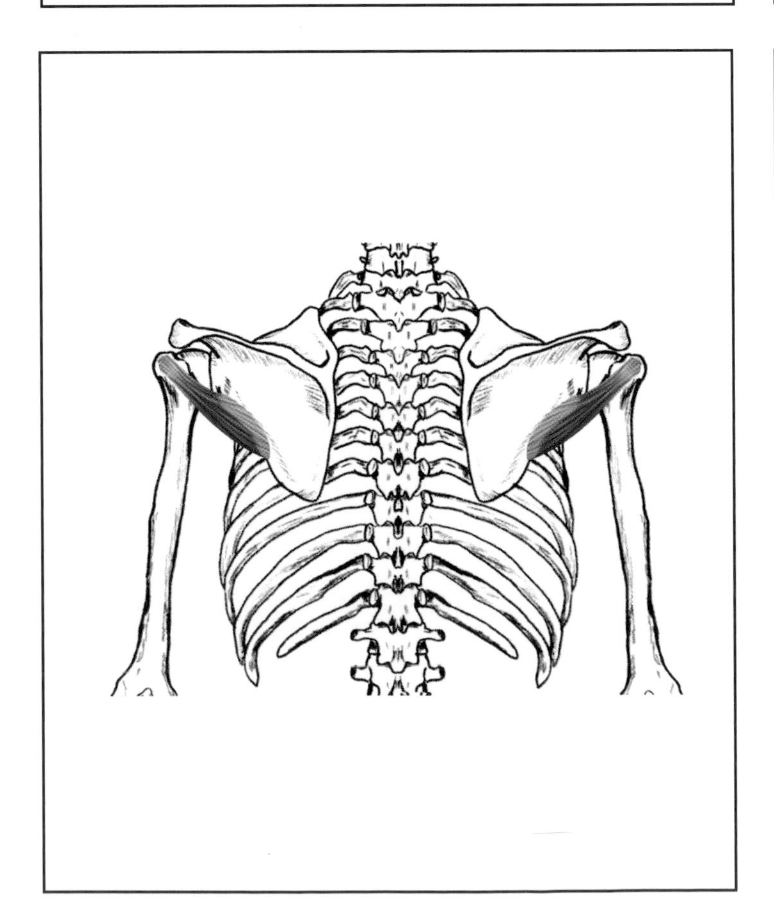

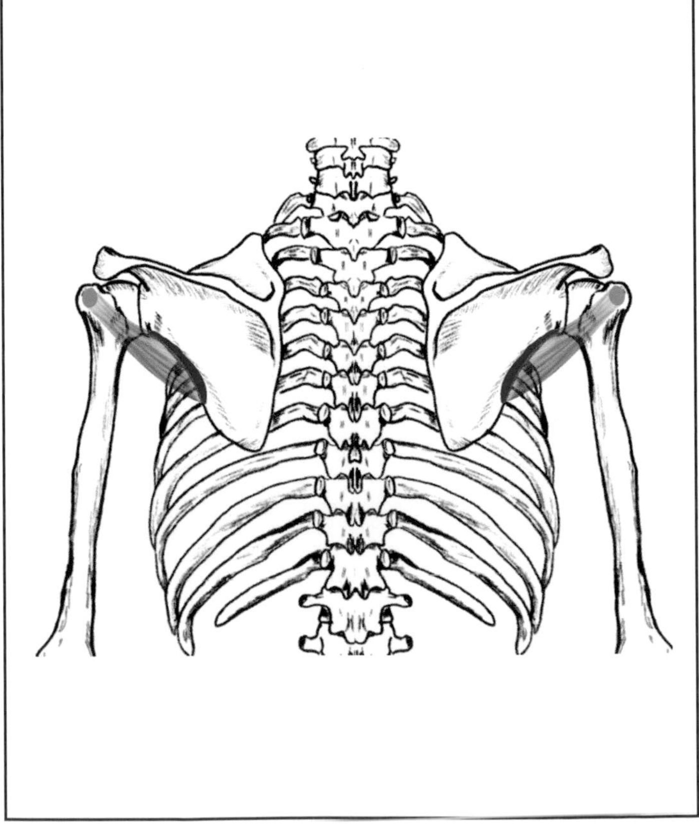

LEVATOR SCAPULAE

Description:
A back muscle that mostly elevates the scapula but also assists with downward rotation.

Joint Crossings:
N/A

Rank/Body Action:

1	Scapula elevation
2	Scapula downward rotation
3	Scapula adduction (slight)

Blood Supply:
Transverse cervical artery

Insertion:
Vertebral border of the scapula, from the spine to the superior angle

Nerves:
- Branches of C3,4
- Dorsal scapular nerve C5

Origin:
Transverse processes of the first four cervical vertebrae

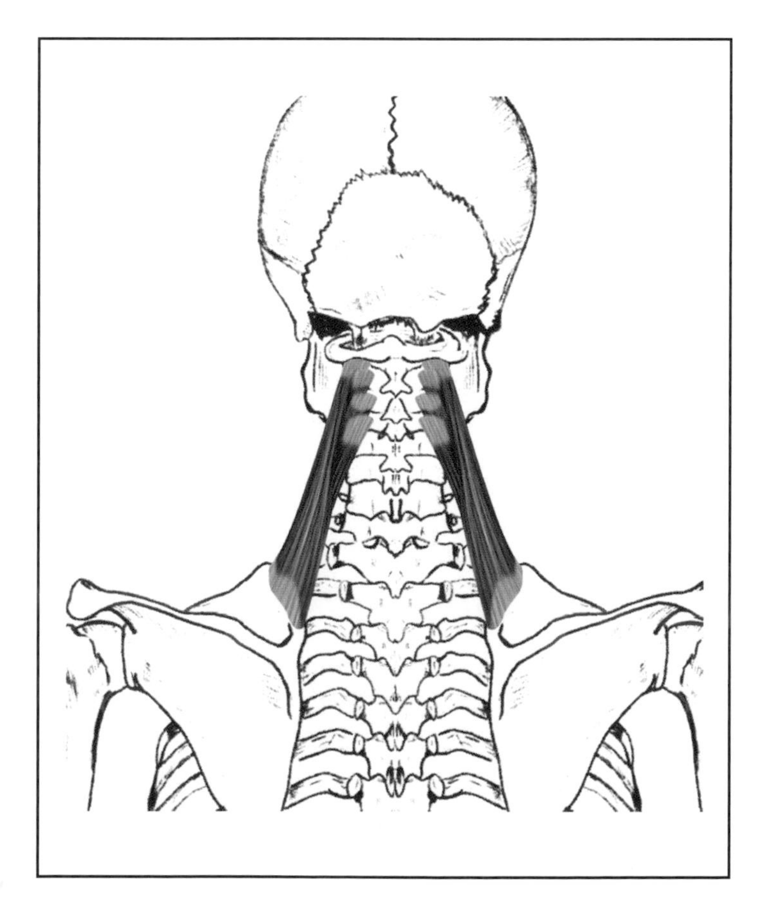

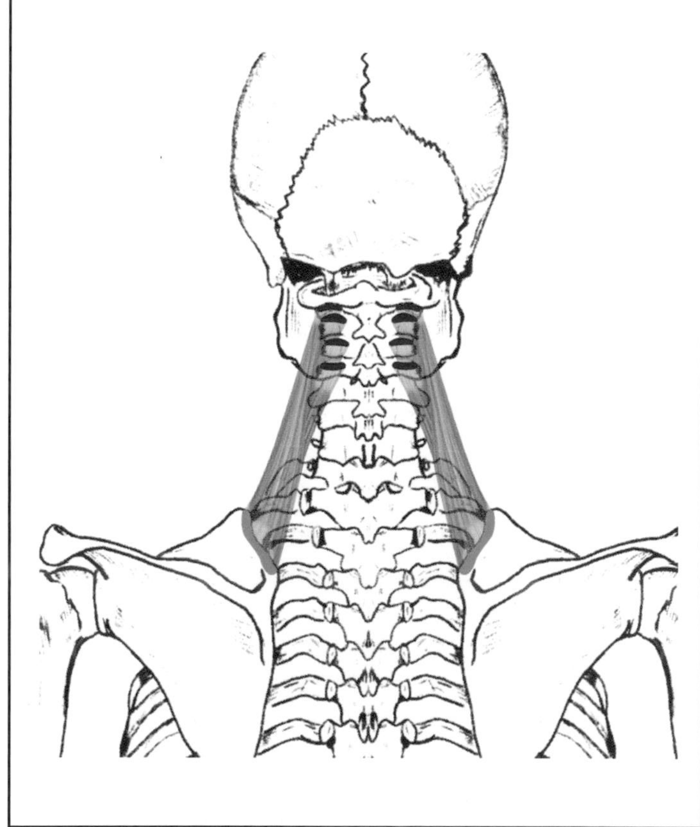

Pectoralis Minor

Description:
A non-locomotor muscle that is used in forced inspiration. It lies beneath the pectoralis major.

Joint Crossings:
N/A

Rank/Body Action:
1 Rib elevation
2 Scapula depression

Blood Supply:
Lateral thoracic artery

Insertion:
Coracoid process of scapula

Nerves:
Medial pectoral nerve C8, T1

Origin:
Anterior surface of third, fourth, and fifth ribs

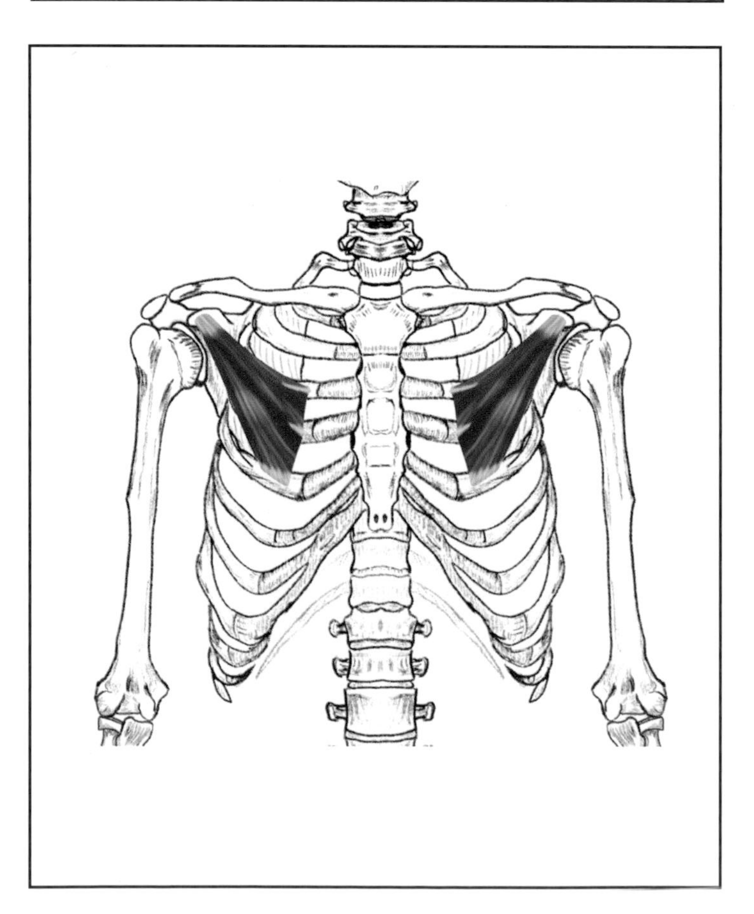

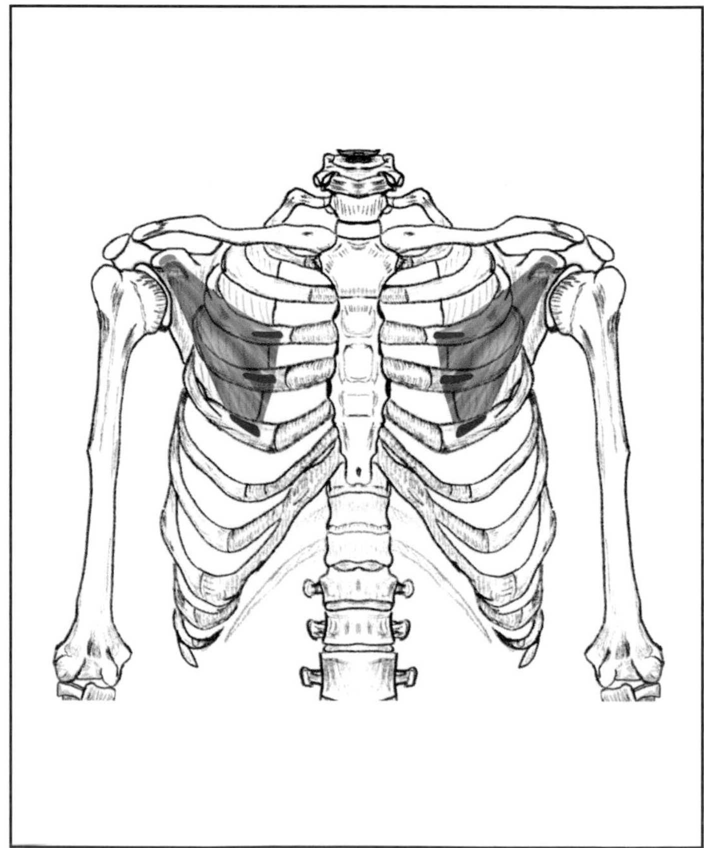

Rhomboid Major & Minor

Description:
A muscle that acts to adduct and downwardly rotate the scapula. The movements of the scapula affect the movements of the shoulder girdle.

Joint Crossings:
N/A

Rank/Body Action:

Rank	Body Action
1	Scapula adduction
2	Scapula downward rotation
3	Scapula elevation

Blood Supply:
Deep branch of transverse cervical artery (or dorsal scapular artery)

Insertion:
Vertebral border of the scapula, from the spine to the inferior angle

Nerves:
Dorsal scapular nerve C5

Origin:
- Lower part of the ligamentum nuchae
- Spine of the seventh cervical vertebra
- Spines of the first five thoracic vertebrae

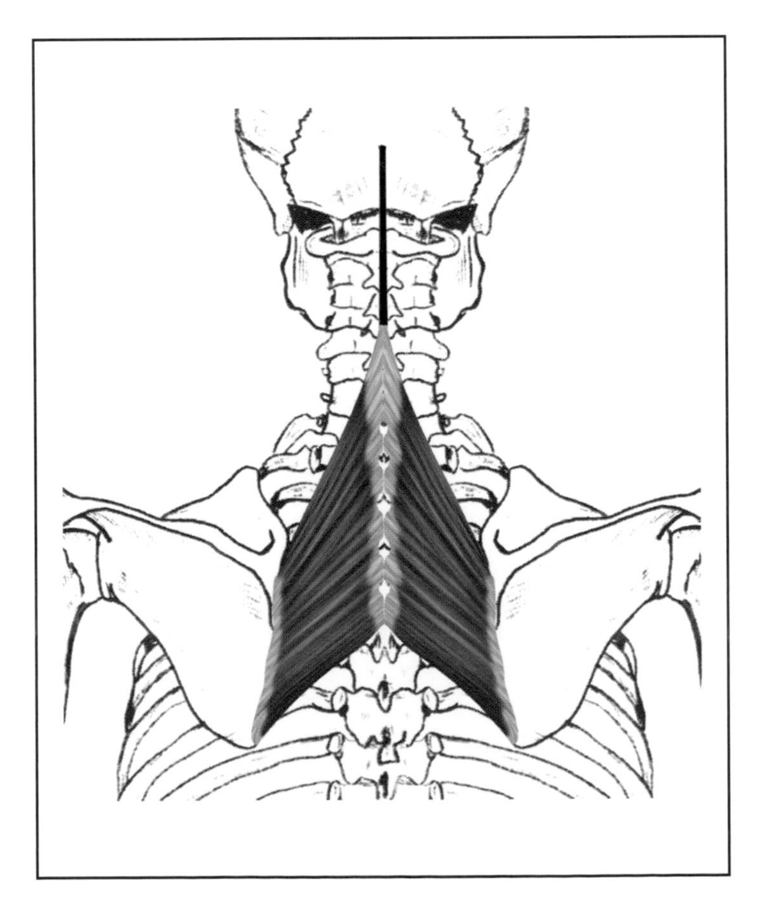

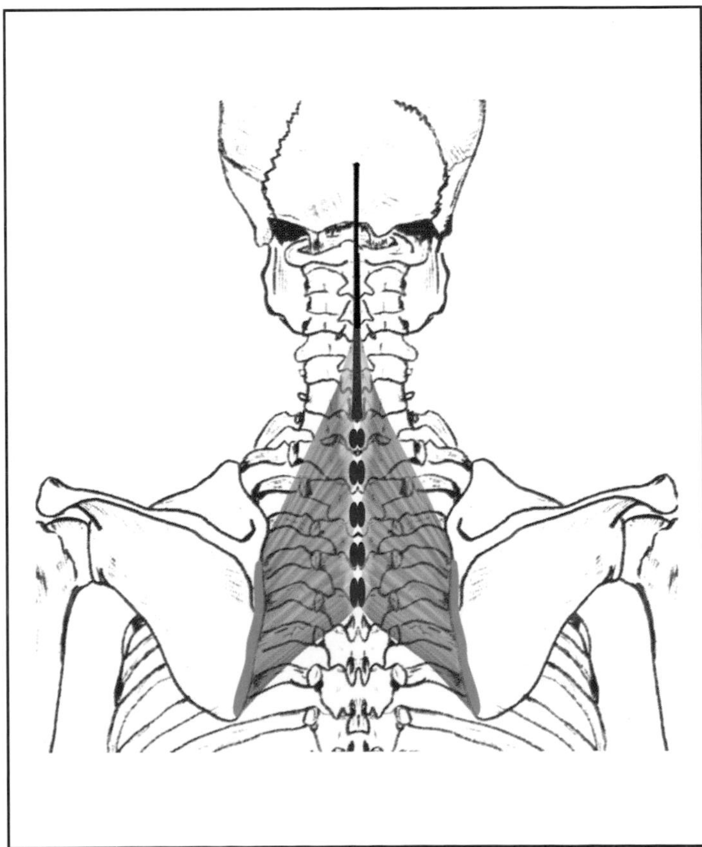

Serratus Anterior

Description:
A broad muscle that wraps around the thorax and dovetails with the external abdominal oblique (in a serrated junction). It acts to stabilize the scapula by holding it against the back thoracic wall and abducts and upwardly rotates it. Also called the serratus magnus.

Joint Crossings:
N/A

Rank/Body Action:

Rank	Body Action
1	Scapula abduction
2	Scapula upward rotation
3	Scapula depression (lower fibers)

Blood Supply:
- Lateral thoracic artery (upper portion)
- Thoracodorsal artery (lower portion)

Insertion:
Costal surface of vertebral border of scapula

Nerves:
Long thoracic nerve C5, 6, 7

Origin:
Lateral, anterior surfaces of first eight or nine ribs

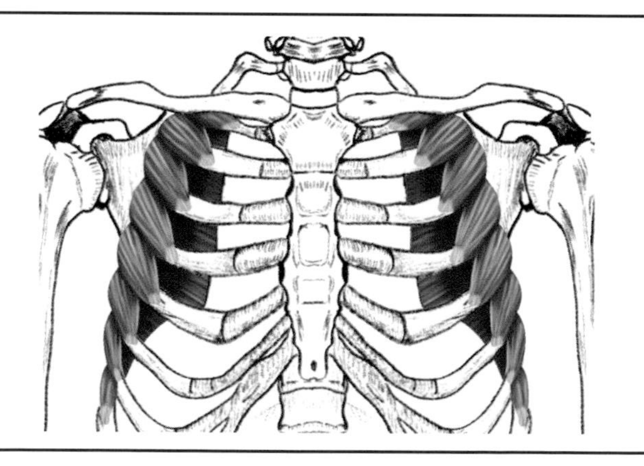

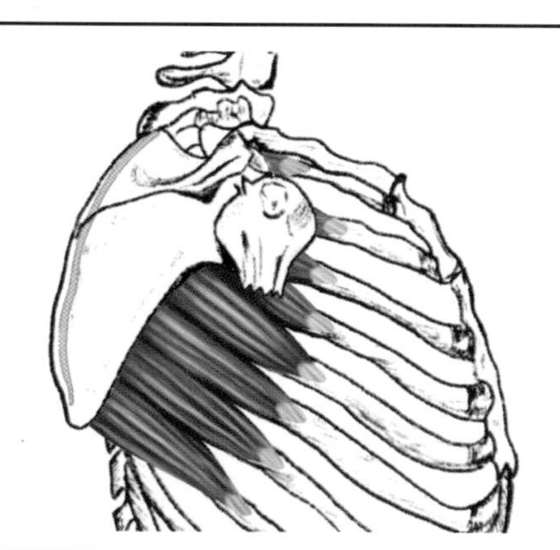

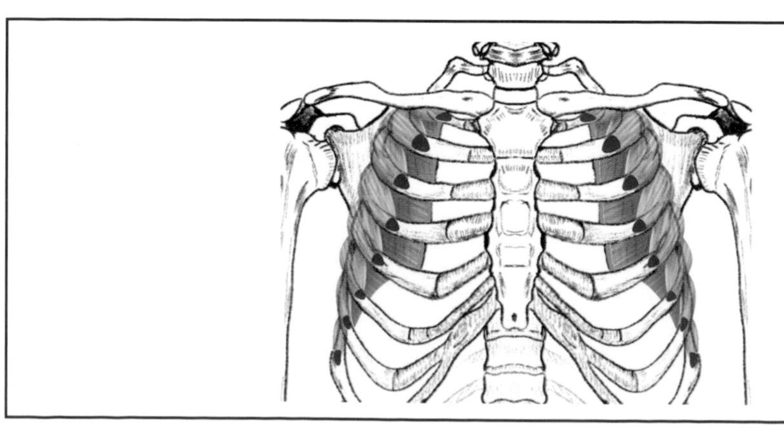

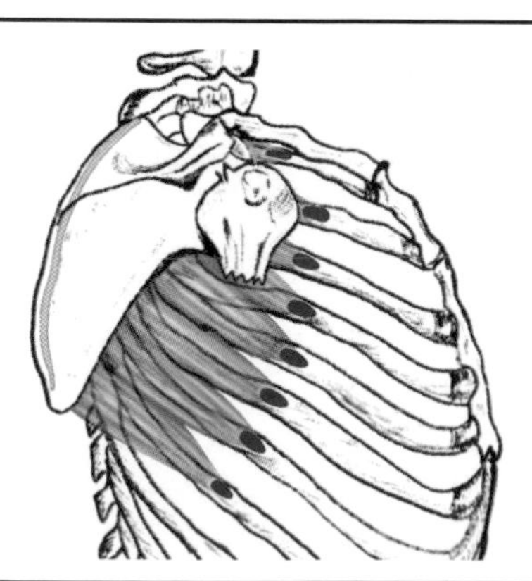

Subclavius

Description:
A small muscle on each side of the body that acts to stabilize the clavicle by depressing and drawing forward its lateral end during movements of the shoulder.

Joint Crossings:
N/A

Body Action:
Scapula depression (sternoclavicular joint stabilization and protection)

Blood Supply:
Clavicular branch of thoracoacromial artery

Insertion:
Inferior groove in the mid-portion of the clavicle

Nerves:
Nerve fibers from C5, 6

Origin:
Superior aspect of first rib at its junction with its costal cartilage

TRAPEZIUS

Description:
A large flat triangular superficial muscle on each side of the upper back that serves mainly to elevate the scapula.

Joint Crossings:
One

Rank/Body Action:

Rank	Body Action
1	Scapula adduction (middle fibers)
2	Scapula elevation (upper fibers)
3	Scapula depression (lower fibers)
4	Scapula upward rotation
5	Head extension
6	Head hyperextension
7	Head rotation
8	Head lateral flexion

Blood Supply:
Transverse cervical artery

Insertion:
- Lateral one-third of the clavicle
- Acromion process of the scapula
- Spine of the scapula

Nerves:
- Spinal accessory nerve
- Branches of C3, 4

Origin:
- Superior nuchal line
- Occipital protuberance
- Ligamentum nuchae
- Spines of 7th cervical and all 12 thoracic vertebrae

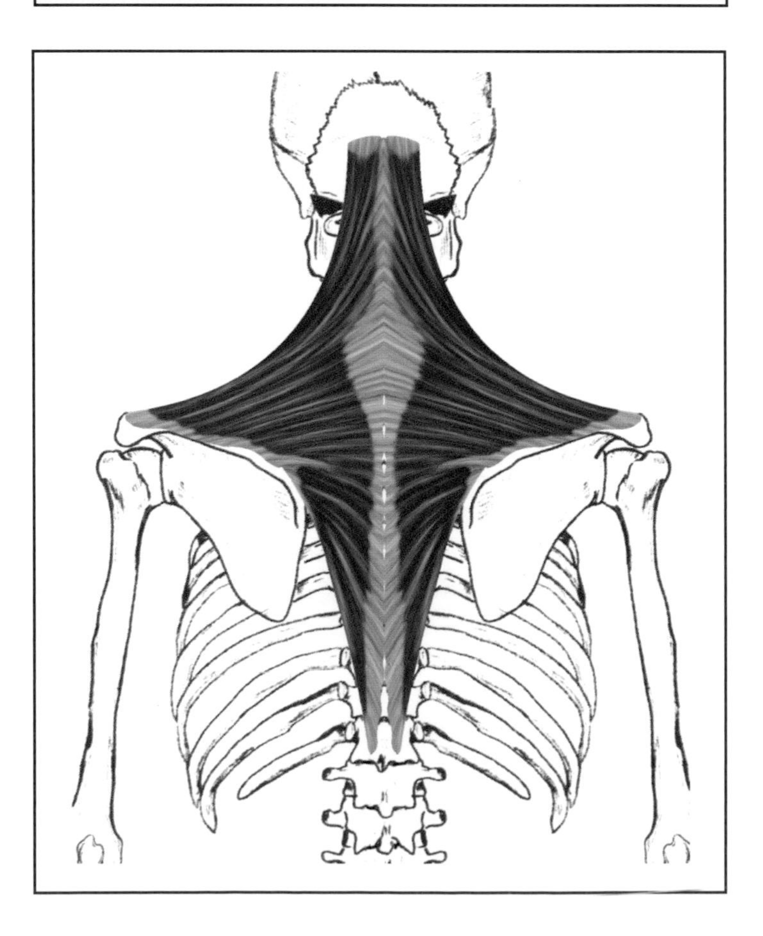

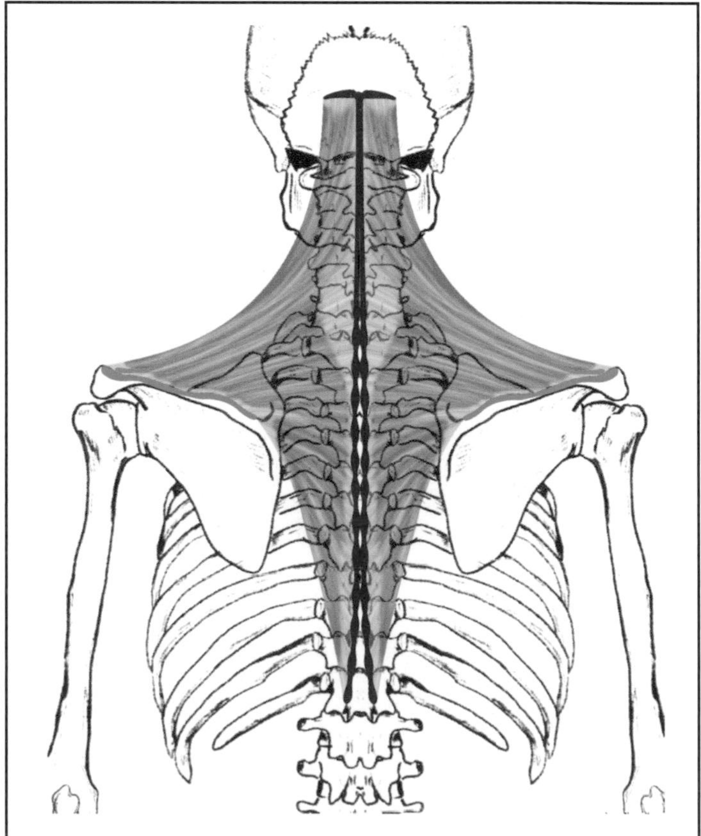

INTERSPINALIS

Description:
Any of the various short muscles that are located between each vertebra to extend the spine.

Joint Crossings:
One

Rank/Body Action:
1 Spine extension
2 Spine hyperextension

Blood Supply:
Muscular branches of the aorta

Insertion:
Spinous process of next vertebra

Nerves:
Posterior primary ramus of spinal nerves

Origin:
Spinous processes of each vertebra

INTERTRANSVERSARII

Description:
A deep set of muscles that connect adjacent vertebrae and laterally flex the spine. This set of muscles includes four parts: intertransversarii anteriores, posteriores, laterales, and mediales.

Joint Crossings:
One

Body Action:
Spine lateral flexion

Blood Supply:
Muscular branches of the aorta

Insertion:
Tubercles of transverse processes of next vertebra

Nerves:
Anterior primary ramus of spinal nerves

Origin:
Tubercles of transverse processes of each vertebra

Longus Colli (Inferior Oblique)

Description:
A group of three muscles (longus colli superior oblique, inferior oblique, and vertical) that are located posteriorly on the cervical vertebrae, and assist in spine flexion.

Joint Crossings:
Spine

Body Action:
Spine flexion (cervical spine)

Blood Supply:
Muscular branches of the aorta

Insertion:
Transverse processes of C5 - C6

Nerves:
Ventral rami C2-C7

Origin:
Bodies of T1 - T3

Longus Colli (Superior Oblique)

Description:
A group of three muscles (longus colli superior oblique, inferior oblique, and vertical) that are located posteriorly on the cervical vertebrae, and assist in spine flexion.

Joint Crossings:
Spine

Body Action:
Spine flexion (cervical spine)

Blood Supply:
Muscular branches of the aorta

Insertion:
Anterior arch of atlas

Nerves:
Ventral rami C2-C7

Origin:
Transverse processes of C3-C5

Longus Colli (Vertical)

Description:
A group of three muscles (longus colli superior oblique, inferior oblique, and vertical) that are located posteriorly on the cervical vertebrae, and assist in spine flexion.

Joint Crossings:
Spine

Body Action:
Spine flexion (cervical spine)

Blood Supply:
Muscular branches of the aorta

Insertion:
Anterior surface of C2-C4

Nerves:
Ventral rami C2-C7

Origin:
Bodies of C5-C7 and T1-T3

Multifidus

Description:
A deep muscle of the back, filling up the groove on each side of the spinous processes of the vertebrae from the sacrum to the skull and consisting of many fasciculi that pass upward and inward to the spinous processes and help extend and rotate the spine.

Joint Crossings:
One

Rank/Body Action:

1	Spine extension
2	Spine hyperextension
3	Spine rotation (contralateral)

Blood Supply:
Muscular branches of the aorta

Insertion:
Spinous processes of 2nd, 3rd, or 4th vertebrae above the origin

Nerves:
Posterior primary ramus of spinal nerves

Origin:
- Sacrum
- Iliac spine
- Transverse processes of lumbar, thoracic, and lower four cervical vertebrae

Rotatores

Description:
Any of several small muscles in the dorsal region of the spine that acts to rotate and extend the spine.

Blood Supply:
Muscular branches of the aorta

Insertion:
Base of spinous process of next vertebra above

Joint Crossings:
One

Nerves:
Posterior primary ramus of spinal nerves

Rank/Body Action:
1 Spine extension
2 Spine hyperextension
3 Spine rotation (contralateral rotation)

Origin:
Transverse process of each vertebra

Semispinalis Cervicis

Description:
A deep longitudinal muscle of the back that, with the semispinalis thoracis, acts to extend the spinal column and rotate it toward the opposite side. Also called the semispinalis colli.

Blood Supply:
Muscular branches of the aorta

Insertion:
Spinous processes of C2 - C5

Joint Crossings:
Spine

Nerves:
Posterior primary divisions of spinal nerves, all divisions

Rank/Body Action:
1 Spine extension
2 Spine hyperextension
3 Spine rotation (contralateral)

Origin:
Transverse processes of T1 - T6 vertebrae

SEMISPINALIS THORACIS

Description:
A deep longitudinal muscle of the back that, with the semispinalis cervicis, acts to extend the spinal column and rotate it toward the opposite side.

Joint Crossings:
One

Rank/Body Action:

Rank	Body Action
1	Spine extension
2	Spine hyperextension
3	Spine rotation (contralateral)

Blood Supply:
Muscular branches of the aorta

Insertion:
Spinous processes of C6 - C7 and T1 - T4

Nerves:
Posterior primary ramus of spinal nerves

Origin:
Transverse processes of T6 - T10

Iliocostalis Cervicis

Description:
A muscle that extends from the ribs to the cervical vertebrae and extends and laterally flexes the upper cervical vertebrae. Also known as cervicalis ascendens.

Joint Crossings:
Spine

Rank/Body Action:
1 Spine extension
2 Spine hyperextension
3 Spine lateral flexion

Blood Supply:
Muscular branches of the aorta

Insertion:
Transverse processes of the 4th, 5th, and 6th cervical vertebrae

Nerves:
Dorsal rami of spinal nerves

Origin:
3rd, 4th, 5th and 6th ribs

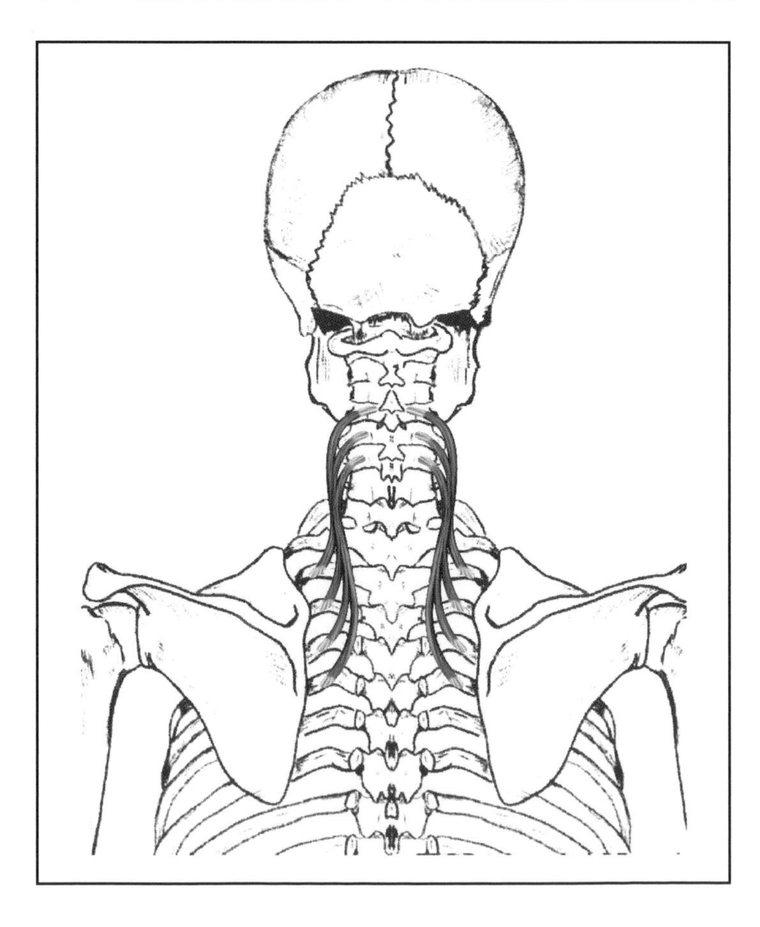

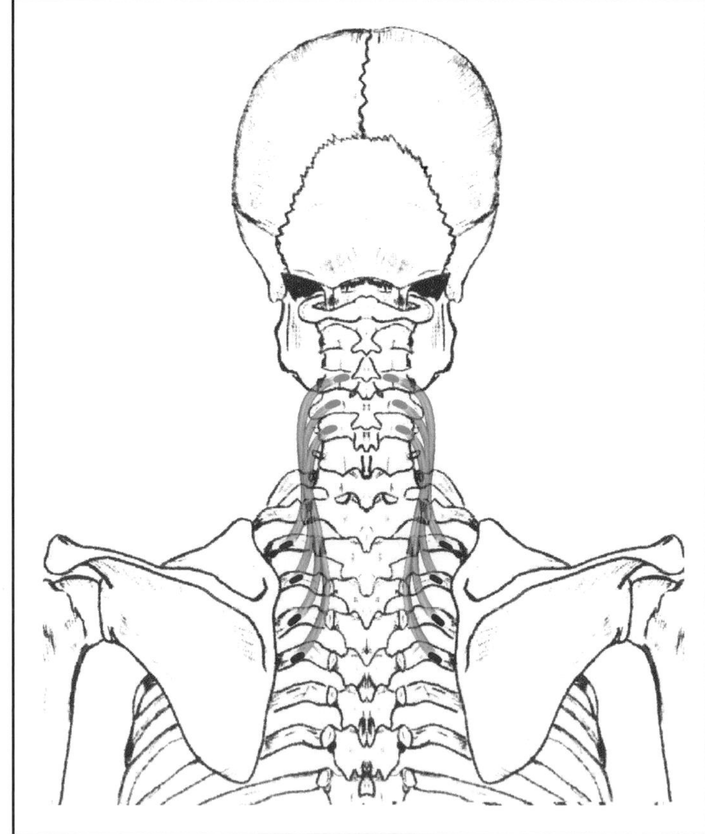

Iliocostalis Lumborum

Description:
A muscle that extends from the lumbodorsal fascia to the lower ribs and individually acts to draw the spine to the same side and to extend the spine when working together. Also known as the sacrolumbalis.

Joint Crossings:
Spine

Rank/Body Action:
1 Spine extension (both)
2 Spine hyperextension (both)
3 Spine lateral flexion (each)

Blood Supply:
Muscular branches of the aorta

Insertion:
Lower six ribs

Nerves:
Posterior branches of spinal nerves

Origin:
Lumbodorsal fascia – (posterior iliac crest, posterior surface of sacrum, lumbar vertebrae)

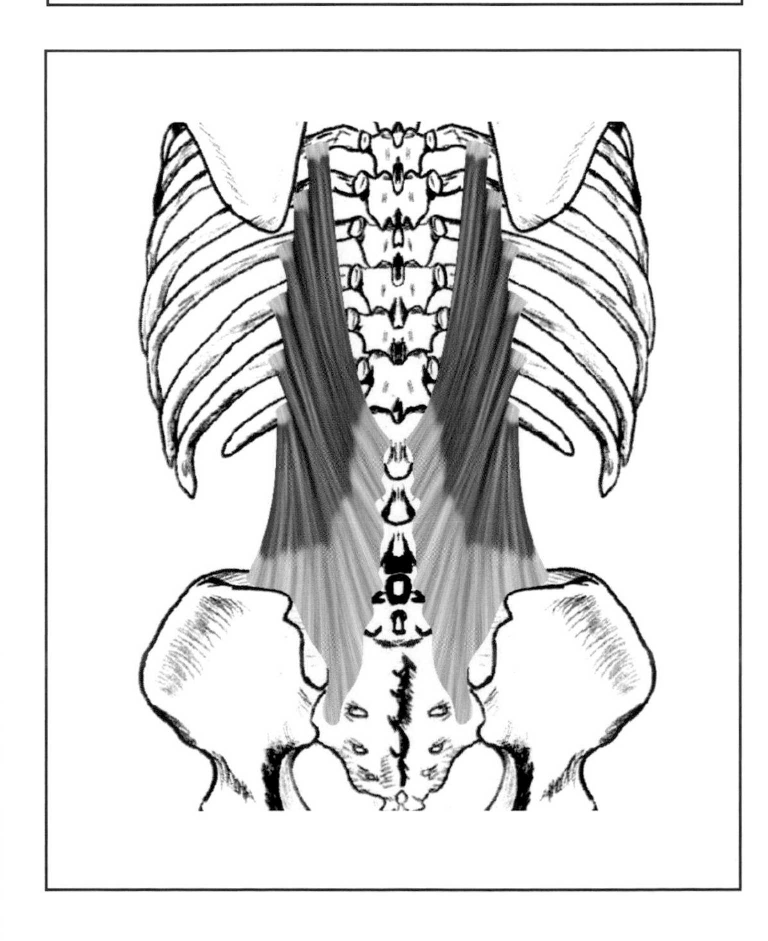

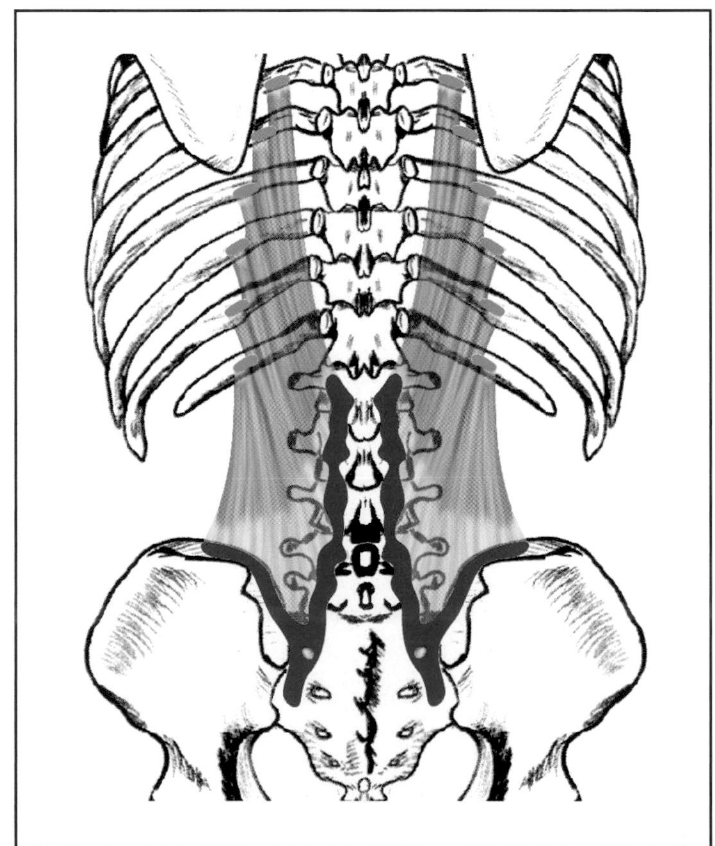

ILIOCOSTALIS THORACIS

Description:
A muscle that extends from the lower to the upper ribs and acts to laterally flex the spine to the same side. Also known as the musculus accessorius.

Joint Crossings:
Spine

Rank/Body Action:
1 Spine extension
2 Spine hyperextension
3 Spine lateral flexion

Blood Supply:
Muscular branches of the aorta

Insertion:
Upper six ribs

Nerves:
Dorsal rami of spinal nerves

Origin:
Lower six ribs

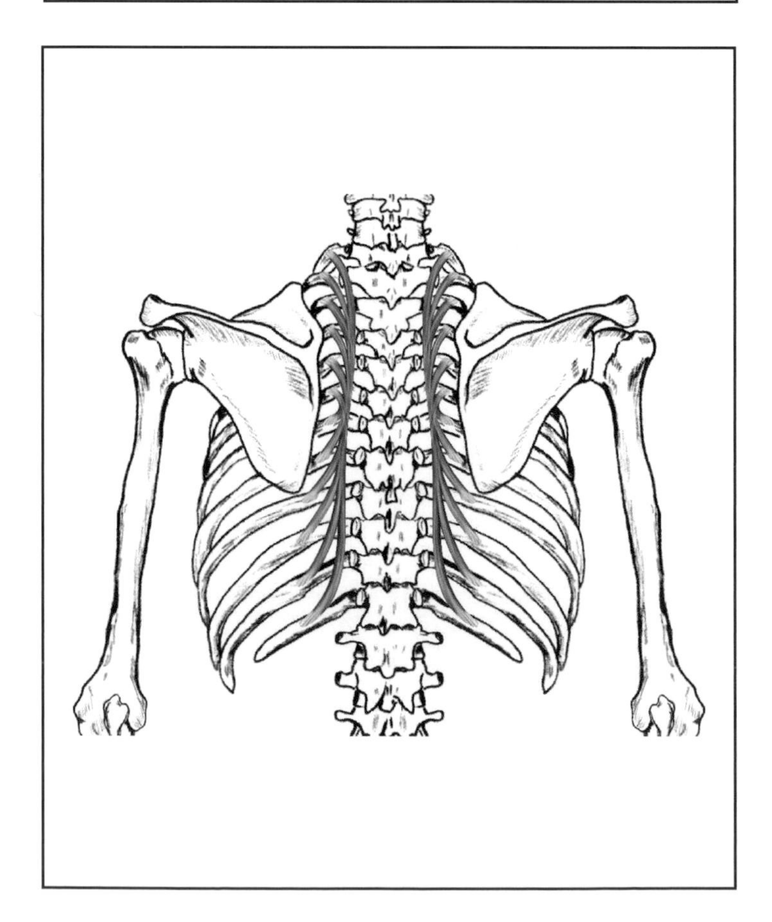

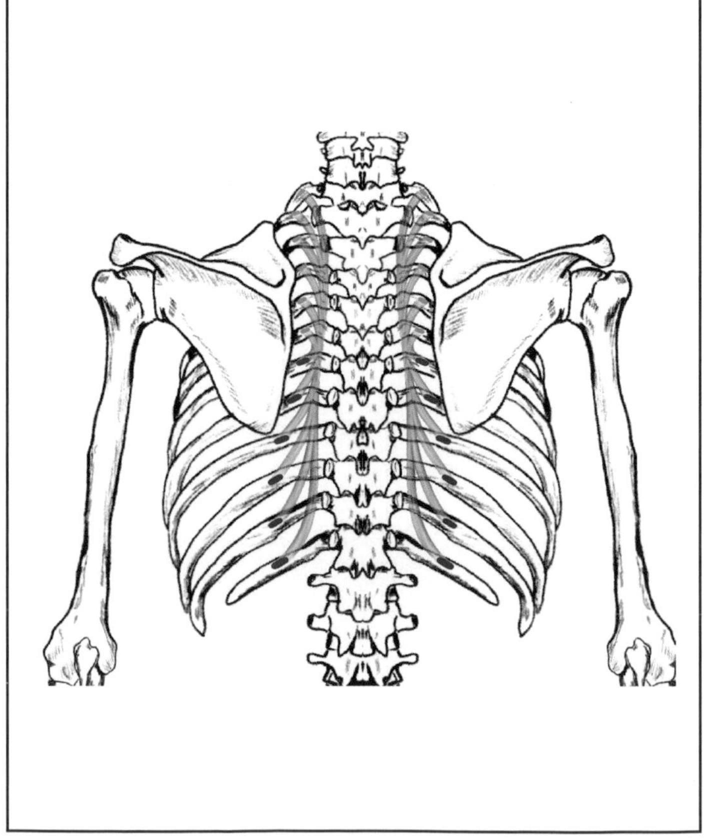

Longissimus Capitis

Description:
A long slender muscle between the longissimus cervicis and the semispinalis capitis that extends the head and spine. Also called the trachelomastoid.

Joint Crossings:
Two

Rank/Body Action:
1 Spine extension
2 Spine hyperextension
3 Head extension
4 Head hyperextension
5 Head lateral flexion
6 Head rotation

Blood Supply:
Muscular branches of the aorta

Insertion:
Mastoid process of the temporal bone

Nerves:
Dorsal rami of spinal nerves

Origin:
Lower four cervical vertebrae and upper four thoracic vertebrae (eight vertebrae total)

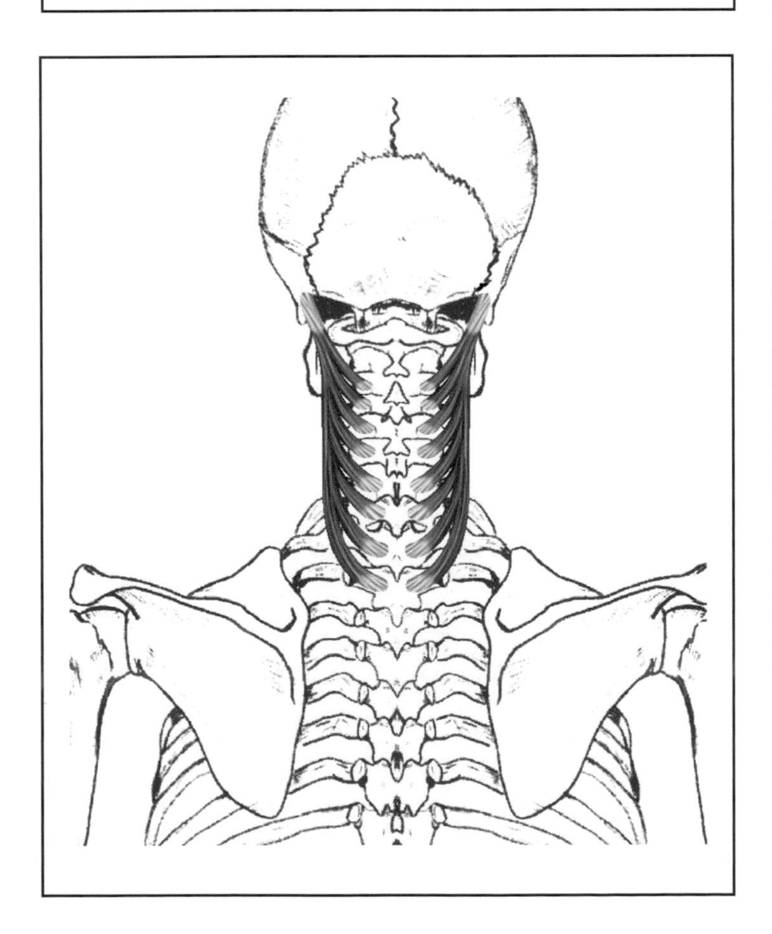

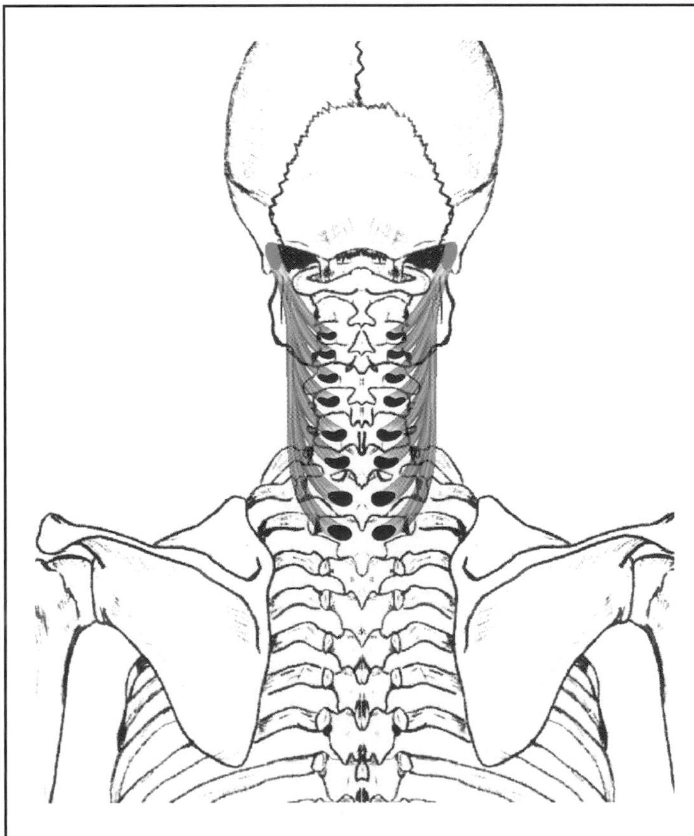

Longissimus Cervicis

Description:
A slender muscle medial to the longissimus thoracis that extends the spine.

Joint Crossings:
Spine

Rank/Body Action:
1 Spine extension
2 Spine hyperextension

Blood Supply:
Muscular branches of the aorta

Insertion:
Second through the sixth cervical vertebrae (all but C1 and C7)

Nerves:
Dorsal rami of spinal nerves

Origin:
Transverse processes of the upper four or five thoracic vertebrae

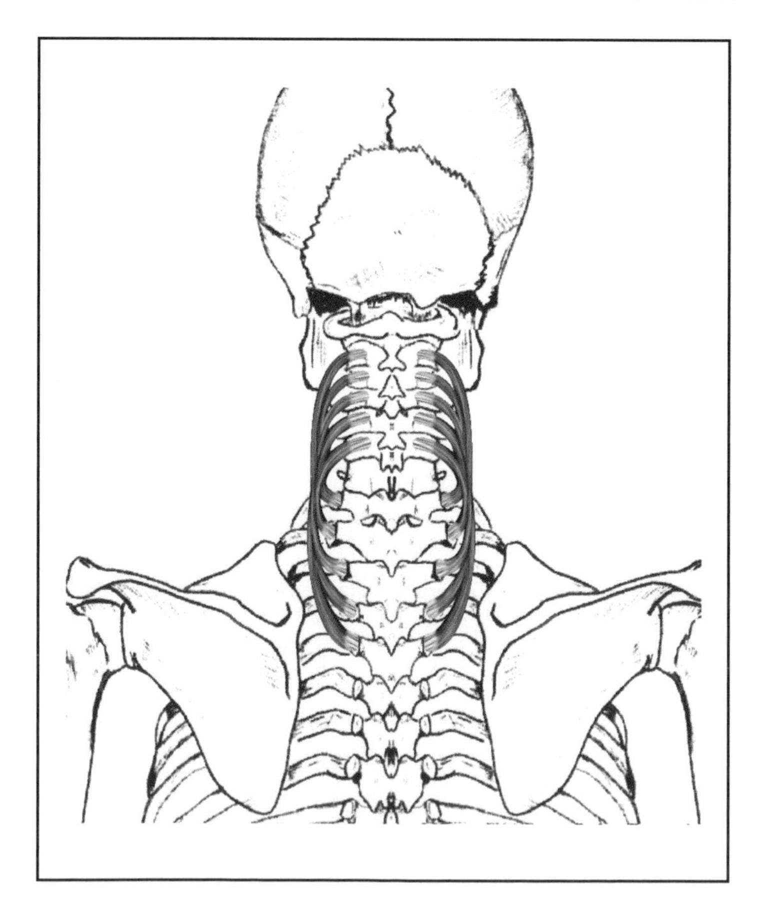

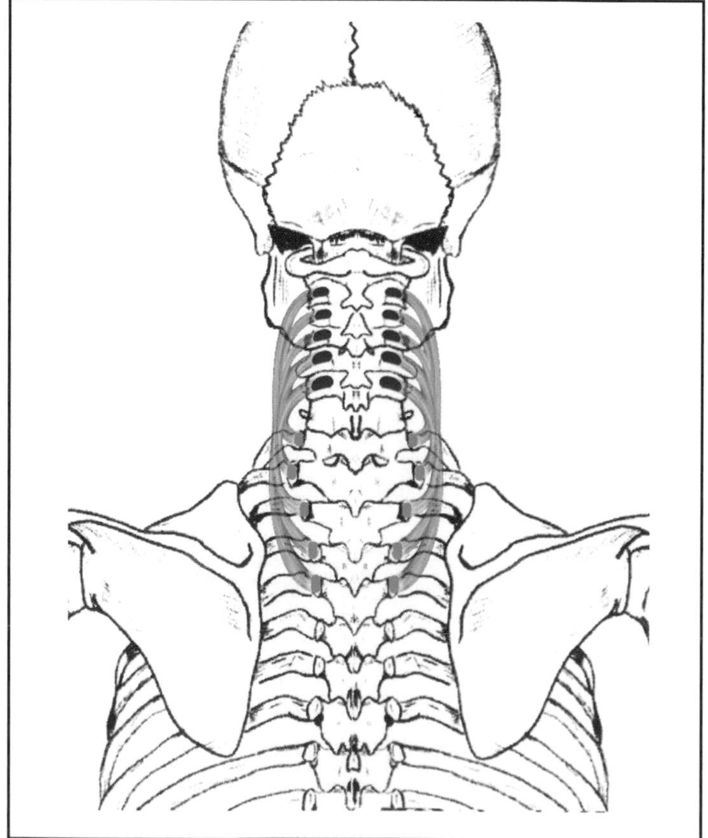

Longissimus Thoracis

Description:
A muscle whose origin blends in with the iliocostalis lumborum.

Joint Crossings:
Spine

Rank/Body Action:
1 Spine extension
2 Spine hyperextension

Blood Supply:
Muscular branches of the aorta

Insertion:
Transverse processes of all of the thoracic vertebrae and adjacent rib surface

Nerves:
Dorsal rami of spinal nerves

Origin:
Lumbodorsal fascia, which includes the posterior iliac crest, posterior surface of sacrum, lumbar vertebrae

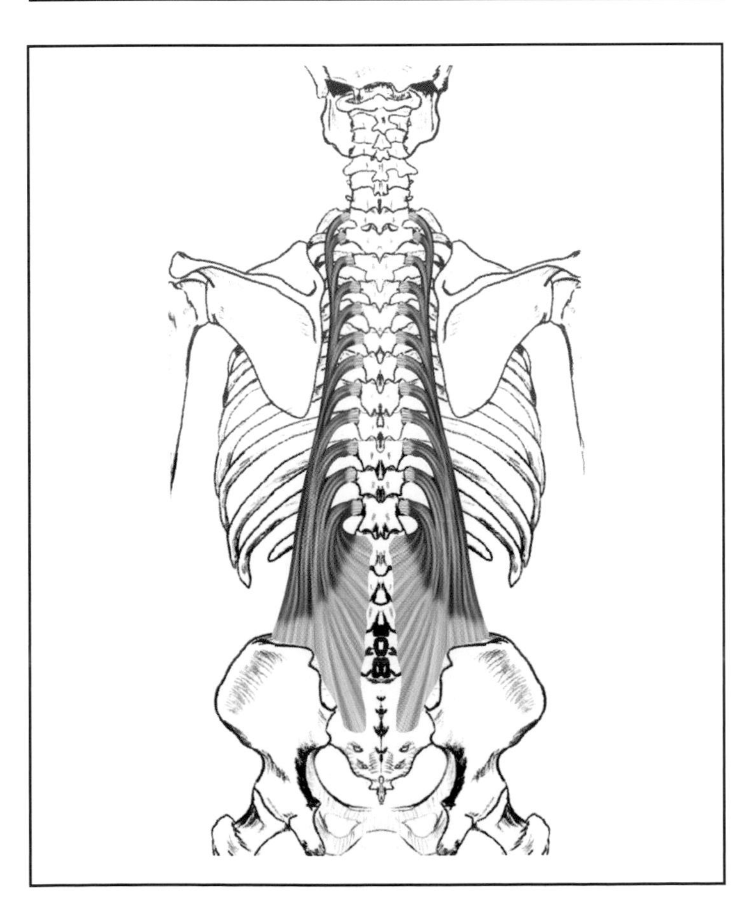

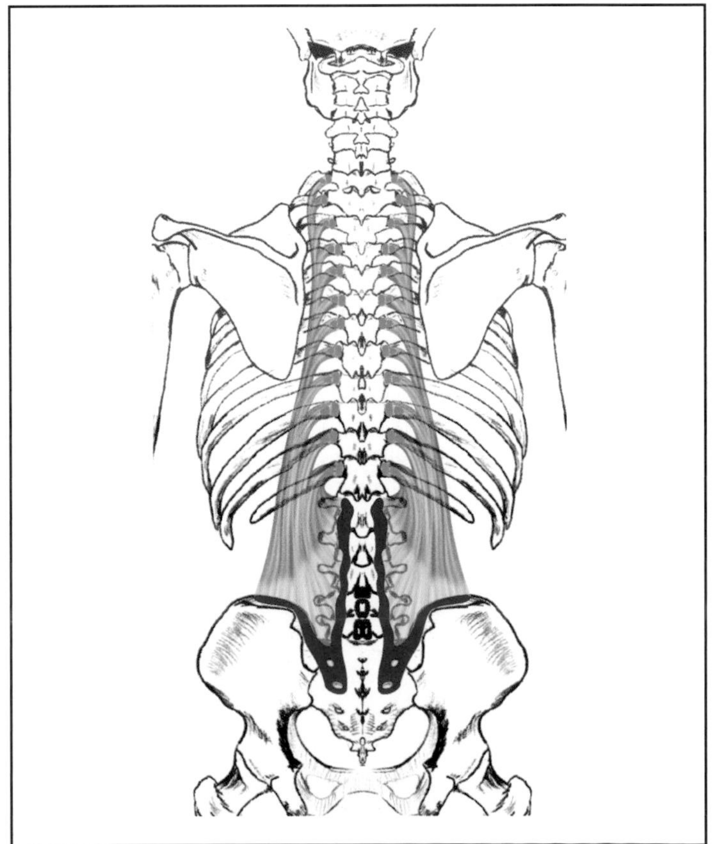

Spinalis Capitis

Description:
Also know as the semispinalis capitis and the biventer cervicis.

Joint Crossings:
Two

Rank/Body Action:

Rank	Body Action
1	Spine extension (both – upper spine)
2	Spine hyperextension (both – upper spine)
3	Head extension (both)
4	Head hyperextension (both)
5	Head lateral flexion (each – slight)
6	Head rotation (each – to the opposite side)

Blood Supply:
Muscular branches of the aorta

Insertion:
Occipital bone between the superior and inferior nuchal line

Nerves:
Dorsal rami of spinal nerves

Origin:
Transverse processes of the first six or seven thoracic vertebrae and the C4-C7 vertebrae (10-11 vertebrae in all)

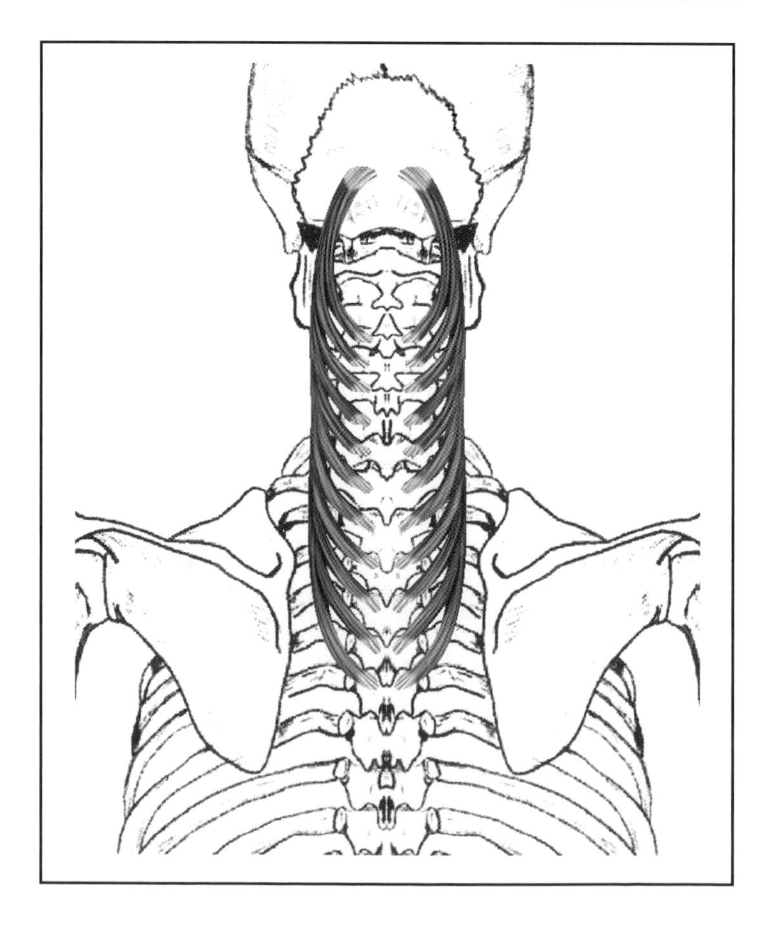

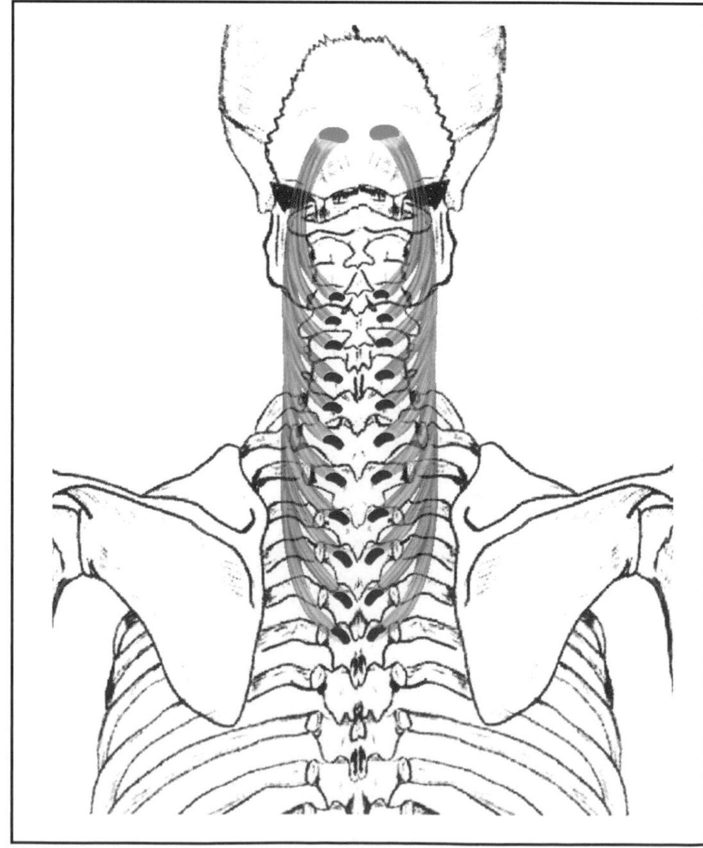

Spinalis Cervicis

Description:
An irregular shaped small muscle that extends the spine and laterally flexes the upper one or two vertebrae.

Joint Crossings:
Spine

Rank/Body Action:
1 Spine extension
2 Spine hyperextension
3 Spine lateral flexion (upper spine – weak)

Blood Supply:
Muscular branches of the aorta

Insertion:
Spine of the axis (second cervical vertebrae)

Nerves:
Dorsal rami of spinal nerves

Origin:
- Lower part of nuchal ligament
- Seventh cervical vertebra
- first and second thoracic vertebrae

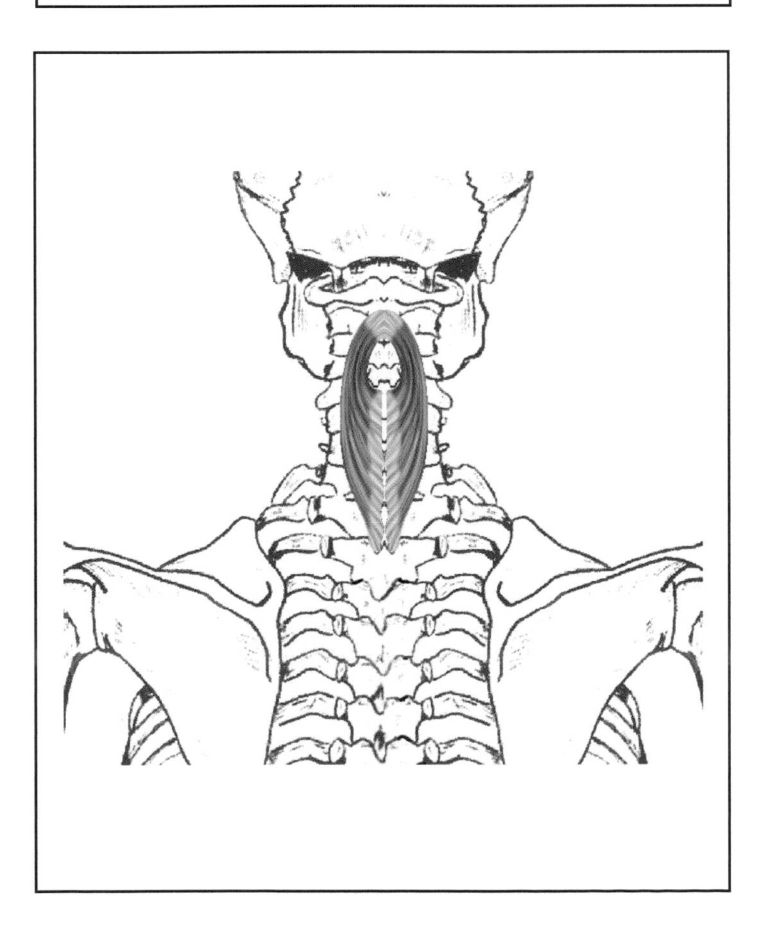

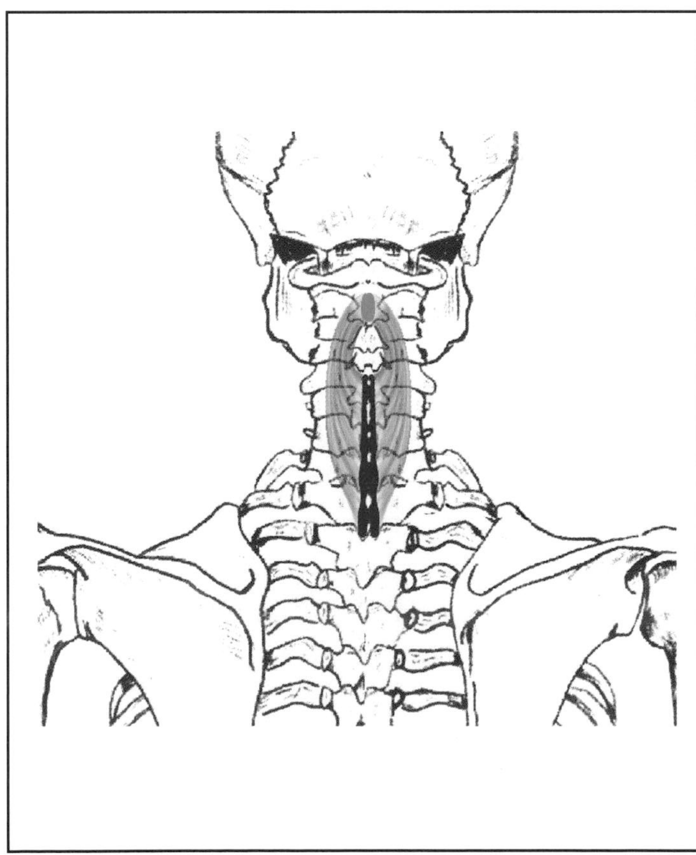

Spinalis Thoracis

Description:
An upward continuation of the sacrospinalis that is situated medially to and blends with the longissimus thoracis. It is a spine extensor.

Joint Crossings:
Spine

Rank/Body Action:
1 Spine extension
2 Spine hyperextension

Blood Supply:
Muscular branches of the aorta

Insertion:
Spines of the upper eight thoracic vertebrae

Nerves:
Dorsal rami of spinal nerves

Origin:
Spines of the 11th and 12th thoracic vertebrae and first and second lumbar vertebrae

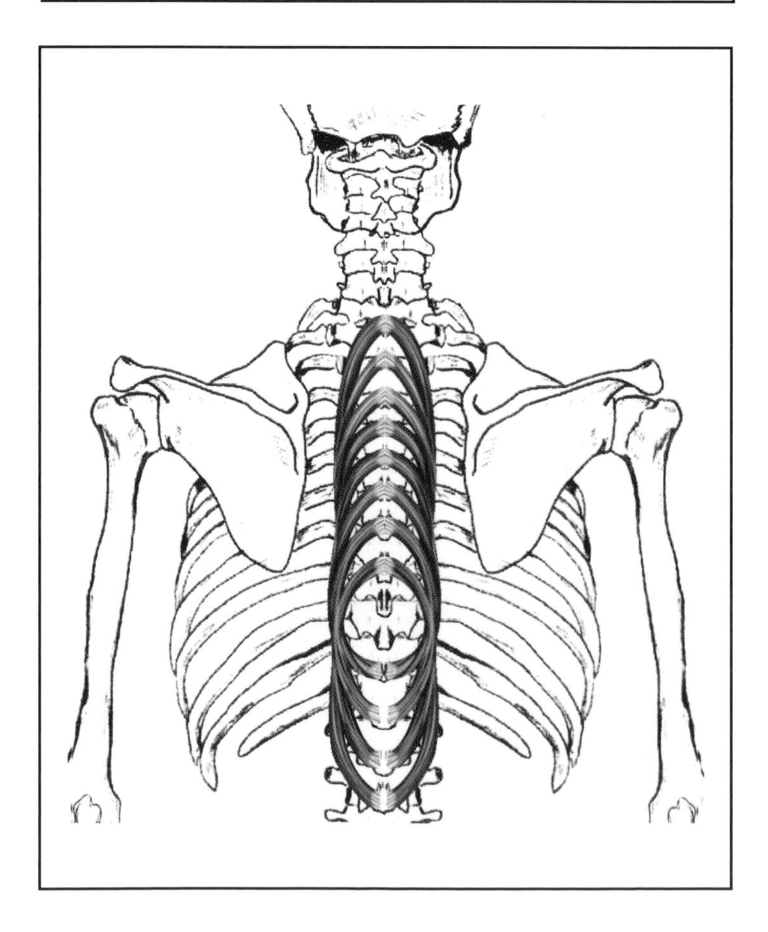

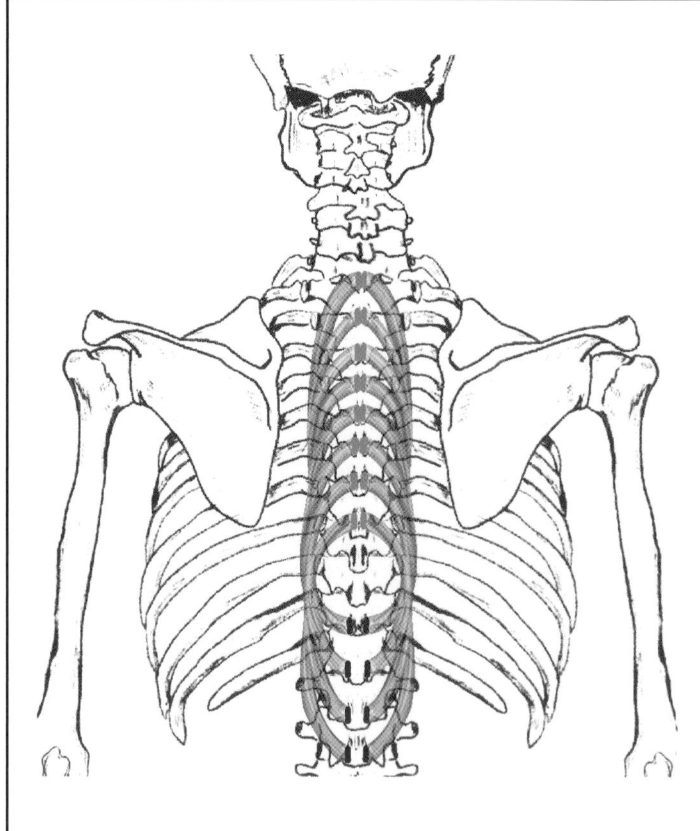

External Intercostals

Description:
These muscles elevate the ribs to increase thoracic volume to assist with inspiration. Also known as the intercostales externi.

Joint Crossings:
One

Body Action:
Rib elevation

Blood Supply:
Intercostal arteries

Insertion:
Superior border of the rib below

Nerves:
Intercostal branches of T1-11

Origin:
Eleven in number; the inferior border of each rib

Internal Intercostals

Description:
Muscles that depress the ribs during exhalation, and help elevate the upper ribs during inhalation. Also known as the intercostales interni.

Joint Crossings:
N/A

Rank/Body Action:

1 Rib elevation (upper four ribs – during inhalation)

2 Rib depression (during exhalation)

Blood Supply:
Intercostal arteries

Insertion:
They extend dorsally to the angle of the ribs

Nerves:
Intercostal branches of T1-11

Origin:
Eleven in number; in the space between the cartilages of the ribs in the sternum

Levator Costarum

Description:
A series of 12 muscles on each side of the spine that raise the ribs increasing the volume of the thoracic cavity and laterally flex the spine.

Joint Crossings:
One

Rank/Body Action:
1 Rib elevation
2 Spine lateral flexion (thoracic spine)

Blood Supply:
Deep cervical artery

Insertion:
They pass obliquely and downward and out to insert of the rib immediately below

Nerves:
Intercostal nerves

Origin:
Twelve in number: from the ends of transverse processes of C7, T2-T11

Scalenus Anterior

Description:
One of usually three deeply situated muscles on each side of the neck that functions to flex the spine forward and laterally, and to rotate it to the side. Also called the anterior scalene or scalenus anticus.

Joint Crossings:
Spine

Rank/Body Action:
1 Rib elevation (First rib)
2 Spine flexion (cervical spine)
3 Spine lateral flexion (cervical spine)
4 Spine rotation (contralateral - cervical spine)

Blood Supply:
Inferior thyroid artery (branch of the thyrocervical trunk)

Insertion:
Inner border and upper surface of 1st rib

Nerves:
Ventral rami C5, 6

Origin:
Transverse processes of C3 - C6

Scalenus Medius

Description:
Any of usually three deeply situated muscles on each side of the neck that functions to flex the spine forward and laterally, and to rotate it to the side. Also called the middle scalene.

Joint Crossings:
One

Rank/Body Action:
1 Rib elevation (first rib)
2 Spine lateral flexion (cervical spine)
3 Spine rotation (cervical spine)

Blood Supply:
Ascending cervical artery

Insertion:
Superior surface of 1st rib

Nerves:
Ventral rami C3-C8

Origin:
Transverse processes of C2 - C7

Scalenus Posterior

Description:
Any of usually three deeply situated muscles on each side of the neck that functions to raise the second rib and to flex and slightly rotate the spine. Also called the posterior scalene.

Joint Crossings:
One

Rank/Body Action:
1 Rib elevation (second rib)
2 Spine flexion (cervical spine)
3 Spine lateral flexion (cervical spine)
4 Spine rotation (contralateral - cervical spine - slight)

Blood Supply:
Ascending cervical artery

Insertion:
Outer surface of second rib

Nerves:
Ventral rami C6, 7, 8

Origin:
Transverse processes of C5 - C7

Serratus Posterior (Inferior)

Description:
A thin quadrilateral muscle at the junction of the thoracic and lumbar regions that acts to counteract the pull of the diaphragm on the ribs to which it is attached.

Blood Supply:
- Lower posterior intercostal artery
- Subcostal artery

Joint Crossings:
One

Insertion:
Inferior borders lateral to angles of ribs 8-12

Body Action:
Rib depression (depress ribs 8-12 downward and outward; counteracts the inward pull of the diaphragm)

Nerves:
Branches from anterior primary rami of T9-T12

Origin:
Spinous processes of T10 - T12 and L1 - L3

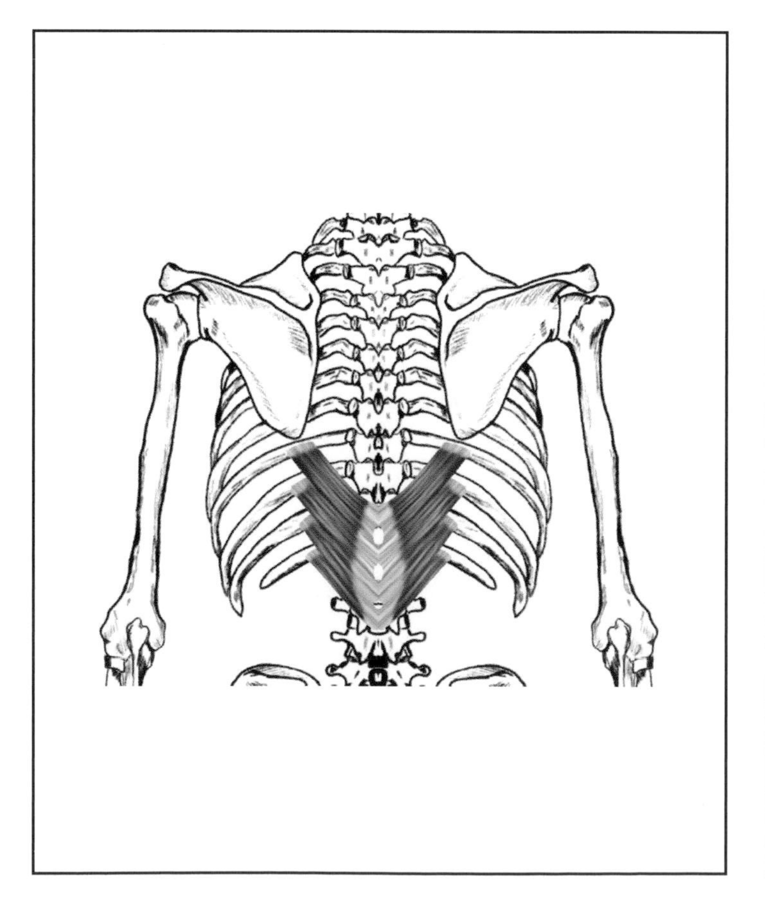

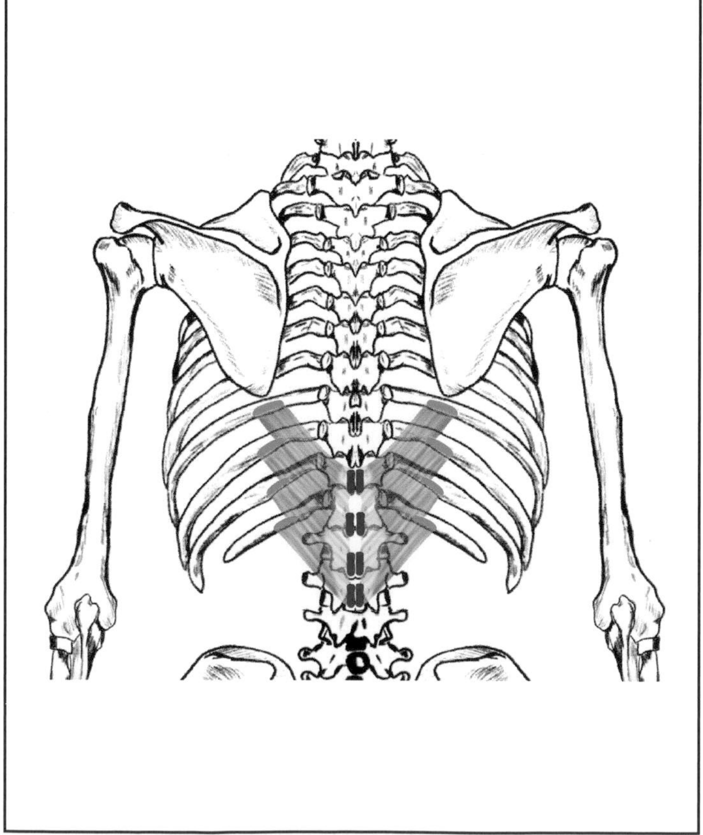

Serratus Posterior (Superior)

Description:
A thin quadrilateral muscle of the upper and dorsal part of the thorax that acts to elevate the upper ribs.

Joint Crossings:
One

Body Action:
Rib elevation (upper ribs)

Blood Supply:
Posterior intercostal arteries

Insertion:
Superior borders lateral to angles of ribs 2-5

Nerves:
Branches from anterior primary rami of T1-T4

Origin:
- Ligamentum nuchae
- Spinous processes of C7, T1, and T2 or T3

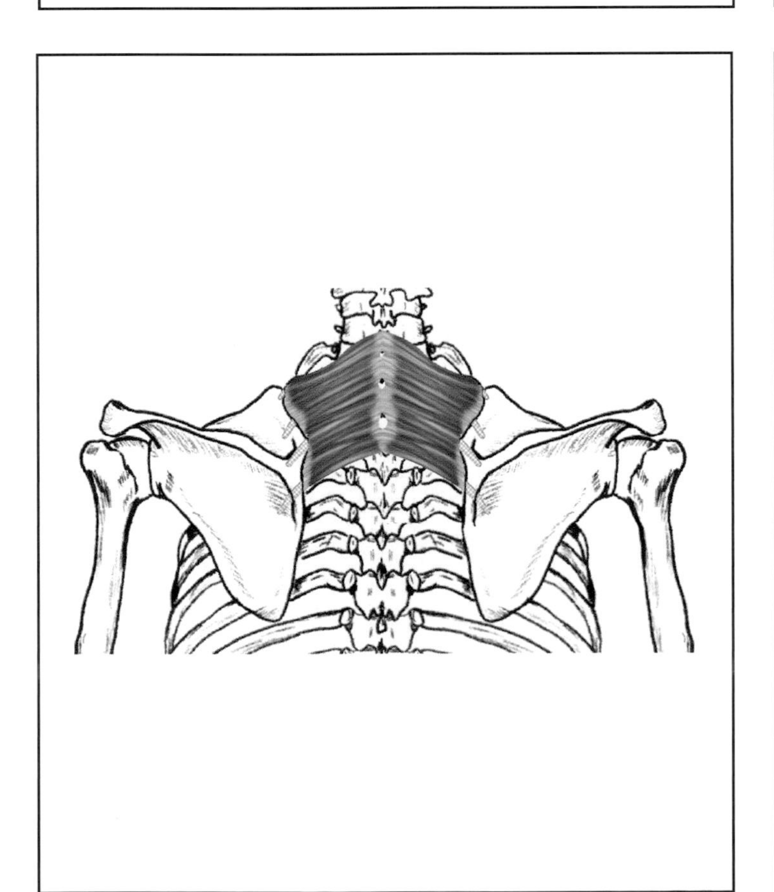

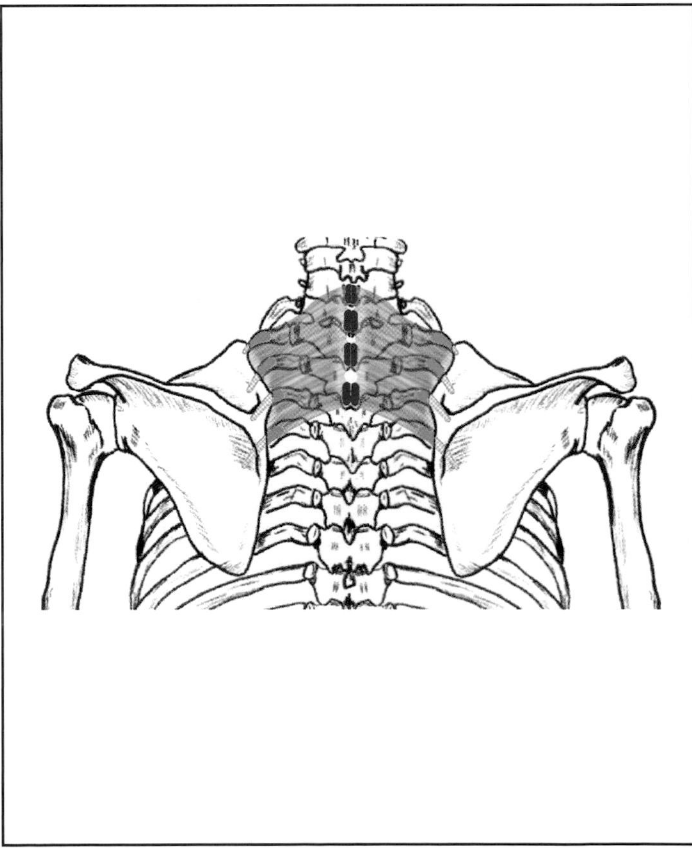

SUBCOSTALES

Description:
Any of a variable number of small muscles that function to depress the ribs by drawing them together. Also known as the intracostales.

Joint Crossings:
N/A

Body Action:
Rib depression (decreasing thoracic cavity volume)

Blood Supply:
Intercostal arteries

Insertion:
Medially on the inner surface of second or third rib below

Nerves:
Intercostal nerves

Origin:
Inner surface of each rib near its angle

TRANSVERSUS THORACIS

Description:
A thin flat sheet of muscle and tendon fibers of the anterior wall of the chest that acts to draw the ribs downward. Also called the triangularis sterni.

Joint Crossings:
One

Body Action:
Rib depression (ribs 3-6)

Blood Supply:
Intercostal arteries

Insertion:
Inner surfaces and inferior borders of costal cartilages 2 - 6

Nerves:
Intercostal branches of T3-T6

Origin:
- Inner surface of the sternum
- Xiphoid process
- Sternal ends of costal cartilages of the last three or four true ribs

Abductor Pollicis Longus

Description:
A muscle of the forearm that abducts the thumb and radially flexes the wrist. Also known as the extensor ossis metacarpi pollicis.

Joint Crossings:
One

Rank/Body Action:
1 Thumb abduction
2 Wrist radial flexion

Blood Supply:
Radial recurrent artery

Insertion:
Dorsal surface of the base of the first metacarpal (thumb)

Nerves:
Deep branch of radial nerve C6, 7

Origin:
- The middle one third of the medial posterior surface of the radius
- The upper lateral surface of the middle of the ulna

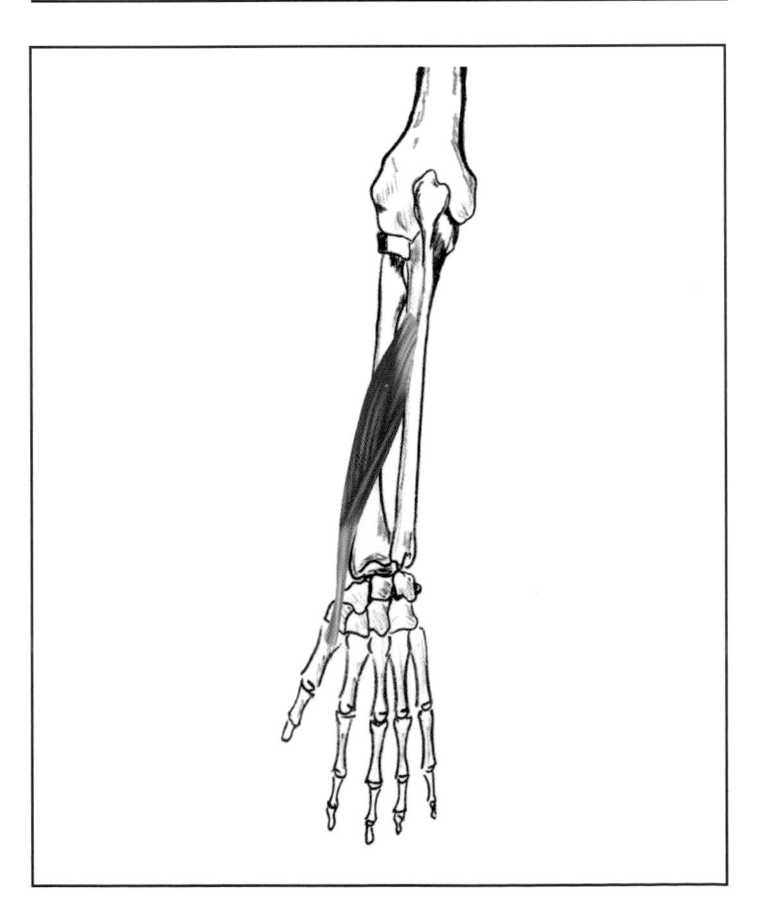

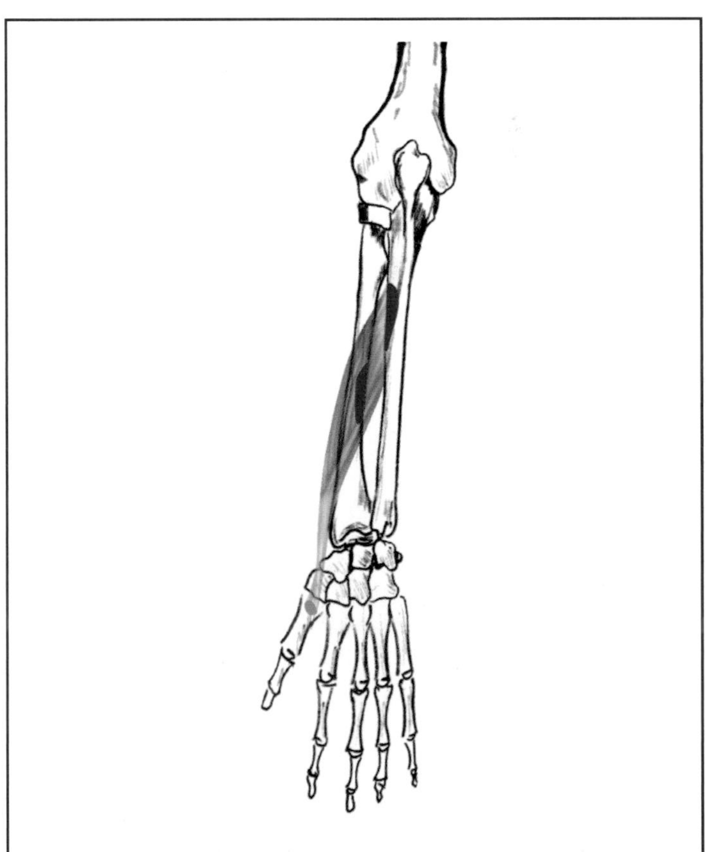

EXTENSOR INDICIS

Description:
A thin muscle that extends the index finger, especially used when pointing. Also known as the extensor indicis proprius.

Joint Crossings:
One

Rank/Body Action:

1	Finger extension (index finger)
2	Finger hyperextension (index finger)
3	Wrist extension (weak)
4	Wrist hyperextension (weak)

Blood Supply:
Posterior interosseous artery

Insertion:
The tendon of the extensor digitorum at about the region of the second metacarpal of the index finger

Nerves:
Posterior interosseous nerve of radial nerve C6, 7, 8

Origin:
Dorsal surface of the lower shaft of the ulna

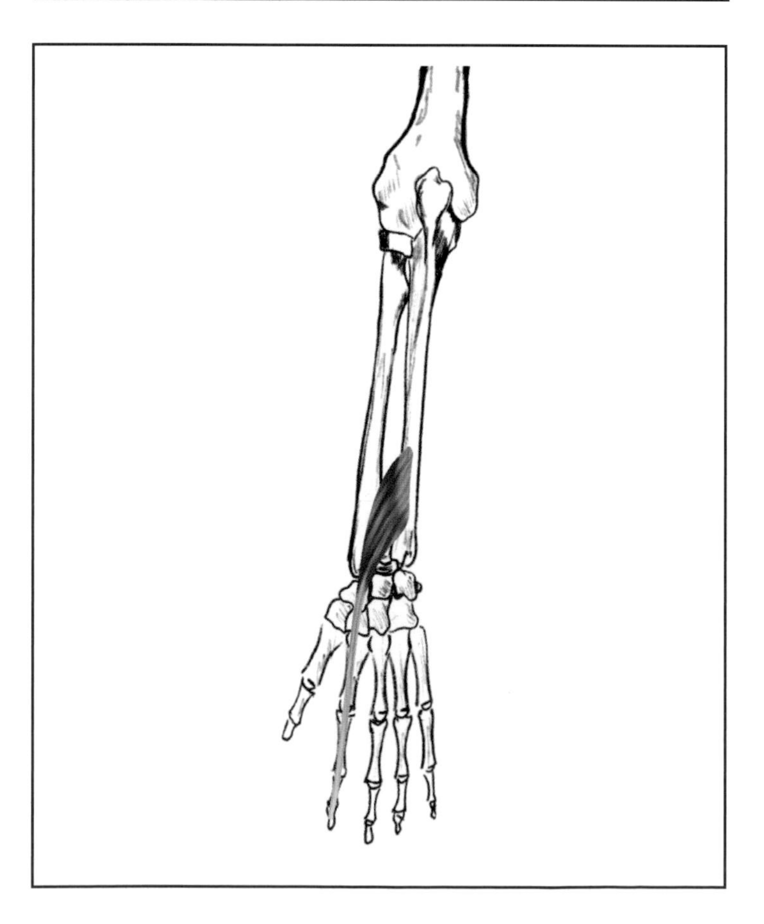

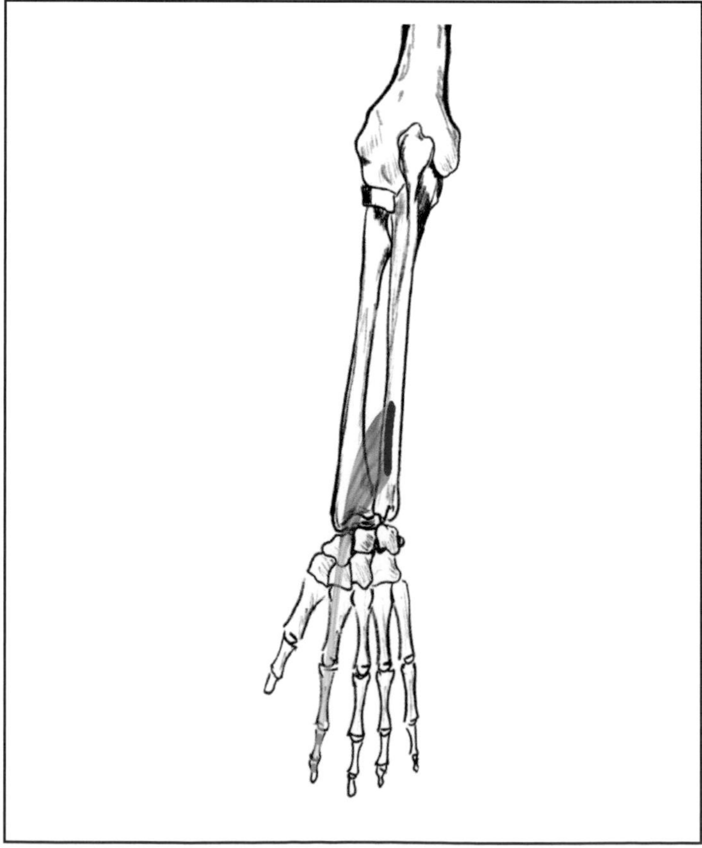

Extensor Pollicis Brevis

Description:
A muscle that extends the first phalange of the thumb, and slightly extends the wrist.

Joint Crossings:
Two

Rank/Body Action:
1 Thumb extension
2 Wrist extension (slight)
3 Wrist hyperextension (slight)

Blood Supply:
Posterior interosseous artery

Insertion:
Dorsal surface of the first phalange of the thumb

Nerves:
Posterior interosseous nerve of radial nerve C6, 7

Origin:
Posterior surface of the lower one-third of the radius

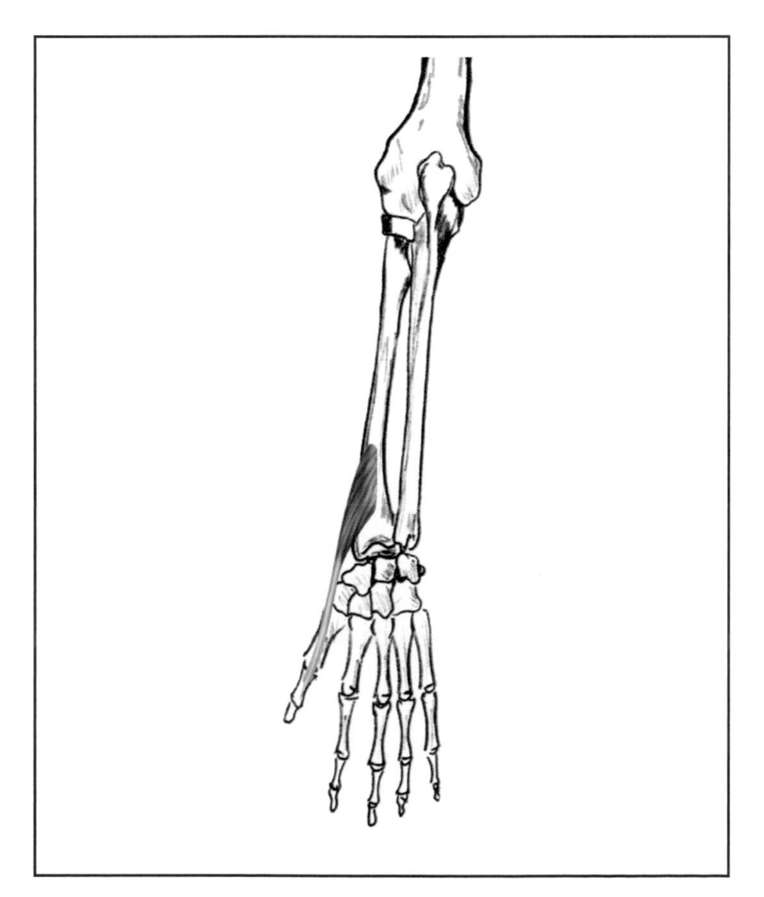

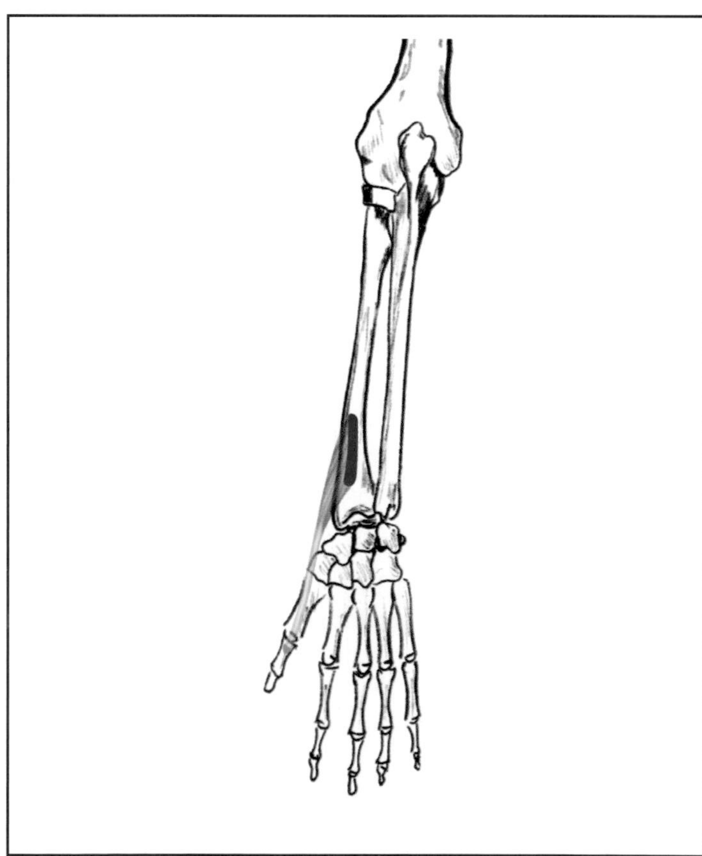

Extensor Pollicis Longus

Description:
A muscle that extends the second phalanx of the thumb, and extends the wrist. Also known as the extensor secundi internodii pollicis.

Joint Crossings:
Two

Rank/Body Action:

1	Thumb extension
2	Wrist extension
3	Wrist hyperextension

Blood Supply:
Posterior interosseous artery

Insertion:
Dorsal surface of the base of the terminal phalange of the thumb

Nerves:
Posterior interosseous nerve of radial nerve C6, 7, 8

Origin:
Lateral part of the middle third of the posterior surface of the shaft of the ulna

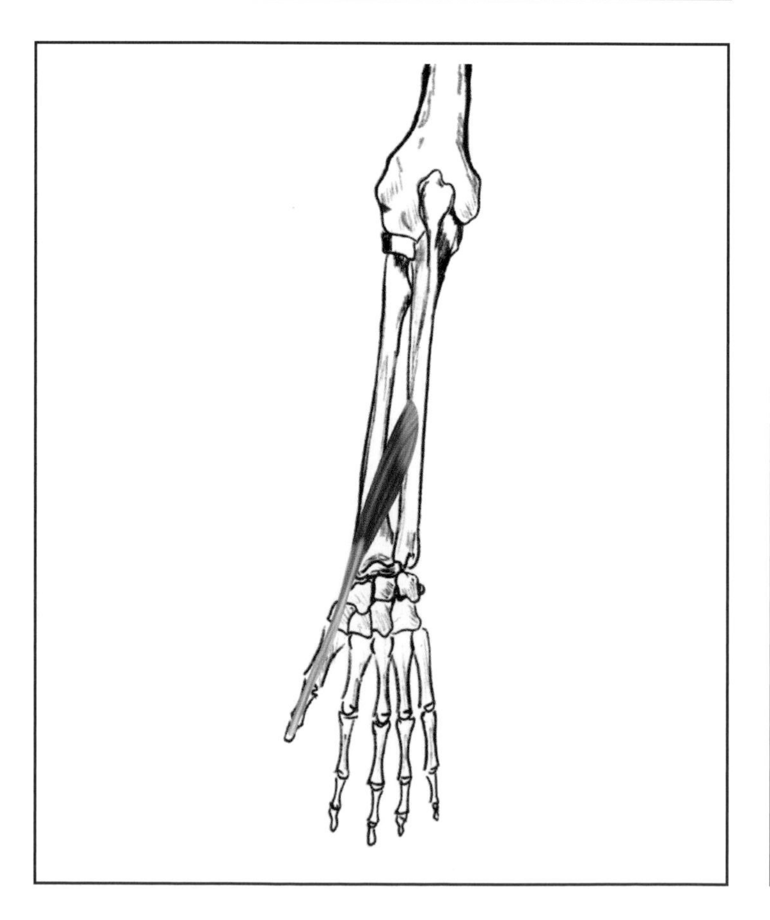

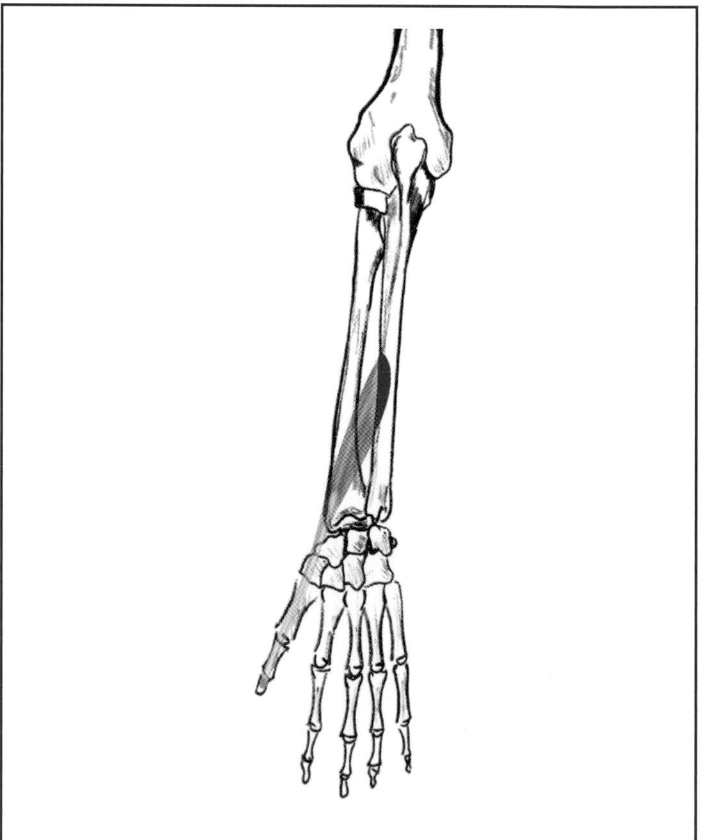

Flexor Digitorum Profundus

Description:
A deep muscle of the outside of the forearm that flexes the terminal phalanges of the four fingers - used in any type of gripping action. With the flexor digitorum superficialis, it is one of only two muscles that flex all four fingers.

Joint Crossings:
Two

Rank/Body Action:
1 Finger flexion
2 Wrist flexion

Blood Supply:
- Muscular branches of the ulnar artery
- Muscular branches of the radial artery
- Anterior interosseous artery

Insertion:
By split tendons to the terminal phalanges of the four fingers

Nerves:
- Fingers 2,3: medial nerve C8, T1
- Fingers 4,5: ulnar nerve C8, T1

Origin:
Upper anterior three-fourths of the ulna

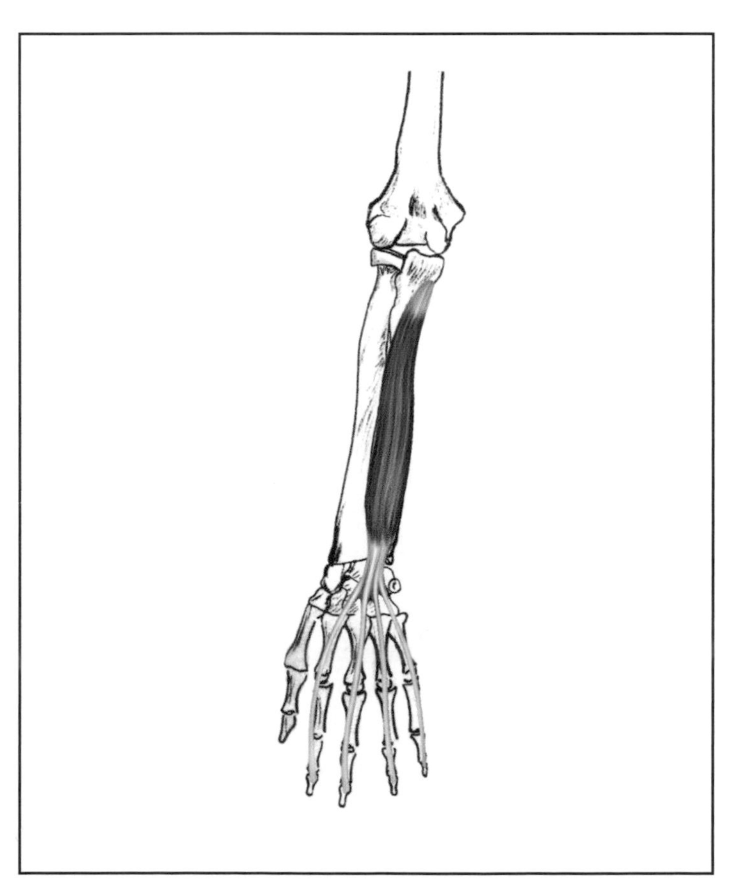

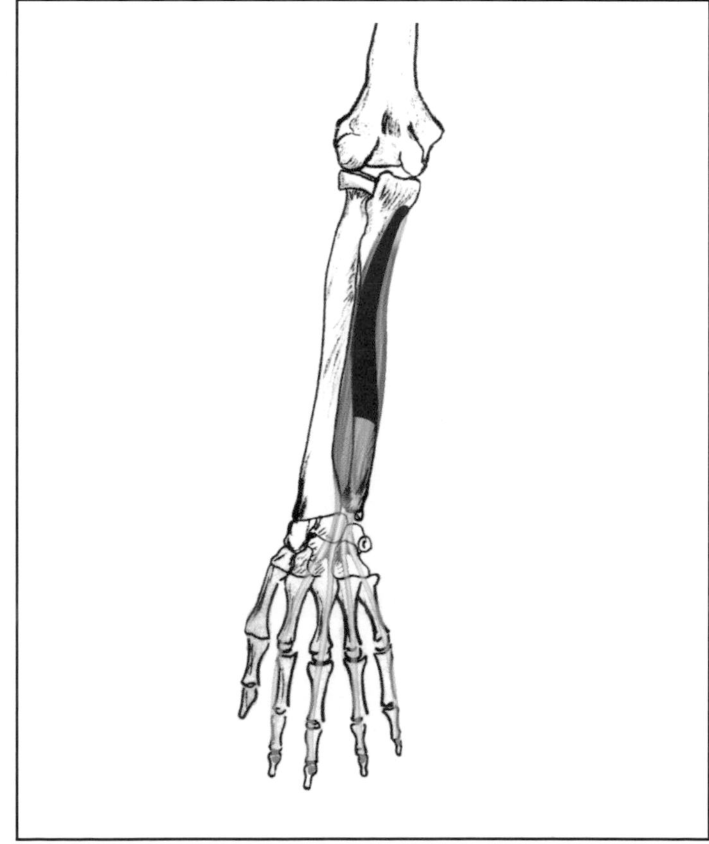

Flexor Pollicis Longus

Description:
A muscle of the radial side of the forearm that flexes the terminal phalange of the thumb.

Joint Crossings:
Two

Rank/Body Action:
1 Thumb flexion
2 Wrist flexion

Blood Supply:
- Muscular branches of radial artery
- Anterior interosseous artery

Insertion:
Base of the terminal phalange of the thumb (palmar surface)

Nerves:
Palmar interosseous branch of median nerve C8, T1

Origin:
- Middle anterior surface of the radius
- Anterior medial border of the ulna just distal to the coronoid process

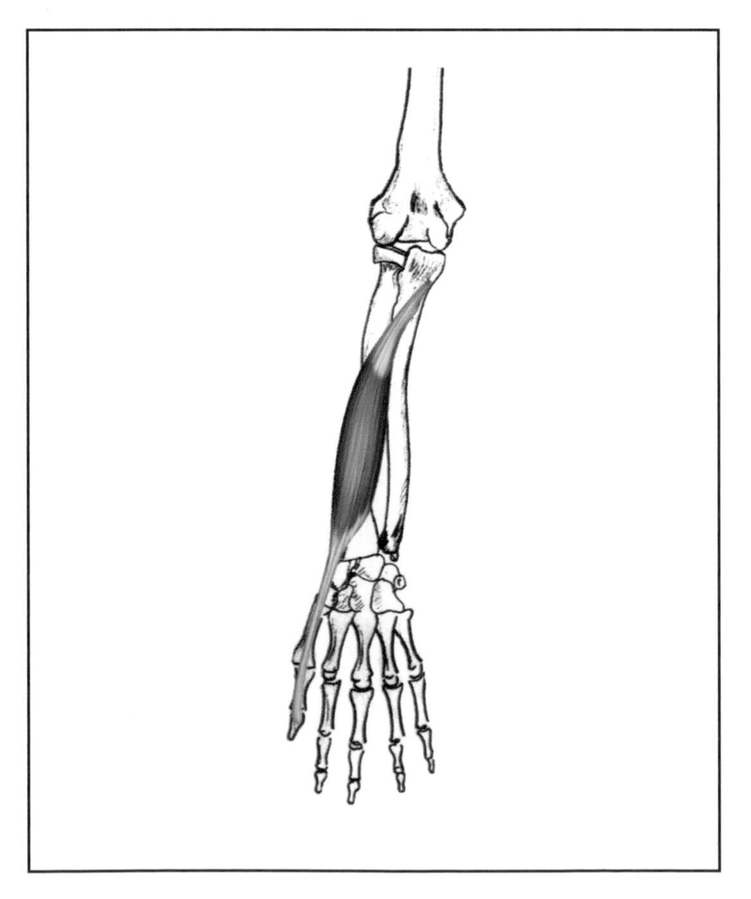

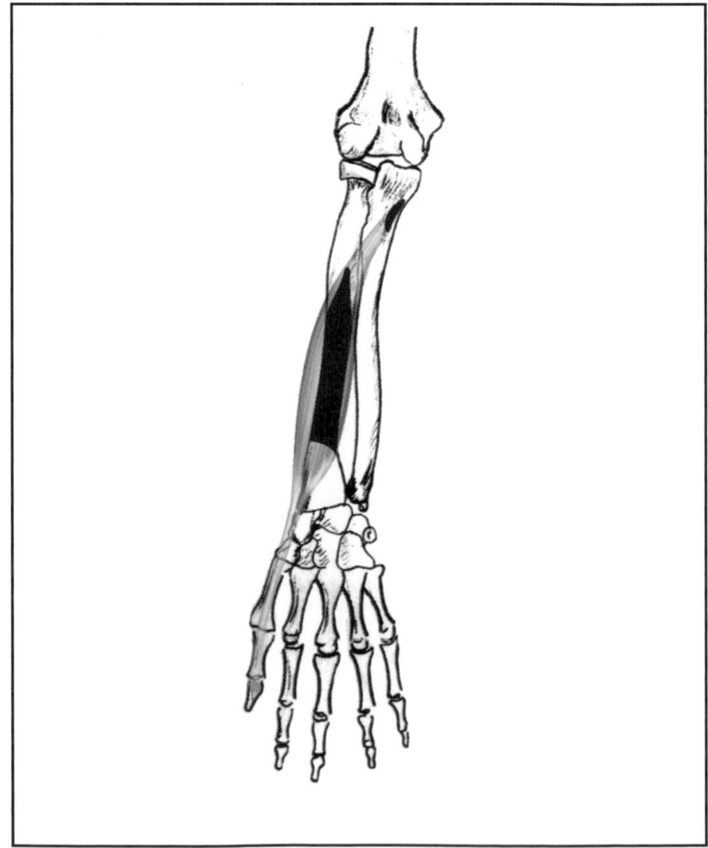

Extensor Carpi Radialis Brevis

Description:
A muscle on the radial side of the back of the forearm that extends and radially flexes the wrist; a strong wrist extender.

Joint Crossings:
Two

Rank/Body Action:
1 Wrist extension
2 Wrist hyperextension
3 Wrist radial flexion
4 Elbow extension (slight)

Blood Supply:
Radial recurrent artery

Insertion:
Base of the third metacarpal (dorsal surface)

Nerves:
Deep branch of radial nerve C6, 7

Origin:
Posterior lateral epicondyle of the humerus

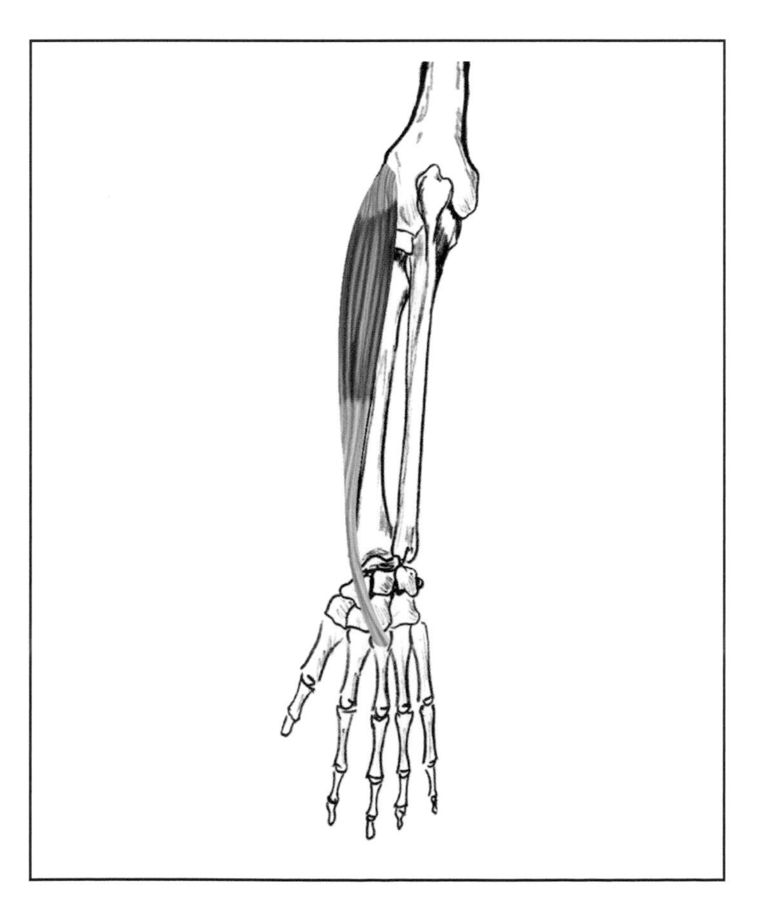

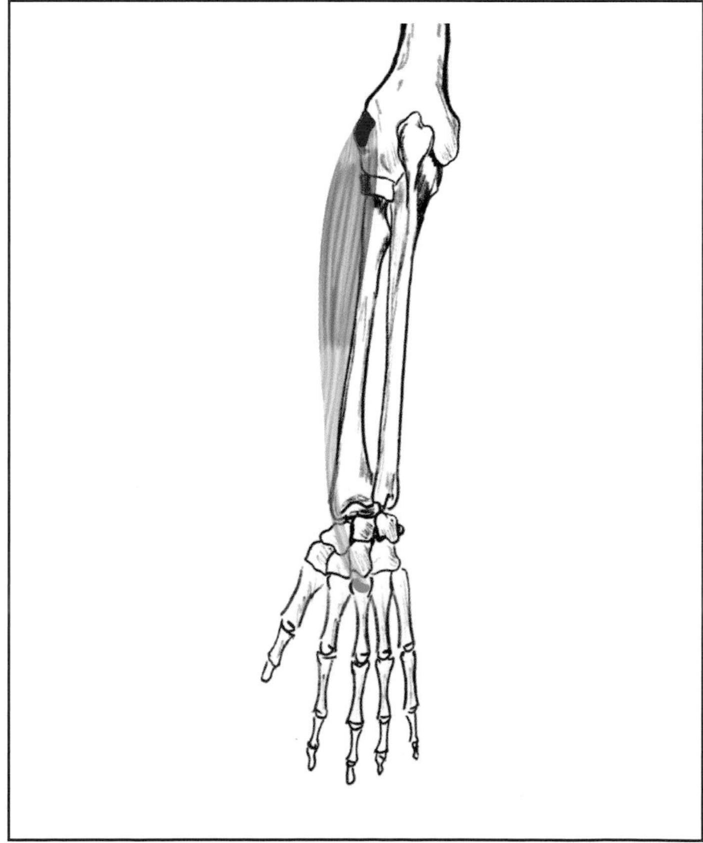

Extensor Carpi Radialis Longus

Description:
A long muscle on the radial side of the back of the forearm that extends and radially flexes the wrist.

Joint Crossings:
Two

Rank/Body Action:
1 Wrist extension
2 Wrist hyperextension
3 Wrist radial flexion
4 Elbow extension (slight)

Blood Supply:
Radial recurrent artery

Insertion:
Dorsal surface of the base of the second metacarpal

Nerves:
Radial nerve C6, 7

Origin:
- Distal one-third of the lateral supracondylar ridge of the humerus
- Lateral epicondyle of the humerus

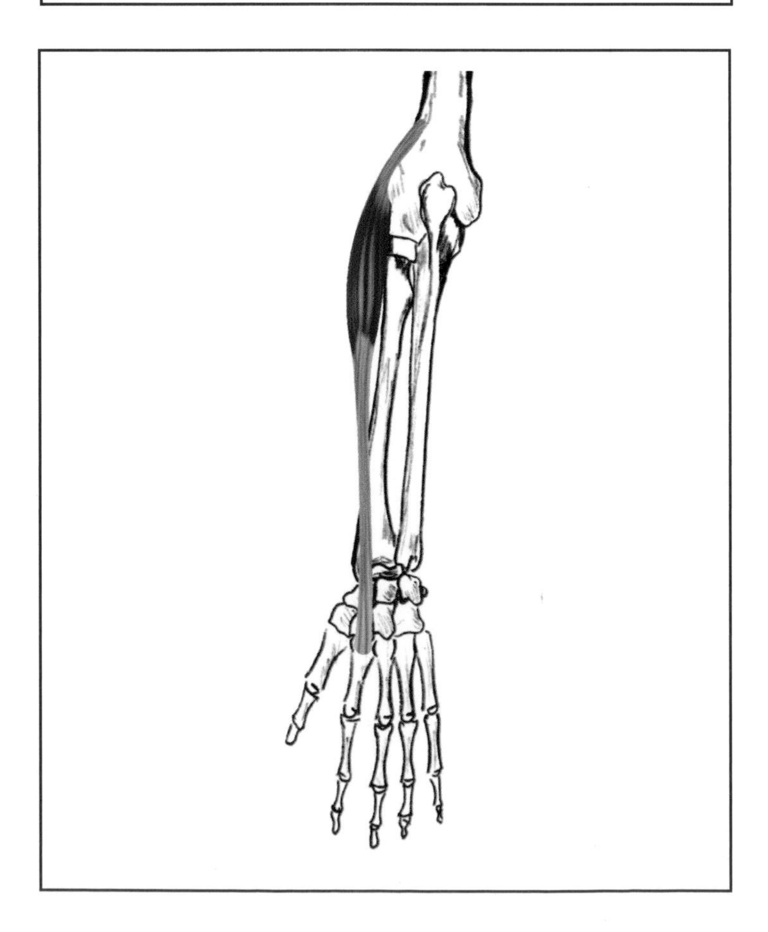

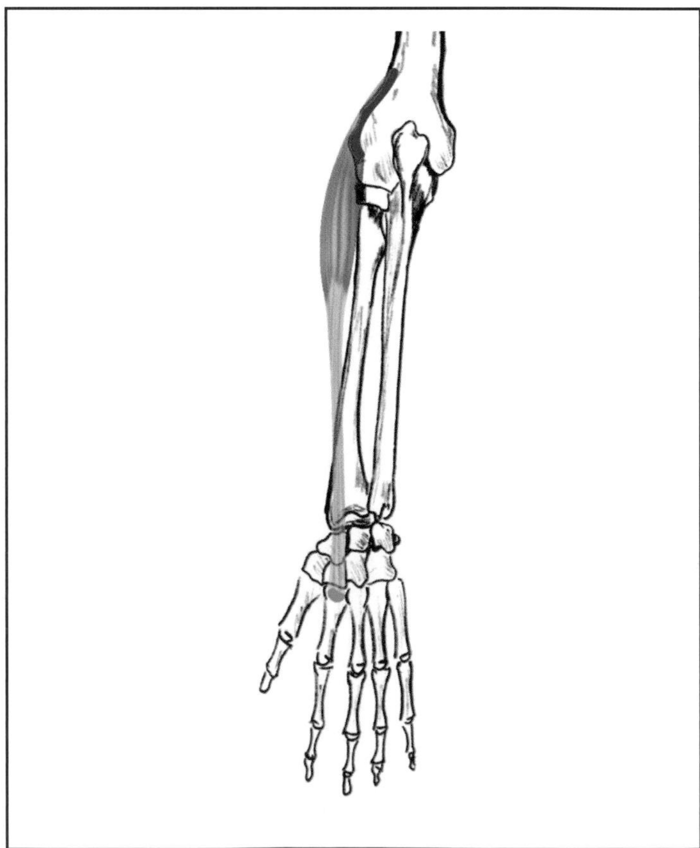

Extensor Carpi Ulnaris

Description:
A muscle on the ulnar side of the back of the forearm that extends and ulnar flexes the wrist. With the flexor carpi ulnaris, it is one of the two muscles that ulnar flex the wrist.

Joint Crossings:
Two

Rank/Body Action:
1 Wrist extension
2 Wrist hyperextension
3 Wrist ulnar flexion
4 Elbow extension (slight)

Blood Supply:
Posterior interosseous artery

Insertion:
Base of the 5th metacarpal (dorsal surface)

Nerves:
Posterior interosseous nerve of radial nerve C6, 7, 8

Origin:
- Posterior lateral epicondyle of the humerus
- Middle half of the posterior ulna

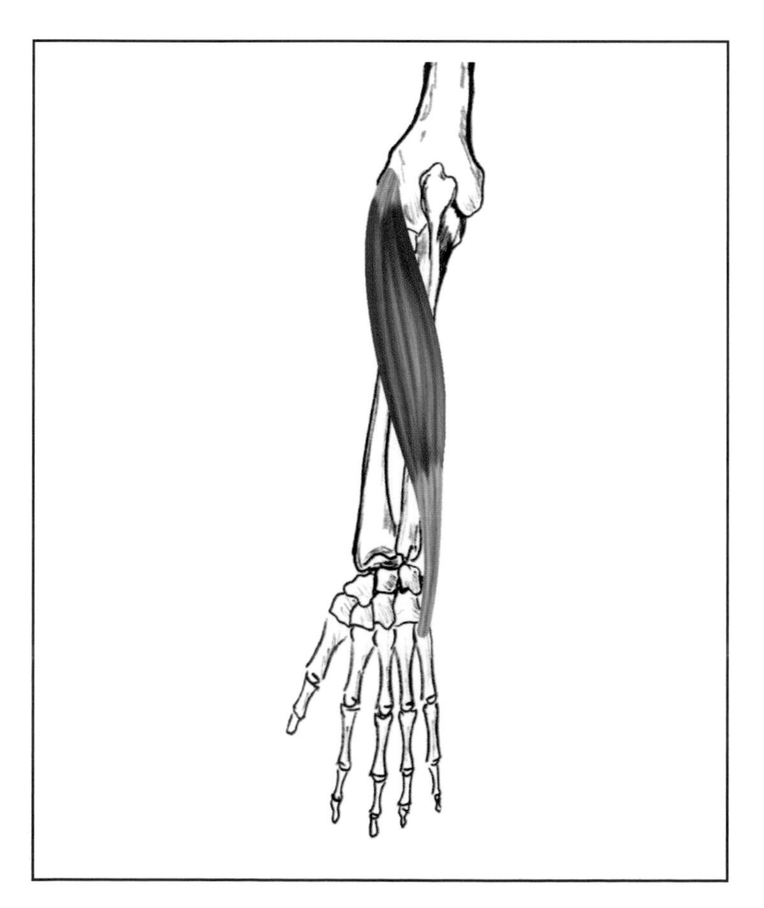

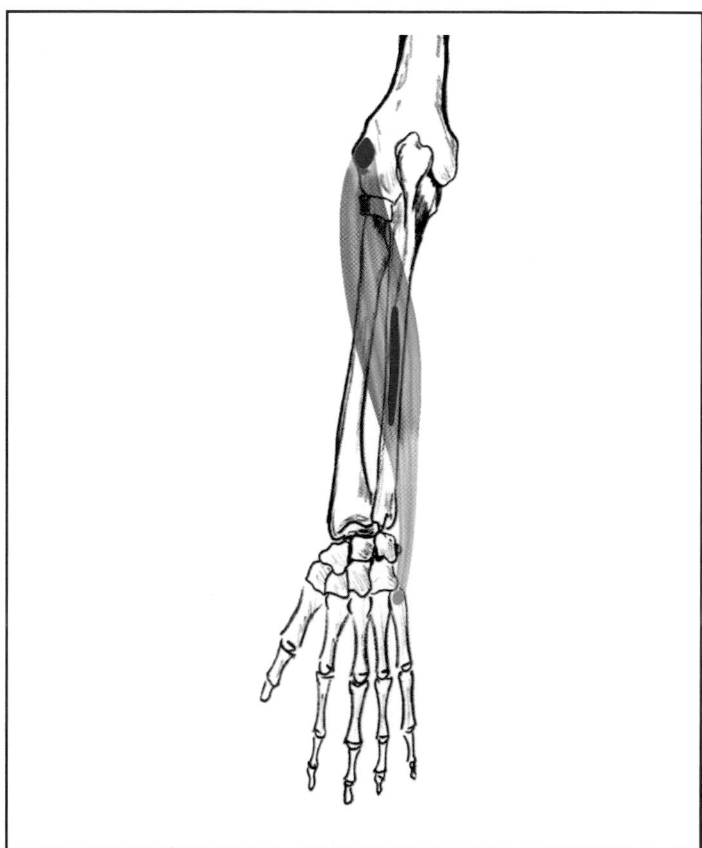

Extensor Digiti Minimi

Description:
A slender muscle on the medial side of the extensor digitorum that assists to extend the little finger.

Joint Crossings:
Three

Rank/Body Action:

1. Finger extension (little finger)
2. Finger hyperextension (little finger)
3. Wrist extension (slight)
4. Wrist hyperextension (slight)
5. Elbow extension (weak)

Blood Supply:
Posterior interosseous artery

Insertion:
Base of second and third phalange of the little finger (dorsal surface)

Nerves:
Posterior interosseous nerve of radial nerve C6, 7, 8

Origin:
Posterior lateral epicondyle of the humerus

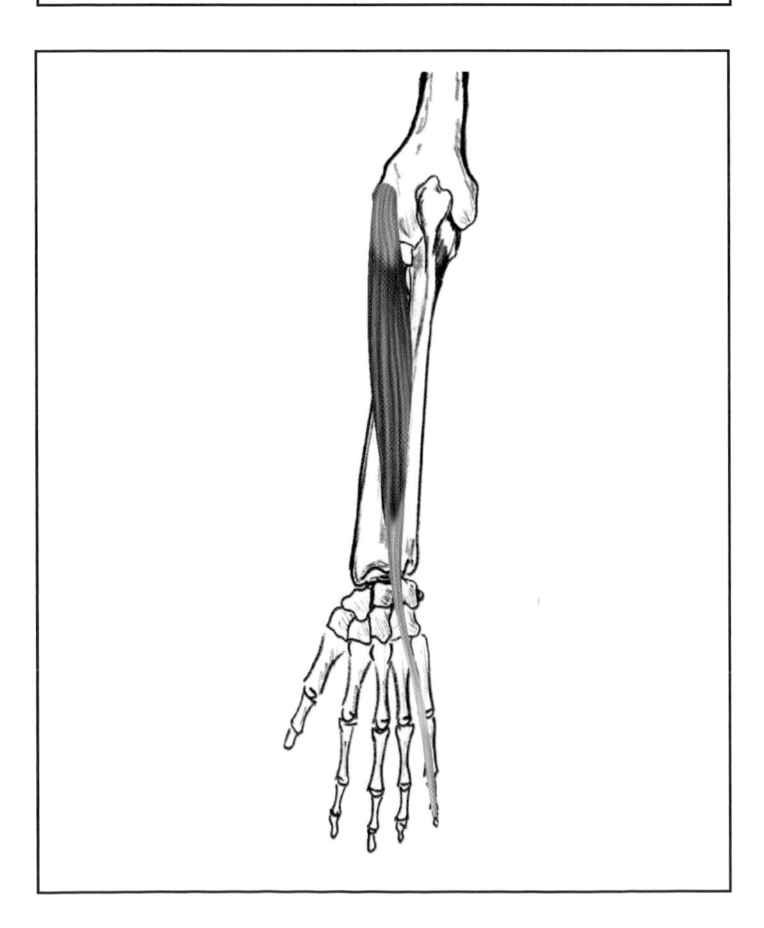

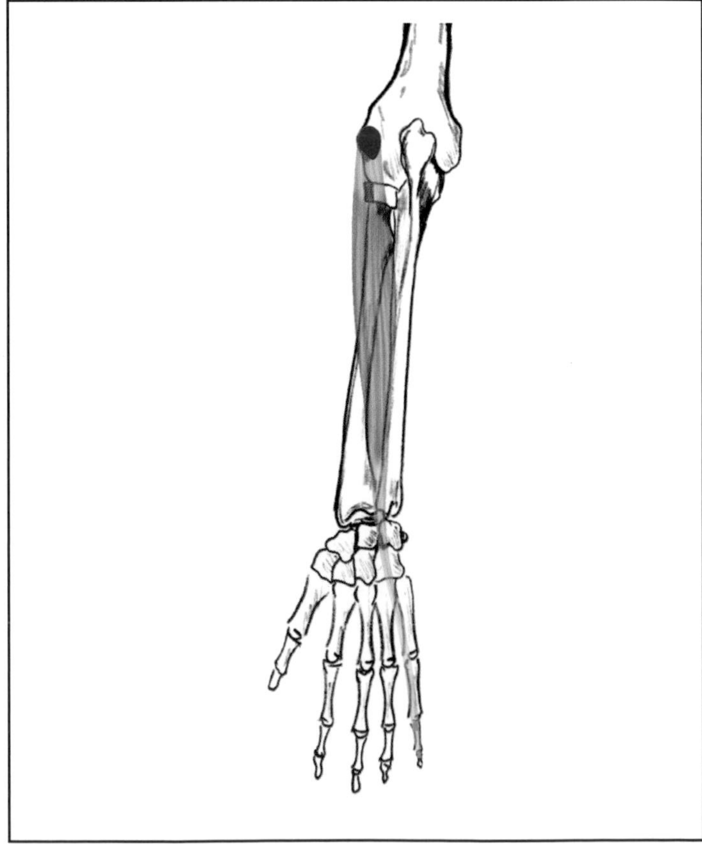

EXTENSOR DIGITORUM

Description:
The only muscle involved in the extension of all four fingers. Also known as the extensor digitorum communis.

Joint Crossings:
Three

Rank/Body Action:

1	Finger extension
2	Finger hyperextension
3	Wrist extension
4	Wrist hyperextension
5	Elbow extension (slight)

Blood Supply:
Posterior interosseous artery

Insertion:
Split tendons to the second and third phalange of the four fingers

Nerves:
Posterior interosseous nerve of radial nerve C6, 7, 8

Origin:
Posterior surface of the lateral epicondyle of the humerus

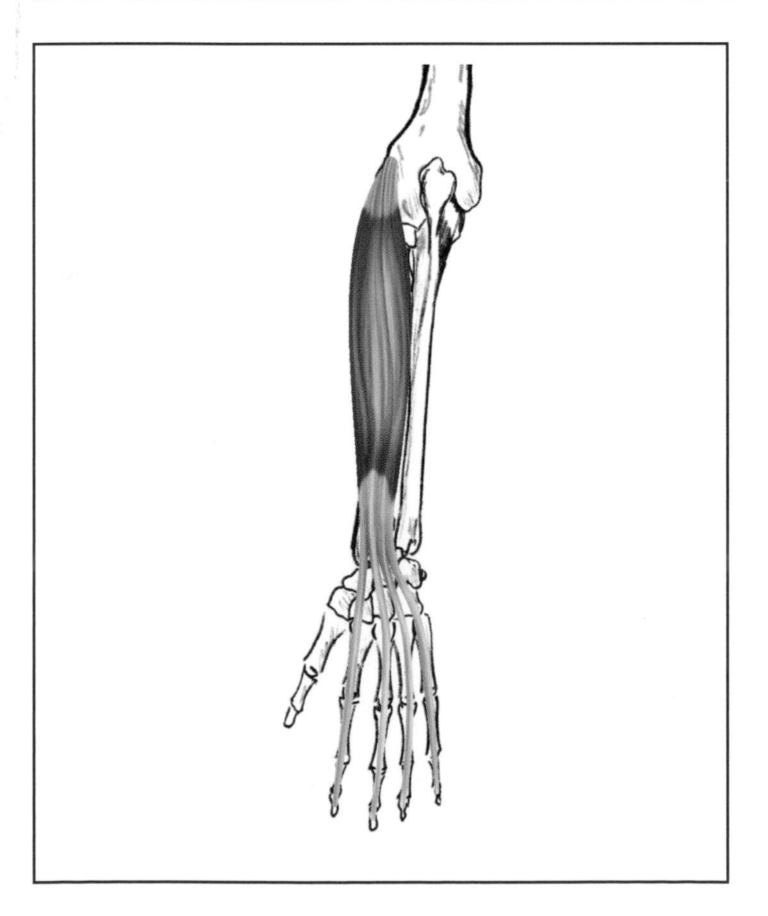

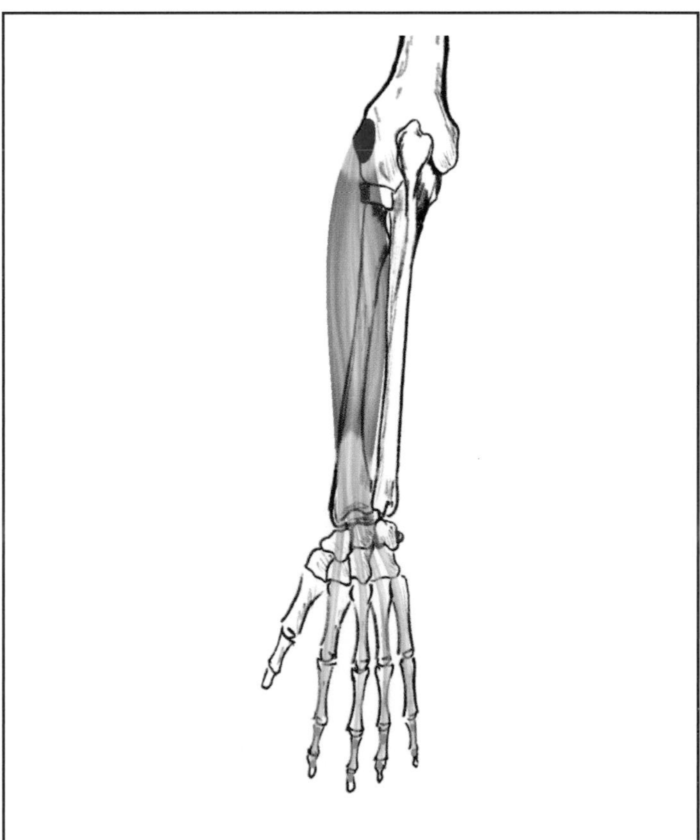

Flexor Carpi Radialis

Description:
A superficial muscle of the radial side of the forearm that flexes the wrist and assists in radially flexing it. This muscle may be absent in some individuals.

Joint Crossings:
Two

Rank/Body Action:
1 Wrist flexion
2 Wrist radial flexion
3 Elbow flexion (slight)

Blood Supply:
Muscular branches of radial artery

Insertion:
Base of the second and third metacarpals, on the palmar surface

Nerves:
Median nerve C6, 7

Origin:
Medial epicondyle of the humerus

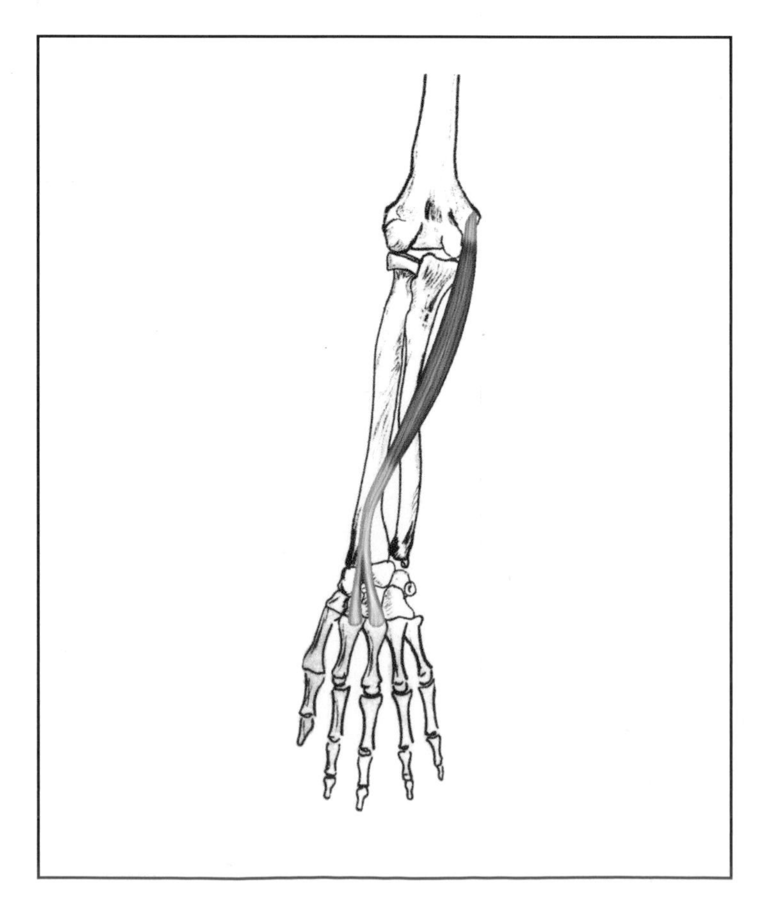

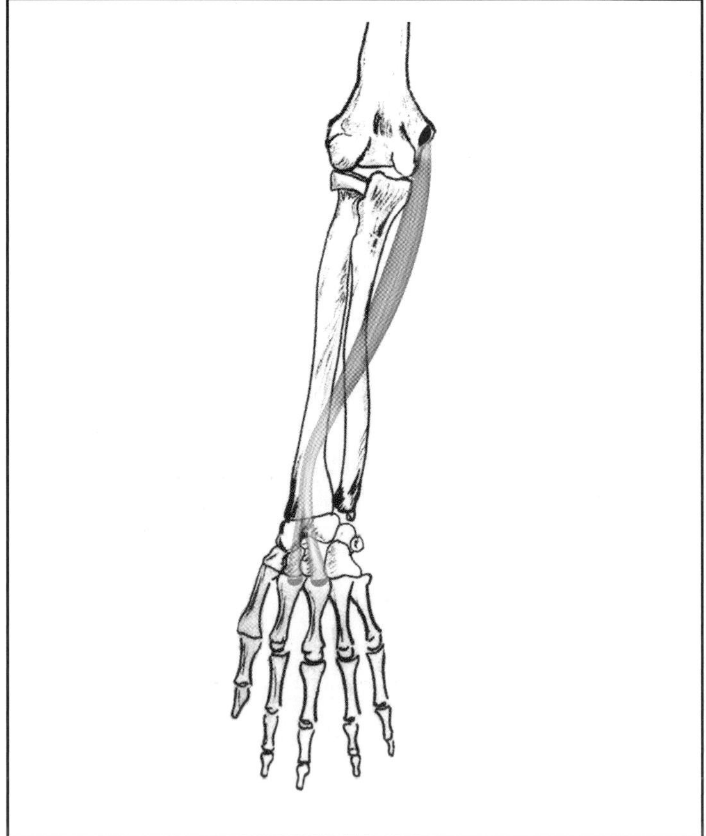

Flexor Carpi Ulnaris

Description:
A superficial muscle on the ulnar side of the forearm that flexes the wrist. With the extensor carpi ulnaris, it is one of the two muscles that ulnar flex the wrist.

Joint Crossings:
Two

Rank/Body Action:
1 Wrist flexion
2 Wrist ulnar flexion
3 Elbow flexion (slight)

Blood Supply:
Muscular branches of ulnar artery

Insertion:
- Anterior - palmar surface:
- Pisiform
- Hamate
- Base of the fifth metacarpal

Nerves:
Ulnar nerve C8, T1

Origin:
- Medial epicondyle of the humerus
- Medial border of the olecranon process and the upper two-thirds of the posterior ulna

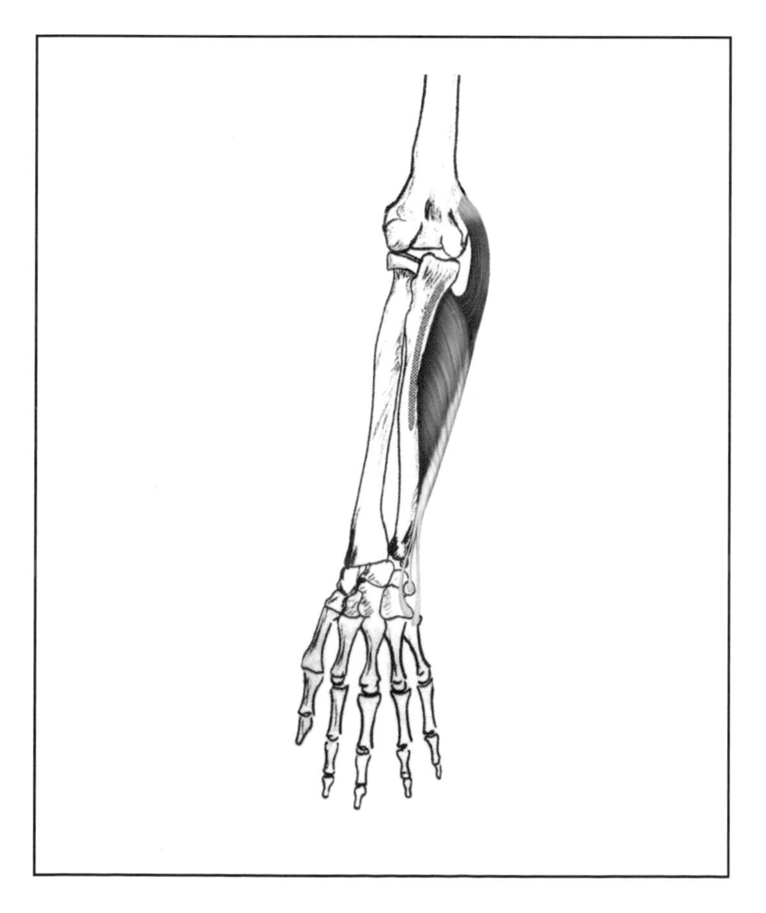

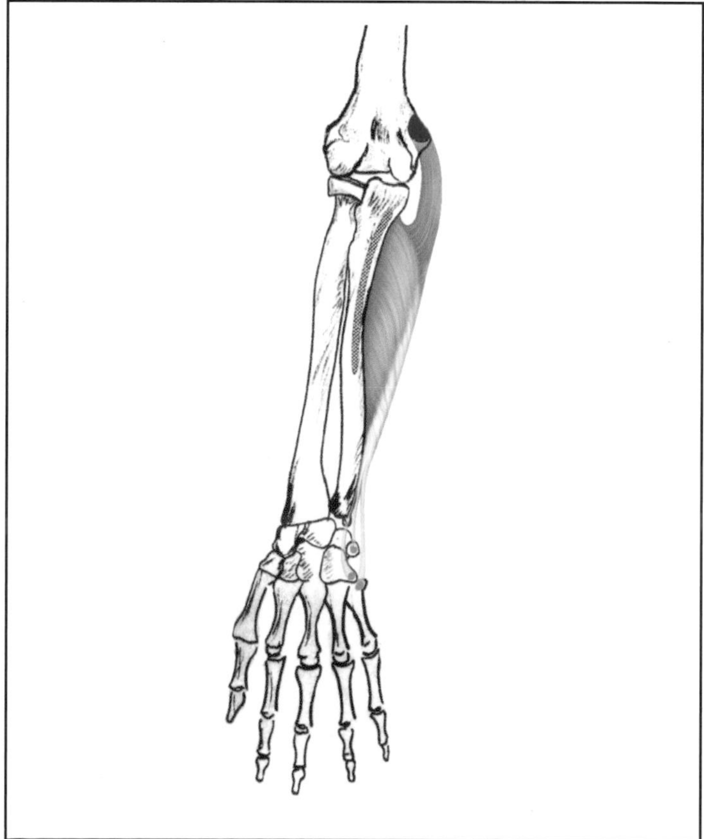

Flexor Digitorum Superficialis

Description:
A superficial muscle of the palmar side of the forearm that flexes the second phalanges of the four fingers. Also called the flexor digitorum sublimis. With the flexor digitorum profundus, it is one of only two muscles that flex all four fingers.

Joint Crossings:
Three

Rank/Body Action:

1	Finger flexion
2	Wrist flexion
3	Elbow flexion (slight)

Blood Supply:
- Muscular branches of ulnar artery
- Muscular branches of radial artery

Insertion:
The muscle splits into four tendons running down each of the four fingers again splitting and inserting on each side of the second phalanges of the four fingers

Nerves:
Median nerve C7, 8, T1

Origin:
- Medial epicondyle of the humerus
- Ulnar head – medial coronoid process
- Radial head – diagonal line cross the upper radius

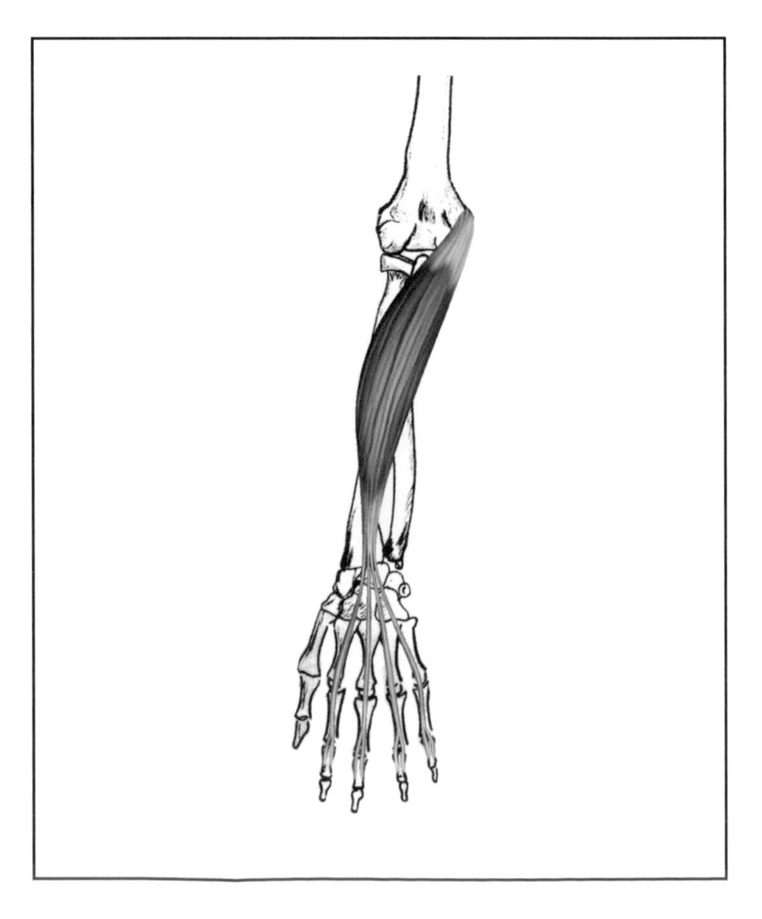

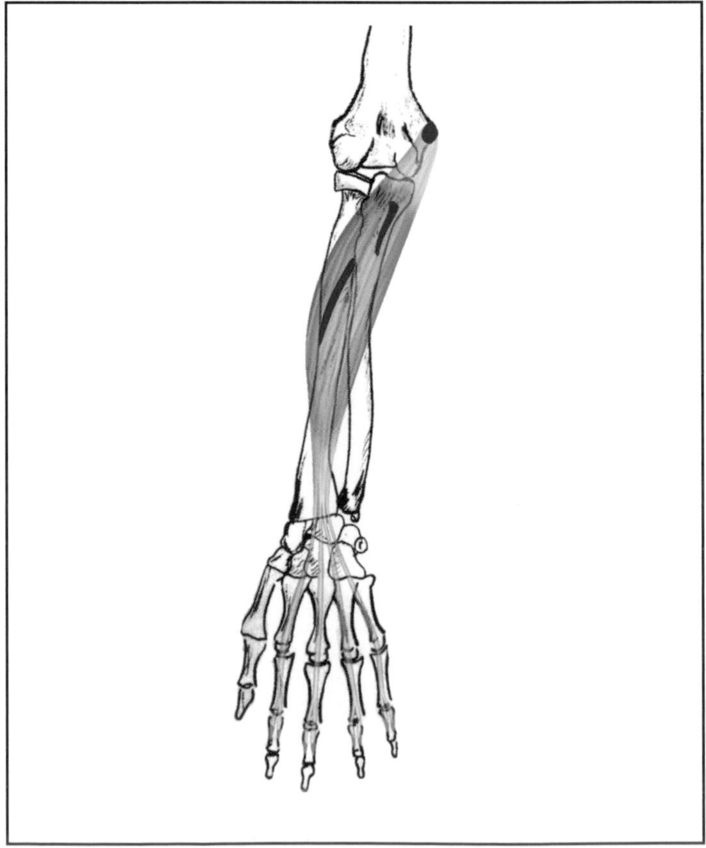

Palmaris Longus

Description:
A superficial muscle of the forearm, lying on the medial side of the flexor carpi radialis, that acts to flex the wrist. One of the most variable muscles in the body, this muscle is absent in 10% of individuals.

Joint Crossings:
Two

Rank/Body Action:
1 Wrist flexion
2 Elbow flexion (slight)

Blood Supply:
Muscular branches of ulnar artery

Insertion:
Palmar aponeurosis

Nerves:
Median nerve C6, 7

Origin:
Medial epicondyle of the humerus

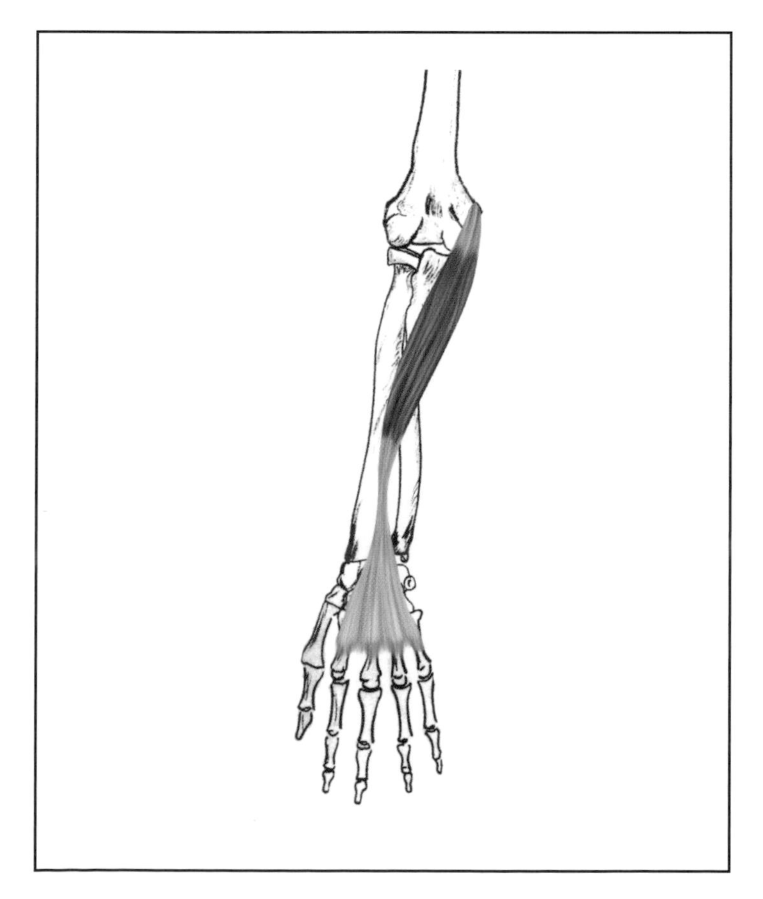

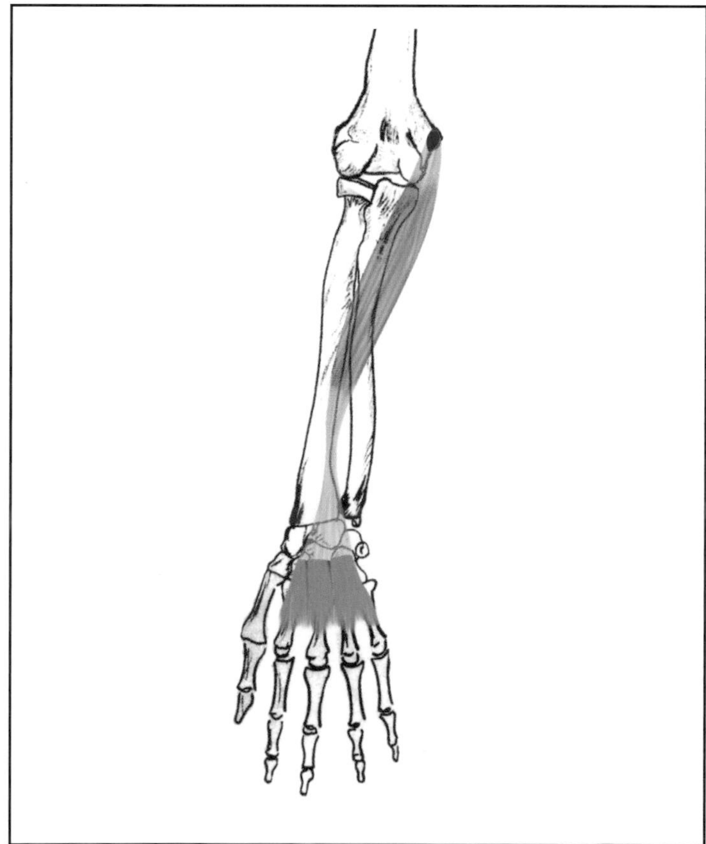

Chapter 5 • Movement Analysis

This chapter deals with an analysis of human movements. The chapter includes a brief description of the three roles that muscles take in a particular movement (prime movers, stabilizers, and neutralizers), a list of every body action with the muscles that contribute to that action, and finally a detailed movement analysis of selected exercises and an overview of the body actions involved in specific activities.

Analysis of movements divides a movement into a list of body actions. A movement may be an exercise, sport, or some other activity. Initially, movement analysis breaks up a particular movement into specific motions. For example, with archery (shooting an arrow), the movement involved could be broken into the following motions: holding the bow out in front of you; pulling the string back and aiming; and releasing the string. The identified motions are then ranked as to the most important from a strength perspective. Subsequently, each motion is then broken up into a ranked list of body actions that comprise that motion. In other words, movement analysis involves breaking down a particular movement into ranked motions, which are then broken into ranked body actions. In this way, a concise, complete description of the original movement is derived by a list of ranked body actions.

Role of Muscles in Movement

When analyzing the involvement of muscles in an activity or an exercise, the following three terms are used to describe the role that muscles take in the movement – prime movers, stabilizers, and neutralizers.

• Prime Movers

A prime mover is the muscle or muscle group that does the main movement. In the push-up, for example, the prime movers are the elbow extensors and the arm flexors. These are the muscles that get the body from the "floor" position to the "up" position. The elbow extensors straighten the elbow, and the arm flexors bring the arm into the final "up" position for the push-up. In pull-ups, the prime movers are the elbow flexors and the shoulder extensors; both sets of muscles must work together to pull you to the "up" position for a pull-up.

Prime movers typically imply concentric contraction. In the movement analysis of the various exercises and activities that appears in subsequent sections of this chapter, the prime mover actions are typically for the first motion listed.

• Stabilizers

Using the previously discussed push-up example as a reference point, if you lay on the floor in the "down" position for the push-up, and then only did the actions of the prime movers, you would be in the "up" position, but your body would be sagging toward the floor (hyperextending the spine and hips). A good push-up involves keeping your body straight or rigid when pushing to the "up" position. If the body is to be kept straight, then muscles have to keep it rigid. The muscles that make this happen are called stabilizers. They stabilize the body, while the prime movers do their job. In the push-up, since no stabilizers cause the spine and hips to be hyperextended, the stabilizers are the spine and hip flexors (to prevent hyperextension). Depending on the activity or movement, muscles can be either prime movers or stabilizers.

Stabilizers typically imply isometric contractions. In the analysis of the various exercises and activities that appears in subsequent sections of this chapter, the actions listed as isometric contractions usually involve stabilizers.

• Neutralizers

The third role of muscles in movement is as neutralizers. Many muscles have more than one action. For example, the biceps is an elbow flexor, an arm flexor, and a supinator of the forearm. When it contracts, all of these actions are initiated. If the movement being analyzed only involves supination (for example, driving a screw into a piece of wood), then the biceps will be a prime mover (as one of the forearm supinators). However, of its three possible actions that the biceps could undertake, only supination is needed. To prevent the other two actions, muscles with the opposite actions have to be innervated to "neutralize" the elbow and arm flexion, (in other words, the elbow and arm extensors). In this instance, these muscles act as neutralizers.

Muscles for Each Body Action

This section presents a list of every body action, followed by the muscles that contribute to that action. The actions are listed in alphabetical order. The muscles associated with each action are also listed. The indented muscles have a smaller contribution to the action.

- ❑ Abdomen Compressors:
  - External oblique abdominal
  - Internal oblique abdominal
  - Rectus abdominis
  - Transversus abdominis

❑ Ankle Dorsiflexors:
- Extensor digitorum longus
- Peroneus tertius
- Tibialis anterior
  - extensor hallucis longus

❑ Ankle Everters:
- Extensor digitorum longus
- Peroneus brevis
- Peroneus longus
- Peroneus tertius

❑ Ankle Inverters:
- Flexor hallucis longus
- Tibialis anterior
- Tibialis posterior
  - extensor hallucis longus
  - flexor digitorum longus

❑ Ankle Plantar Flexors
- Flexor digitorum longus
- Gastrocnemius
- Peroneus longus
- Plantaris
- Soleus
- Tibialis posterior
  - flexor hallucis longus
  - peroneus brevis

❑ Arm Abductors:
- Deltoid anterior (upper fibers)
- Deltoid middle
- Deltoid posterior (upper fibers)
- Supraspinatus
  - infraspinatus (upper fibers)

❑ Arm Adductors:
- Latissimus dorsi
- Teres major
- Teres minor
  - coracobrachialis
  - deltoid anterior (lower fibers)
  - deltoid posterior (lower fibers)
  - infraspinatus (lower fibers)
  - pectoralis major
  - subscapularis
  - triceps brachii

❑ Arm Extensors:
- Deltoid posterior
- Latissimus dorsi
- Subscapularis
- Teres major
- Teres minor
  - triceps brachii

❑ Arm Flexors:
- Coracobrachialis
- Deltoid anterior
- Pectoralis major
  - biceps brachii

❑ Arm Horizontal Abductors:
- Infraspinatus
  - deltoid posterior
  - teres minor

❑ Arm Horizontal Adductors:
- Coracobrachialis
- Pectoralis major
  - deltoid anterior
  - latissimus dorsi

❑ Arm Hyperextensors:
- Deltoid posterior
- Latissimus dorsi
- Subscapularis
- Teres major
- Teres minor
  - triceps brachii

❑ Arm Internal Rotators:
- Deltoid anterior
- Latissimus dorsi
- Pectoralis major
- Subscapularis
  - teres major

❑ Arm External Rotators:
- Deltoid posterior
- Infraspinatus
- Teres minor
  - coracobrachialis

❑ Elbow Extensors:
- Anconeus
- Triceps brachii
  - extensor carpi radialis brevis
  - extensor carpi radialis longus
  - extensor carpi ulnaris
  - extensor digiti minimi

√ extensor digitorum

❑ Elbow Flexors:
- Biceps brachii
- Brachialis
- Brachioradialis
- Pronator teres
  - flexor carpi radialis
  - flexor carpi ulnaris
  - flexor digitorum superficialis
  - palmaris longus
  - supinator

❑ Finger Abductors:
- Abductor digiti minimi
- Dorsal interossei
  - opponens digiti minimi

❑ Finger Adductors:
- Palmar interossei

❑ Finger Extensors:
- Dorsal interossei
- Extensor digiti minimi
- Extensor digitorum
- Extensor indicis
- Lumbricals
  - palmar interossei

❑ Finger Flexors:
- Abductor digiti minimi
- Flexor digiti minimi brevis
- Flexor digitorum profundus
- Flexor digitorum superficialis
- Lumbricals
- Palmar interossei
  - opponens digiti minimi

❑ Finger Hyperextensors:
- Dorsal interossei
- Extensor digiti minimi
- Extensor digitorum
- Extensor indicis
- Lumbricals
  - palmar interossei

❑ Forearm Pronators:
- Pronator quadratus
- Pronator teres

❑ Forearm Supinators:
- Biceps brachii
- Brachioradialis
- Supinator

❑ Head Extensors:
- Longissimus capitis
- Obliquus capitis superior
- Rectus capitis posterior (major)
- Rectus capitis posterior (minor)
- Spinalis capitis
- Splenius capitis
- Splenius cervicis
  - trapezius

❑ Head Flexors:
- Longus capitis
- Rectus capitis anterior
- Sternocleidomastoid

❑ Head Hyperextensors:
- Longissimus capitis
- Obliquus capitis superior
- Rectus capitis posterior (major)
- Rectus capitis posterior (minor)
- Spinalis capitis
- Splenius capitis
- Splenius cervicis
  - trapezius

❑ Head Lateral Flexors:
- Longissimus capitis
- Obliquus capitis superior
- Rectus capitis lateralis
- Splenius capitis
- Splenius cervicis
- Sternocleidomastoid
  - spinalis capitis
    - √ trapezius

❑ Head Rotators:
- Obliquus capitis inferior
- Rectus capitis posterior (major)
- Sternocleidomastoid
  - longissimus capitis
  - spinalis capitis
  - splenius capitis
  - splenius cervicis
    - √ trapezius

❑ Hip-Leg Abductors:
- Gluteus medius
- Gluteus minimus
- Tensor fasciae latae
  - sartorius

❑ Hip-Leg Adductors:
- Adductor brevis
- Adductor longus
- Adductor magnus
- Gracilis
- Pectineus
  - biceps femoris (hamstrings)
  - gluteus maximus
  - iliacus
  - semimembranosus (hamstrings)
  - semitendinosus (hamstrings)

❑ Hip-Leg Extensors:
- Biceps femoris (hamstrings)
- Gluteus maximus
- Semimembranosus (hamstrings)
- Semitendinosus (hamstrings)
  - adductor magnus (posterior fibers)
  - gluteus medius (posterior fibers)

❑ Hip-Leg Flexors:
- Gracilis
- Iliacus
- Pectineus
- Psoas major
- Psoas minor
- Sartorius
- Tensor fasciae latae
  - adductor brevis
  - adductor longus
  - adductor magnus
  - gluteus medius
  - gluteus minimus
  - rectus femoris (quads)

❑ Hip-Leg Hyperextensors
- Biceps femoris (hamstrings)
- Gluteus maximus
- Semimembranosus (hamstrings)
- Semitendinosus (hamstrings)
  - adductor magnus
  - gluteus medius

❑ Hip-Leg Internal Rotators:
- Gluteus medius (anterior fibers)
- Gluteus minimus (anterior fibers)
  - adductor magnus
  - semimembranosus (hamstrings)
  - semitendinosus (hamstrings)
  - tensor fasciae latae

❑ Hip-Leg External Rotators:
- Adductor magnus
- Biceps femoris (hamstrings)
- Gemellus inferior
- Gemellus superior
- Gluteus maximus
- Gluteus medius
- Obturator externus
- Obturator internus
- Piriformis
- Quadratus femoris
- Sartorius
  - adductor brevis
  - adductor longus
  - iliacus
  - pectineus
  - psoas major
  - psoas minor
    - √ gracilis

❑ Knee Extensors:
- Rectus femoris (quads)
- Vastus intermedius (quads)
- Vastus lateralis (quads)
- Vastus medialis (quads)

❑ Knee Flexors:
- Biceps femoris (hamstrings)
- Popliteus
- Sartorius
- Semimembranosus (hamstrings)
- Semitendinosus (hamstrings)
  - gastrocnemius
  - gracilis
  - plantaris

❑ Rib Depressors:
- Internal intercostals
- Serratus posterior (inferior)
- Subcostales
- Transversus thoracis

❑ Rib Elevators:
- External intercostals
- Internal intercostals
- Levator costarum
- Pectoralis minor
- Scalenus anterior
- Scalenus medius
- Scalenus posterior
- Serratus posterior (superior)

❑ Scapula Abductors:
- Serratus anterior

❑ Scapula Adductors:
- Rhomboid major & minor
- Trapezius
  - levator scapulae

❑ Scapula Depressors:
- Pectoralis minor
- Subclavius
- Trapezius
  - serratus anterior

❑ Scapula Elevators:
- Levator scapulae
- Rhomboid major & minor
- Trapezius

❑ Scapula Downward Rotators:
- Levator scapulae
- Rhomboid major & minor

❑ Scapula Upward Rotators:
- Serratus anterior
- Trapezius

❑ Spine Extensors:
- Iliocostalis cervicis
- Iliocostalis lumborum
- Iliocostalis thoracis
- Interspinalis
- Longissimus capitis
- Longissimus cervicis
- Longissimus thoracis
- Multifidus
- Quadratus lumborum
- Rotatores
- Semispinalis cervicis
- Semispinalis thoracis
- Spinalis capitis
- Spinalis cervicis
- Spinalis thoracis

❑ Spine Flexors:
- External oblique abdominal
- Internal oblique abdominal
- Longus colli (inferior oblique)
- Longus colli (superior oblique)
- Longus colli (vertical)
- Rectus abdominis
  - scalenus anterior
  - scalenus posterior

❑ Spine Hyperextensors:
- Iliocostalis cervicis
- Iliocostalis lumborum
- Iliocostalis thoracis
- Interspinalis
- Longissimus capitis
- Longissimus cervicis
- Longissimus thoracis
- Multifidus
- Quadratus lumborum
- Rotatores
- Semispinalis cervicis
- Semispinalis thoracis
- Spinalis capitis
- Spinalis cervicis
- Spinalis thoracis

❑ Spine Lateral Flexors:
- Iliocostalis cervicis
- Iliocostalis lumborum
- Iliocostalis thoracis
- Intertransversarii
- Levator costarum
- Quadratus lumborum
- Scalenus anterior
- Scalenus medius
- Scalenus posterior
- Spinalis cervicis
  - external oblique abdominal

❑ Spine Rotators:
- External oblique abdominal
- Internal oblique abdominal
- Multifidus
- Rotatores
- Semispinalis cervicis
- Semispinalis thoracis
  - scalenus anterior
  - scalenus medius
  - scalenus posterior

❑ Thumb Abductors:
- Abductor pollicis brevis
- Abductor pollicis longus
- Opponens pollicis

❑ Thumb Adductors:
- Adductor pollicis
- Flexor pollicis brevis

❑ Thumb Extensors:
- Extensor pollicis brevis
- Extensor pollicis longus

❑ Thumb Flexors:
- Flexor pollicis brevis
- Flexor pollicis longus
- Opponens pollicis

❑ Thumb-Finger Opposers:
- Opponens digiti minimi
- Opponens pollicis
  - abductor pollicis brevis

❑ Toe Abductors:
- Abductor digiti minimi (quinti)
- Abductor hallucis
- Dorsal interossei (four muscles)

❑ Toe Adductors:
- Adductor hallucis
- Plantar interossei (three muscles)

❑ Toe Extensors:
- Extensor digitorum brevis
- Extensor digitorum longus
- Extensor hallucis longus
- Lumbricals (four muscles)
- Plantar interossei (three muscles)

❑ Toe Flexors
- Dorsal interossei (four muscles)
- Flexor digiti minimi (quinti) brevis
- Flexor digitorum brevis
- Flexor digitorum longus
- Flexor Hallucis brevis
- Flexor Hallucis longus
- Lumbricals (four muscles)
- Quadratus plantae
  - adductor hallucis

❑ Toe Hyperextensors:
- Extensor digitorum brevis
- Extensor digitorum longus
- Extensor hallucis longus
- Lumbricals (four muscles)
- Plantar interossei (three muscles)

❑ Wrist Extensors:
- Extensor carpi radialis brevis
- Extensor carpi radialis longus
- Extensor carpi ulnaris
  - extensor digiti minimi
  - extensor digitorum
  - extensor indicis
  - extensor pollicis brevis
  - extensor pollicis longus

❑ Wrist Flexors:
- Flexor carpi radialis
- Flexor carpi ulnaris
- Palmaris longus
  - flexor digitorum profundus
  - flexor digitorum superficialis
  - flexor pollicis longus

- ❑ Wrist Hyperextensors:
  - Extensor carpi radialis brevis
  - Extensor carpi radialis longus
  - Extensor carpi ulnaris
    - extensor digiti minimi
    - extensor digitorum
    - extensor indicis
    - extensor pollicis brevis
    - extensor pollicis longus
- ❑ Wrist Radial Flexors:
  - Abductor pollicis longus
  - Extensor carpi radialis longus
  - Flexor carpi radialis
    - abductor pollicis brevis
    - extensor carpi radialis brevis
- ❑ Wrist Ulnar Flexors:
  - Extensor carpi ulnaris
  - Flexor carpi ulnaris

Exercise Examples

In this section, movement analyses are provided for 13 exercises. Each analysis is first divided into the major movements involved in the lifting and lowering phases of the exercise. Then, each movement is divided into the motions that occur at the moving and primary stabilizing joints. The joint motions are ranked numerically according to the importance they play in performing the actual exercise. Next to each joint motion is the muscle group primarily responsible for the motion, and the type of contraction that the muscle group listed performs during the motion occurring at that particular joint. The companion CD software includes a very similar analysis for over 300 exercises in its database.

To get the names of the individual muscles that are part of a muscle group listed for a joint motion, refer to the prior section of this chapter, which lists all the muscles involved for each body action. For example, in the dumbbell fly shown in this section, the main muscle group listed for arm horizontal adduction is the arm horizontal adductors. Referring to arm horizontal adductors in the previous section, it can be seen that the muscles that are prime movers for this motion are the coracobrachialis, pectoralis major, deltoid anterior, and latissimus dorsi (during initiation of the joint movement).

Dumbbell Fly (Supine on Bench)

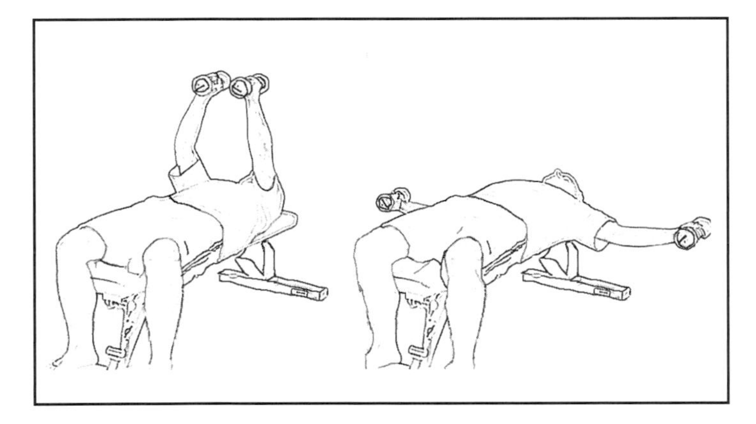

❑ Fly—Lifting Movement (Arms Up and In)

Rank	*Motion*	*Muscle Group Involved*	*Contraction Type*
1	Arm horizontal adduction	Arm horizontal adductors	Concentric
2	Elbow stabilization	Elbow flexors	Isometric
3	Scapula abduction	Scapula abductors	Concentric

❑ Fly—Lowering Movement (Arms Down and Out)

Rank	*Motion*	*Muscle Group Involved*	*Contraction Type*
1	Arm horizontal abduction	Arm horizontal adductors	Eccentric
2	Elbow stabilization	Elbow flexors	Isometric
3	Scapula adduction	Scapula abductors	Eccentric

Bench Press (Barbell)

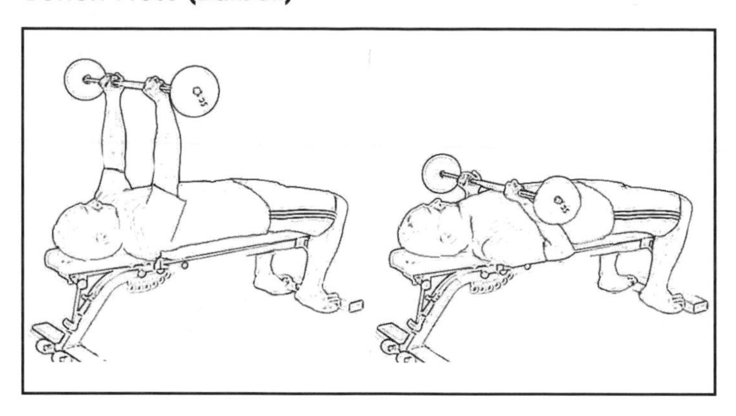

❑ Press—Lifting Weight (Upward Movement)

Rank	*Motion*	*Muscle Group Involved*	*Contraction Type*
1	Elbow extension	Elbow extensors	Concentric
2	Arm flexion	Arm flexors	Concentric
3	Arm horizontal adduction	Arm horizontal adductors	Concentric
4	Scapula abduction	Scapula abductors	Concentric

❑ Press—Lowering Weight (Downward Movement)

Rank	*Motion*	*Muscle Group Involved*	*Contraction Type*
1	Elbow flexion	Elbow extensors	Eccentric
2	Arm extension	Arm flexors	Eccentric
3	Arm horizontal abduction	Arm horizontal adductors	Eccentric
4	Scapula adduction	Scapula abductors	Eccentric

Lunge (Dumbbells)

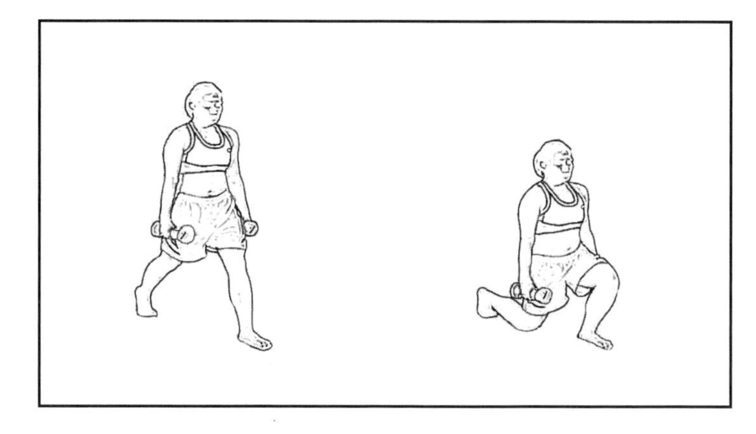

❑ Lunge—Upward Movement (Front Leg)

Rank	*Motion*	*Muscle Group Involved*	*Contraction Type*
1	Knee extension	Knee extensors (quads)	Concentric
2	Hip-leg extension	Hip-leg extensors	Concentric
3	Ankle plantar flexion	Ankle plantar flexors	Concentric

❑ Lunge—Downward Movement (Front Leg)

Rank	*Motion*	*Muscle Group Involved*	*Contraction Type*
1	Knee flexion	Knee extensors (quads)	Eccentric
2	Hip-leg flexion	Hip-leg extensors	Eccentric
3	Ankle dorsiflexion	Ankle plantar flexors	Eccentric

Bicep Hammer Curl (Dumbbells)

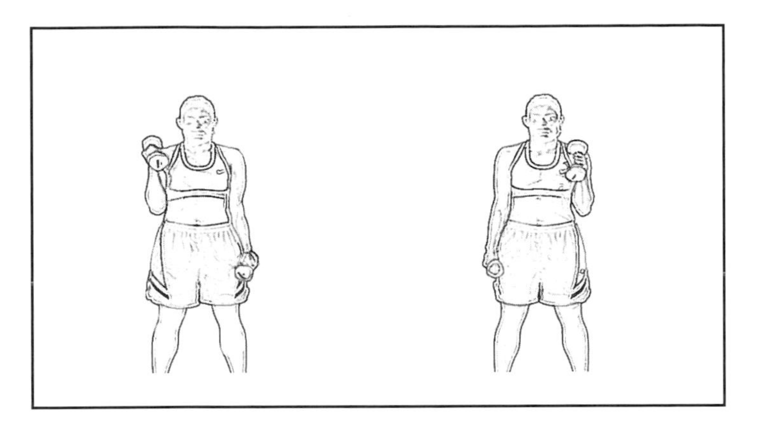

❑ Curl—Lifting (Upward) Movement

Rank	*Motion*	*Muscle Group Involved*	*Contraction Type*
1	Elbow flexion	Elbow flexors	Concentric

❑ Curl—Lowering (Downward) Movement

Rank	*Motion*	*Muscle Group Involved*	*Contraction Type*
1	Elbow extension	Elbow flexors	Eccentric

Bent Over Lateral Raise—Reverse Fly (Dumbbells)

❑ Reverse Fly—Lifting Movement (Arms Up and Out)

Rank	*Motion*	*Muscle Group Involved*	*Contraction Type*
1	Arm horizontal abduction	Arm horizontal abductors	Concentric
2	Elbow stabilization	Elbow extensors	Isometric
2	Scapula adduction	Scapula adductors	Concentric

❑ Reverse Fly—Lowering Movement (Arms Down and In)

Rank	*Motion*	*Muscle Group Involved*	*Contraction Type*
1	Arm horizontal adduction	Arm horizontal abductors	Eccentric
2	Elbow stabilization	Elbow extensors	Isometric
2	Scapula abduction	Scapula adductors	Eccentric

❑ Back and Hip Stabilization (Lifting and Lowering Movements)

Rank	*Motion*	*Muscle Group Involved*	*Contraction Type*
1	Spinal stabilization	Spine extensors	Isometric
1	Knee stabilization	Knee extensors (quads)	Isometric
1	Hip-leg stabilization	Hip-leg extensors	Isometric
1	Head stabilization	Head extensors	Isometric

Bent Over Row (Barbell)

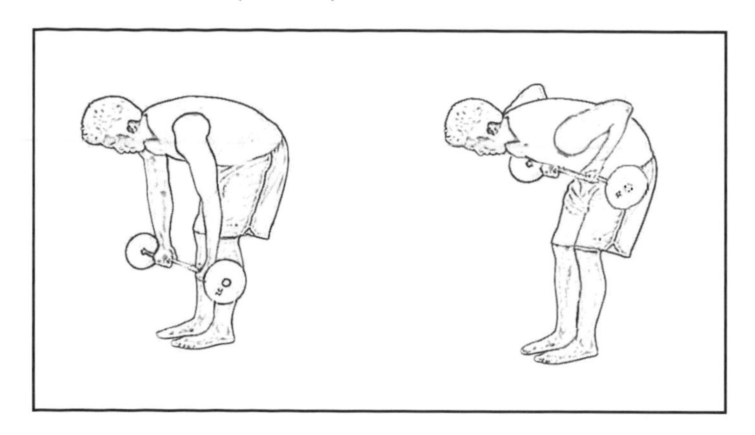

❑ Row—Lifting (Upward) Movement

Rank	*Motion*	*Muscle Group Involved*	*Contraction Type*
1	Arm extension	Arm extensors	Concentric
1	Scapula adduction	Scapula adductors	Concentric
2	Elbow flexion	Elbow flexors	Concentric

❑ Row—Lowering (Downward) Movement

Rank	*Motion*	*Muscle Group Involved*	*Contraction Type*
1	Arm flexion	Arm extensors	Eccentric
1	Scapula abduction	Scapula adductors	Eccentric
2	Elbow extension	Elbow flexors	Eccentric

❑ Back and Hip Stabilization (Lifting and Lowering Movements)

Rank	*Motion*	*Muscle Group Involved*	*Contraction Type*
1	Head stabilization	Head extensors	Isometric
1	Hip-leg stabilization	Hip-leg extensors	Isometric
1	Spinal stabilization	Spine extensors	Isometric
2	Knee stabilization	Knee extensors (quads)	Isometric

Deadlift (Barbell)

❑ Deadlift—Lifting (Upward) Motion

Rank	*Motion*	*Muscle Group Involved*	*Contraction Type*
1	Hip-leg extension	Hip-leg extensors	Concentric
1	Knee extension	Knee extensors (quads)	Concentric
2	Spinal stabilization	Spine extensors	Isometric

❑ Deadlift—Lowering (Downward) Motion

Rank	*Motion*	*Muscle Group Involved*	*Contraction Type*
1	Hip-leg flexion	Hip-leg extensors	Eccentric
1	Knee flexion	Knee extensors (quads)	Eccentric
2	Spinal stabilization	Spine extensors	Isometric

Side Plank (Hip Raise)

❑ Lifting Hips Upward Into Side Plank

Rank	*Motion*	*Muscle Group Involved*	*Contraction Type*
1	Spine lateral flexion	(Lower) spine lateral flexors	Concentric
1	Hip-leg abduction	(Lower) hip-leg abductors	Concentric
1	Hip-leg adduction	(Upper) hip-leg adductors	Concentric

❑ Lowering Hips Downward From Side Plank

Rank	*Motion*	*Muscle Group Involved*	*Contraction Type*
1	Spine lateral extension	(Lower) spine lateral flexors	Eccentric
1	Hip-leg adduction	(Lower) hip-leg abductors	Eccentric
1	Hip-leg abduction	(Upper) hip-leg adductors	Eccentric

❑ Arm and Head Stabilization

Rank	*Motion*	*Muscle Group Involved*	*Contraction Type*
1	Elbow stabilization	Elbow flexors	Isometric
1	Wrist stabilization	Wrist flexors	Isometric
2	Head stabilization	(Upper) head lateral flexors	Isometric

Lateral Raise (Dumbbells)

❑ Lateral Raise—Lifting (Upward) Movement

Rank	*Motion*	*Muscle Group Involved*	*Contraction Type*
1	Arm abduction	Arm abductors	Concentric
2	Elbow stabilization	Elbow extensors	Isometric
2	Scapula upward rotation	Scapula upward rotators	Concentric

❑ Lateral Raise—Lowering (Downward) Movement

Rank	*Motion*	*Muscle Group Involved*	*Contraction Type*
1	Arm adduction	Arm abductors	Eccentric
2	Elbow stabilization	Elbow extensors	Isometric
2	Scapula downward rotation	Scapula upward rotators	Eccentric

Push-Up, Full

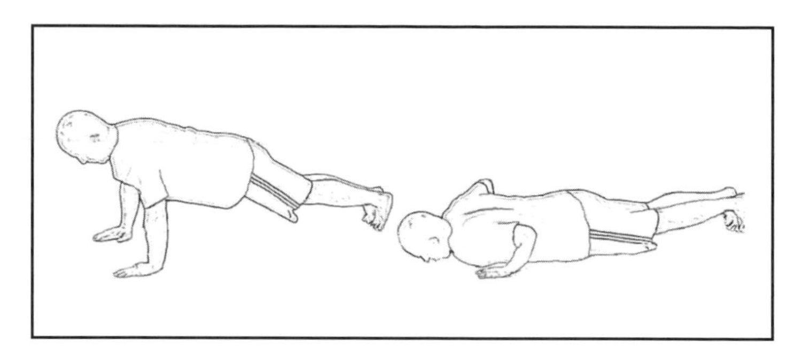

❑ Push-Up—Lifting Movement

Rank	*Motion*	*Muscle Group Involved*	*Contraction Type*
1	Elbow extension	Elbow extensors	Concentric
2	Arm flexion	Arm flexors	Concentric
3	Arm horizontal adduction	Arm horizontal adductors	Concentric
4	Scapula abduction	Scapula abductors	Concentric

❑ Push-Up—Lowering Movement

Rank	*Motion*	*Muscle Group Involved*	*Contraction Type*
1	Elbow flexion	Elbow extensors	Eccentric
2	Arm extension	Arm flexors	Eccentric
3	Arm horizontal abduction	Arm horizontal adductors	Eccentric
4	Scapula adduction	Scapula abductors	Eccentric

❑ Body Stabilization

Rank	*Motion*	*Muscle Group Involved*	*Contraction Type*
1	Hip-leg stabilization	Hip-leg flexors	Isometric
1	Spinal stabilization	Spine flexors	Isometric
1	Head stabilization	Head extensors	Isometric
2	Knee stabilization	Knee extensors (quads)	Isometric

Squat

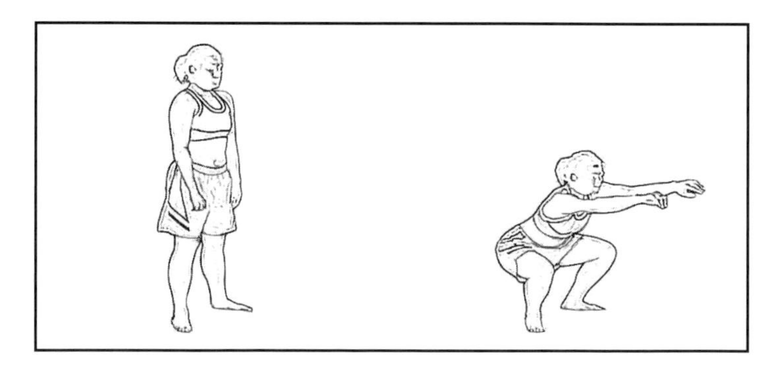

❑ Squat—Lifting (Upward) Movement

Rank	*Motion*	*Muscle Group Involved*	*Contraction Type*
1	Knee extension	Knee extensors (quads)	Concentric
2	Hip-leg extension	Hip-leg extensors	Concentric
3	Ankle plantar flexion	Ankle plantar flexors	Concentric
4	Spinal stabilization	Spine extensors and flexors	Isometric

❑ Squat—Lowering (Downward) Movement

Rank	*Motion*	*Muscle Group Involved*	*Contraction Type*
1	Knee flexion	Knee extensors (quads)	Eccentric
2	Hip-leg flexion	Hip-leg extensors	Eccentric
3	Ankle dorsiflexion	Ankle plantar flexors	Eccentric
4	Spinal stabilization	Spine extensors and flexors	Isometric

Upright Row (Barbell)

Note: The upright row is contraindicated for people with shoulder issues, especially shoulder impingement.

❑ Upright Row—Lifting Movement (Arms Up)

Rank	*Motion*	*Muscle Group Involved*	*Contraction Type*
1	Elbow flexion	Elbow flexors	Concentric
1	Arm abduction	Arm abductors	Concentric
2	Scapula upward rotation	Scapula upward rotators	Concentric

❑ Upright Row—Lowering Movement (Arms Down)

Rank	*Motion*	*Muscle Group Involved*	*Contraction Type*
1	Elbow extension	Elbow flexors	Eccentric
1	Arm adduction	Arm abductors	Eccentric
2	Scapula downward rotation	Scapula upward rotators	Eccentric

Pull-Up

❑ Pull-Up—Lifting Movement

Rank	*Motion*	*Muscle Group Involved*	*Contraction Type*
1	Elbow flexion	Elbow flexors	Concentric
1	Arm extension	Arm extensors	Concentric
1	Arm adduction	Arm adductors	Concentric
2	Scapula downward rotation	Scapula downward rotators	Concentric
2	Scapula depression	Scapula depressors	Concentric

❑ Pull-Up—Lowering Movement

Rank	*Motion*	*Muscle Group Involved*	*Contraction Type*
1	Elbow extension	Elbow flexors	Eccentric
1	Arm flexion	Arm extensors	Eccentric
1	Arm abduction	Arm adductors	Eccentric
2	Scapula upward rotation	Scapula downward rotators	Eccentric
2	Scapula elevation	Scapula depressors	Eccentric

Activity Examples

In this section, 12 activities have been analyzed. Each movement analysis is presented as a series of sequential steps that occur while performing the movement. Each step is divided into the joint motions performed during the movement, with the joint motions presented in the order that they occur during the full movement. The joint motions are ranked numerically, with the muscle group responsible for the motion and the contraction type for the muscle group listed on the same row. The companion CD software includes a similar analysis for over 100 activities in its database.

To get the names of the individual muscles that are part of a muscle group listed for a joint motion, refer to the prior section of this chapter, which lists all the muscles involved for each body action. For example, in the dumbbell fly shown in this section, the main muscle group listed for arm horizontal adduction is the arm horizontal adductors. Referring to arm horizontal adductors in the previous section, it can be seen that the muscles that are prime movers for this motion are the coracobrachialis, pectoralis major, deltoid anterior, and latissimus dorsi (during initiation of the joint movement).

Archery

Holding the bow, pulling the string back, and letting go

❑ Holding the bow and string (bow arm/forward arm)

Rank	*Motion*	*Muscle Group Involved*	*Contraction Type*
1	Wrist stabilization	Wrist extensors	Isometric
1	Elbow stabilization	Elbow extensors	Isometric
2	Finger stabilization	Finger flexors	Isometric
2	Thumb stabilization	Thumb flexors	Isometric

❑ Pulling the string back (shooting arm/back arm)

Rank	*Motion*	*Muscle Group Involved*	*Contraction Type*
1	Arm horizontal abduction	Arm horizontal abductors	Concentric
2	Finger stabilization	Finger flexors	Isometric
2	Scapula adduction	Scapula adductors	Concentric

❑ Letting go of the arrow and bow string (shooting hand/back arm)

Rank	*Motion*	*Muscle Group Involved*	*Contraction Type*
1	Finger extension	Finger extensors	Concentric
1	Finger abduction	Finger abductors	Concentric

Hitting a Baseball

Standing at the plate waiting for a pitch, backswing (or load), swinging at the ball, and follow-through

❑ Gripping the bat

Rank	*Motion*	*Muscle Group Involved*	*Contraction Type*
1	Finger flexion	Finger flexors	Isometric
1	Thumb flexion	Thumb flexors	Isometric
2	Wrist flexion	Wrist flexors	Isometric
2	Wrist extension	Wrist extensors	Isometric

❑ Holding the bat ready (awaiting the pitch)

Rank	*Motion*	*Muscle Group Involved*	*Contraction Type*
1	Arm stabilization	Arm flexors	Isometric
1	Arm stabilization	Arm abductors	Isometric
1	Elbow stabilization	Elbow flexors	Isometric
1	Spine stabilization	Spine rotators	Isometric
2	Hip-leg stabilization	Hip-leg extension	Isometric
2	Head stabilization	Head rotators	Isometric
2	Hip-leg stabilization	Hip-leg internal rotators	Isometric
2	Hip-leg stabilization	Hip-leg external rotators	Isometric
2	Knee stabilization	Knee extensors (quads)	Isometric

❑ Backswing (or load)—small, quick counter-movement just prior to the power swing

Rank	*Motion*	*Muscle Group Involved*	*Contraction Type*
1	Spinal rotation (away from ball)	Spine rotators	Concentric
1	Hip-leg internal rotation (rear leg)	Hip-leg internal rotators	Concentric
1	Hip-leg external rotation (front leg)	Hip-leg external rotators	Concentric
1	Elbow flexion	Elbow flexors	Concentric
1	Arm horizontal adduction (front arm)	Arm horizontal adductors	Concentric
1	Arm horizontal abduction (rear arm)	Arm horizontal abductors	Concentric
2	Wrist radial flexion	Wrist radial flexors	Concentric
2	Hip-leg flexion	Hip extensors	Eccentric
2	Knee flexion	Knee extensors	Eccentric
2	Ankle dorsiflexion	Ankle plantar flexors	Eccentric
3	Head stabilization	Head extensors	Isometric

❑ Power swing

Rank	*Motion*	*Muscle Group Involved*	*Contraction Type*
1	Spinal rotation (towards ball)	Spine rotators	Concentric
1	Hip-leg external rotation (rear leg)	Hip-leg external rotators	Concentric
1	Hip-leg internal rotation (front leg)	Hip-leg internal rotators	Concentric
1	Elbow extension	Elbow extensors	Concentric
1	Arm horizontal abduction (front arm)	Arm horizontal abductors	Concentric
1	Arm horizontal adduction (rear arm)	Arm horizontal adductors	Concentric
2	Wrist ulnar flexion	Wrist ulnar flexors	Concentric
2	Hip-leg extension	Hip-leg extensors	Concentric
2	Knee extension	Knee extensors	Concentric
2	Ankle plantar flexion	Ankle plantar flexors	Concentric
3	Head stabilization	Head extensors	Isometric

❑ Follow-through (slowing the bat down after the swing)

Rank	*Motion*	*Muscle Group Involved*	*Contraction Type*
1	Spinal rotation (towards ball)	Spine rotators (antagonists)	Eccentric
1	Hip-leg external rotation (rear leg)	Hip-leg internal rotators	Eccentric
1	Hip-leg internal rotation (front leg)	Hip-leg external rotators	Eccentric
1	Elbow flexion	Elbow extensors	Eccentric
1	Arm horizontal abduction (front arm)	Arm horizontal adductors	Eccentric
1	Arm horizontal adduction (rear arm)	Arm horizontal abductors	Eccentric
1	Arm flexion	Arm extensors	Eccentric
1	Wrist radial flexion	Wrist ulnar flexors	Eccentric
2	Hip-leg extension	Hip-leg flexors	Eccentric
2	Knee extension	Knee flexors	Eccentric
2	Ankle plantar flexion	Ankle dorsiflexors	Eccentric

Throwing a Baseball

Overhand throwing of the baseball: bringing the arm back, throwing the ball, and follow-through

❑ Letting go of the arrow and bow string (shooting hand/back arm)

Rank	*Motion*	*Muscle Group Involved*	*Contraction Type*
1	Finger extension	Finger extensors	Concentric
1	Finger abduction	Finger abductors	Concentric

❑ Bringing the arm back

Rank	*Motion*	*Muscle Group Involved*	*Contraction Type*
1	Arm horizontal abduction (throwing arm)	Arm horizontal abductors	Concentric
1	Arm external rotation (throwing arm)	Arm external rotators	Concentric
1	Elbow flexion (throwing arm)	Elbow flexors	Concentric
1	Spinal rotation (backwards)	Spine rotators	Concentric
1	Hip-leg internal rotation (rear leg)	Hip-leg internal rotators	Concentric
1	Hip-leg external rotation (front leg)	Hip-leg external rotators	Concentric
1	Arm horizontal adduction (non-throwing arm)	Arm horizontal adductors	Concentric
1	Wrist extension (throwing arm)	Wrist extensors	Concentric
1	Thumb flexion (throwing arm)	Thumb flexors	Isometric
1	Finger flexion (throwing arm)	Finger flexors	Isometric

❑ Throw

Rank	*Motion*	*Muscle Group Involved*	*Contraction Type*
1	Hip-leg external rotation (rear leg)	Hip-leg external rotators	Concentric
1	Hip-leg internal rotation (front leg)	Hip-leg internal rotators	Concentric
1	Spinal rotation (towards target)	Spine rotators	Concentric
1	Hip-leg extension (rear leg)	Hip-leg extensors	Concentric
1	Knee extension (rear leg)	Knee extensors	Concentric
1	Ankle plantar flexion (rear leg)	Ankle plantar flexors	Concentric
1	Hip-leg flexion (front leg)	Hip-leg flexors	Concentric
1	Arm horizontal abduction (non-throwing arm)	Arm horizontal abductors	Concentric
1	Arm horizontal adduction (throwing arm)	Arm horizontal adductors	Concentric
1	Arm internal rotation (throwing arm)	Arm internal rotators	Concentric
1	Elbow extension (throwing arm)	Elbow extensors	Concentric
1	Arm extension (throwing arm)	Arm extensors	Concentric
1	Wrist flexion (throwing arm)	Wrist flexors	Concentric
1	Thumb extension (throwing arm)	Thumb flexors	Eccentric
1	Finger extension (throwing arm)	Finger flexors	Eccentric

❑ Follow-through (slowing the arm down after release of the ball)

Rank	Motion	Muscle Group Involved	Contraction Type
1	Hip-leg external rotation (rear leg)	Hip-leg internal rotators	Eccentric
1	Hip-leg internal rotation (front leg)	Hip-leg external rotators	Eccentric
1	Spinal rotation (towards target)	Spine rotators (antagonists)	Eccentric
1	Hip-leg extension (rear leg)	Hip-leg flexors	Eccentric
1	Knee extension (rear leg)	Knee flexors	Eccentric
1	Ankle plantar flexion (rear leg)	Ankle dorsiflexors	Eccentric
1	Knee flexion (front leg)	Knee extensors	Eccentric
1	Ankle dorsiflexion (front leg)	Ankle plantar flexors	Eccentric
1	Arm extension (throwing arm)	Arm flexors	Eccentric
1	Arm internal rotation (throwing arm)	Arm external rotators	Eccentric
1	Arm horizontal adduction (throwing arm)	Arm horizontal abductors	Eccentric
1	Elbow extension (throwing arm)	Elbow flexors	Eccentric
1	Wrist flexion (throwing arm)	Wrist extensors	Eccentric
1	Arm horizontal abduction (non-throwing arm)	Arm horizontal adductors	Eccentric
1	Thumb extension (throwing arm)	Thumb flexors	Eccentric
1	Finger extension (throwing arm)	Finger flexors	Eccentric

Shooting a Basketball

Shooting the basketball, as when making a foul shot

❑ Bringing the ball into position from waist level

Rank	Motion	Muscle Group Involved	Contraction Type
1	Wrist stabilization	Wrist flexors and extensors	Isometric
1	Arm flexion	Arm flexors	Concentric
1	Thumb flexion	Thumb flexors	Isometric
1	Elbow flexion	Elbow flexors	Concentric
1	Finger flexion	Finger flexors	Isometric
2	Head extension	Head extensors	Concentric
2	Spine extension	Spine extensors	Concentric

❑ Moving from semi-crouched to extended body position when shooting

Rank	Motion	Muscle Group Involved	Contraction Type
1	Spine extension	Spine extensors	Concentric
1	Hip-leg extension	Hip-leg extensors	Concentric
1	Elbow flexion	Elbow flexors	Concentric
1	Ankle plantar flexion	Ankle plantar flexors	Concentric
1	Knee extension	Knee extensors (quads)	Concentric
2	Thumb flexion	Thumb flexors	Concentric
2	Wrist extension	Wrist extensors	Concentric
2	Head extension	Head extensors	Concentric
2	Finger flexion	Finger flexors	Concentric

❑ Extending the arm and flexing the wrist to shoot

Rank	Motion	Muscle Group Involved	Contraction Type
1	Hip-leg extension	Hip extensors	Concentric
1	Knee extension	Knee extensors	Concentric
1	Ankle plantar flexion	Ankle plantar flexors	Concentric
1	Arm extension	Arm extensors	Concentric
1	Elbow extension	Elbow extensors	Concentric
1	Wrist flexion	Wrist flexors	Concentric
1	Finger flexion	Finger flexors	Concentric
2	Spine stabilization	Spine extensors	Isometric
2	Head stabilization	Head extensors	Isometric

❑ Follow-through (after releasing the basketball)

Rank	*Motion*	*Muscle Group Involved*	*Contraction Type*
1	Hip-leg extension	Hip flexors	Eccentric
1	Knee extension	Knee flexors	Eccentric
1	Ankle plantar flexion	Ankle dorsiflexors	Eccentric
1	Arm extension	Arm flexors	Eccentric
1	Elbow extension	Elbow flexors	Eccentric
1	Wrist flexion	Wrist extensors	Eccentric
1	Finger flexion	Finger extensors	Eccentric
2	Spine stabilization	Spine extensors	Isometric
2	Head stabilization	Head extensors	Isometric

Punting a Football

Holding the football and punting it

❑ Moving the kicking leg back (backswing)

Rank	*Motion*	*Muscle Group Involved*	*Contraction Type*
1	Hip-leg extension (kicking leg)	Hip-leg extensors	Concentric
1	Knee flexion (kicking leg)	Knee flexors	Concentric
1	Hip-leg flexion (non-kicking leg)	Hip-leg extensors	Eccentric
1	Knee flexion (non-kicking leg)	Knee extensors	Eccentric
1	Spinal extension	Spine extensors	Concentric
1	Ankle dorsiflexion (non-kicking leg)	Ankle plantar flexors	Eccentric
2	Arm flexion	Arm flexors	Concentric
2	Elbow stabilization	Elbow flexors	Isometric
2	Wrist stabilization	Wrist extensors	Isometric
2	Finger stabilization (holding ball)	Finger flexors	Isometric
2	Thumb extension (holding ball)	Thumb flexors	Isometric

❑ Punting the football

Rank	*Motion*	*Muscle Group Involved*	*Contraction Type*
1	Hip-leg flexion (kicking leg)	Hip-leg flexors	Concentric
1	Knee extension (kicking leg)	Knee extensors	Concentric
1	Hip-leg extension (non-kicking leg)	Hip-leg extensors	Concentric
1	Knee extension (non-kicking leg)	Knee extensors	Concentric
1	Spinal flexion	Spine flexors	Concentric
2	Arm stabilization	Arm flexors	Isometric
2	Elbow stabilization	Elbow flexors	Isometric
2	Wrist stabilization	Wrist extensors	Isometric
2	Finger extension (letting ball go)	Finger extensors	Concentric
2	Thumb extension (letting ball go)	Thumb extensors	Concentric
2	Ankle plantar flexion (kicking leg)	Ankle plantar flexors	Concentric
2	Ankle plantar flexion (non-kicking leg)	Ankle plantar flexors	Concentric

❑ Follow-through (slowing the leg down after striking the ball)

Rank	*Motion*	*Muscle Group Involved*	*Contraction Type*
1	Hip-leg flexion (kicking leg)	Hip-leg extensors	Eccentric
1	Knee extension (kicking leg)	Knee flexors	Eccentric
1	Ankle dorsiflexion (kicking leg)	Ankle plantar flexors	Eccentric
1	Hip-leg extension (non-kicking leg)	Hip-leg flexors	Eccentric
1	Knee extension (non-kicking leg)	Knee flexors	Eccentric
1	Spinal flexion	Spine flexors	Eccentric
2	Arm extension	Arm flexors	Eccentric
2	Elbow extension	Elbow flexors	Eccentric
2	Wrist stabilization	Wrist extensors	Eccentric
2	Ankle plantar flexion (non-kicking leg)	Ankle dorsiflexors	Eccentric

Gardening
Working with a trowel and pulling weeds

❑ Digging with a trowel

Rank	*Motion*	*Muscle Group Involved*	*Contraction Type*
1	Wrist radial flexion	Wrist radial flexors	Concentric
1	Arm flexion	Arm flexors	Concentric
1	Elbow extension	Elbow extensors	Concentric
1	Finger flexion	Finger flexors	Isometric
1	Thumb flexion	Thumb flexors	Isometric

❑ Pulling up weeds (after grasping them)

Rank	*Motion*	*Muscle Group Involved*	*Contraction Type*
1	Elbow flexion	Elbow flexors	Concentric
1	Forearm supination	Forearm supinators	Concentric
1	Arm extension	Arm extensors	Concentric
1	Finger flexion	Finger flexors	Isometric
1	Thumb flexion	Thumb flexors	Isometric

Driving a Golf Ball
The golf swing as performed off a tee

❑ Backswing

Rank	*Motion*	*Muscle Group Involved*	*Contraction Type*
1	Spinal rotation (backward)	Spine rotators	Concentric
1	Wrist radial flexion	Wrist radial flexors	Concentric
1	Elbow extension (front arm)	Elbow extensors	Isometric
1	Elbow flexion (rear arm)	Elbow flexors	Concentric
1	Hip-leg external rotation (front leg)	Hip-leg external rotators	Concentric
1	Arm adduction (front arm)	Arm adductors	Concentric
1	Arm flexion (front arm)	Arm flexors	Concentric
1	Hip-leg internal rotation (rear leg)	Hip-leg internal rotators	Concentric
1	Arm abduction (rear arm)	Arm abductors	Concentric
1	Thumb flexion	Thumb flexors	Isometric
1	Wrist flexion	Wrist flexors	Isometric
1	Finger flexion	Finger flexors	Isometric
2	Hip-leg flexion	Hip-leg extensors	Eccentric
2	Knee flexion	Knee extensors	Eccentric
3	Head stabilization	Head extensors	Isometric

❑ Power swing (hitting the golf ball)

Rank	*Motion*	*Muscle Group Involved*	*Contraction Type*
1	Spinal rotation (forward)	Spine rotators	Concentric
1	Arm abduction (front arm)	Arm abductors	Concentric
1	Wrist ulnar flexion	Wrist ulnar flexors	Concentric
1	Hip-leg external rotation (rear leg)	Hip-leg external rotators	Concentric
1	Hip-leg internal rotation (front leg)	Hip-leg internal rotators (front leg)	Concentric
1	Elbow extension	Elbow extensors	Concentric
1	Arm adduction (rear arm)	Arm adductors	Concentric
1	Thumb flexion	Thumb flexors	Isometric
1	Wrist flexion	Wrist flexors	Isometric
1	Finger flexion	Finger flexors	Isometric
2	Arm flexion	Arm flexors	Concentric
2	Hip-leg extension	Hip-leg extensors	Concentric
2	Knee extension	Knee extensors	Concentric
2	Ankle plantar flexion	Ankle plantar flexors	Concentric
3	Head stabilization	Head extensors	Isometric

❑ Follow-through

Rank	*Motion*	*Muscle Group Involved*	*Contraction Type*
1	Hip-leg internal rotation (front leg)	Hip-leg external rotators	Eccentric
1	Hip-leg external rotation (rear leg)	Hip-leg internal rotators	Eccentric
1	Elbow flexion	Elbow extensors	Eccentric
1	Spinal rotation (forward)	Spine rotators (antagonists)	Eccentric
1	Arm flexion	Arm extensors	Eccentric
1	Arm abduction (front arm)	Arm abductors	Eccentric
1	Arm adduction (rear arm)	Arm adductors	Eccentric
1	Wrist radial flexion	Wrist ulnar flexors	Eccentric
1	Thumb flexion	Thumb flexors	Isometric
1	Wrist flexion	Wrist flexors	Isometric
1	Finger flexion	Finger flexors	Isometric
2	Hip-leg extension	Hip-leg flexors	Eccentric
2	Knee extension	Knee flexors	Eccentric
2	Ankle plantar flexion	Ankle dorsiflexors	Eccentric

Painting a Wall

Brushing the wall in an upward and downward stroke

❑ Brushing the wall in an upward stroke

Rank	*Motion*	*Muscle Group Involved*	*Contraction Type*
1	Wrist extension	Wrist extensors	Concentric
1	Arm flexion	Arm flexors	Concentric
1	Elbow flexion	Elbow flexors	Concentric
1	Finger flexion	Finger flexors	Isometric
1	Thumb flexion	Thumb flexors	Isometric

❑ Brushing the wall in a downward stroke

Rank	*Motion*	*Muscle Group Involved*	*Contraction Type*
1	Arm extension	Arm extensors	Concentric
1	Elbow extension	Elbow extensors	Concentric
1	Wrist flexion	Wrist flexors	Concentric
1	Finger flexion	Finger flexors	Isometric
1	Thumb flexion	Thumb flexors	Isometric

Hitting a Forehand in Racquetball

The typical forehand shot

❑ Gripping the racquet

Rank	*Motion*	*Muscle Group Involved*	*Contraction Type*
1	Thumb flexion	Thumb flexors	Isometric
1	Finger flexion	Finger flexors	Isometric

❑ Backswing (moving the racquet back prior to forehand shot)

Rank	*Motion*	*Muscle Group Involved*	*Contraction Type*
1	Hip-leg external rotation (front leg)	Hip-leg external rotators	Concentric
1	Hip-leg internal rotation (rear leg)	Hip-leg internal rotators	Concentric
1	Spinal rotation (backwards)	Spine rotators	Concentric
1	Knee flexion	Knee extensors	Eccentric
1	Ankle dorsiflexion	Ankle plantar flexors	Eccentric
1	Arm horizontal abduction (racquet arm)	Arm horizontal abductors	Concentric
1	Wrist extension (racquet arm)	Wrist extensors	Concentric
1	Elbow flexion (racquet arm)	Elbow flexors	Concentric
1	Arm horizontal adduction (non-racquet arm)	Arm horizontal adductors	Concentric

❑ Power forehand swing

Rank	*Motion*	*Muscle Group Involved*	*Contraction Type*
1	Hip-leg internal rotation (front leg)	Hip-leg internal rotators	Concentric
1	Hip-leg external rotation (rear leg)	Hip-leg external rotators	Concentric
1	Spinal rotation (forward)	Spine rotators	Concentric
1	Knee extension	Knee extensors	Concentric
1	Ankle plantar flexion	Ankle plantar flexors	Concentric
1	Arm horizontal adduction (racquet arm)	Arm horizontal adductors	Concentric
1	Wrist flexion (racquet arm)	Wrist flexors	Concentric
1	Elbow extension (racquet arm)	Elbow extensors	Concentric
1	Arm horizontal abduction (non-racquet arm)	Arm horizontal abductors	Concentric

❑ Follow-through (slowing the racquet down)

Rank	*Motion*	*Muscle Group Involved*	*Contraction Type*
1	Hip-leg internal rotation (front leg)	Hip-leg external rotators	Eccentric
1	Hip-leg external rotation (rear leg)	Hip-leg internal rotators	Eccentric
1	Spinal rotation (forward)	Spine rotators (antagonists)	Eccentric
1	Knee extension	Knee flexors	Eccentric
1	Ankle plantar flexion	Ankle dorsiflexors	Eccentric
1	Arm horizontal adduction (racquet arm)	Arm horizontal abductors	Eccentric
1	Wrist flexion (racquet arm)	Wrist extensors	Eccentric
1	Elbow flexion (racquet arm)	Elbow extensors	Eccentric
1	Arm horizontal abduction (non-racquet arm)	Arm horizontal adductors	Eccentric

Shoveling (Snow or Sand)

Using a shovel with only the arms (i.e., without a foot on the blade), as when shoveling snow or sand, and placing the load to the side

❑ Gripping the handle

Rank	*Motion*	*Muscle Group Involved*	*Contraction Type*
1	Thumb flexion	Thumb flexors	Isometric
1	Finger flexion	Finger flexors	Isometric

❑ Placing the shovel into the snow or sand

Rank	*Motion*	*Muscle Group Involved*	*Contraction Type*
1	Knee flexion	Knee extensors	Eccentric
1	Hip-leg flexion	Hip-leg extensors	Eccentric
1	Spinal flexion	Spine extensors	Eccentric
1	Arm flexion	Arm flexors	Concentric
1	Elbow extension	Elbow extensors	Concentric
1	Ankle dorsiflexion	Ankle plantar flexors	Eccentric

❑ Lifting the load

Rank	*Motion*	*Muscle Group Involved*	*Contraction Type*
1	Knee extension	Knee extensors	Concentric
1	Hip-leg extension	Hip-leg extensors	Concentric
1	Spinal extension	Spine extensors	Concentric
1	Elbow flexion	Elbow flexors	Concentric
1	Arm extension	Arm extensors	Concentric
1	Ankle plantar flexion	Ankle plantar flexors	Concentric

❑ Hauling the load to the side

Rank	*Motion*	*Muscle Group Involved*	*Contraction Type*
1	Spinal rotation	Spine rotators	Concentric
1	Arm flexion	Arm flexors	Concentric
1	Elbow extension	Elbow extensors	Concentric

Swimming Freestyle

Swimming using the freestyle or crawl stroke (Arm movement analysis focuses only on one arm; kicking cycle is faster than arm cycle in freestyle stroke, so pace of legs is faster.)

❑ Arm pull (from out far in front of head through to side of leg)

Rank	*Motion*	*Muscle Group Involved*	*Contraction Type*
1	Scapula downward rotation	Scapula downward rotators	Concentric
1	Arm extension	Arm extensors	Concentric
1	Arm internal rotation	Arm internal rotators	Concentric
1	Arm adduction	Arm adductors	Concentric
1	Wrist flexion	Wrist flexors	Concentric
1	Finger stabilization	Finger flexors	Isometric
1	Finger stabilization	Finger adductors	Isometric
2	Elbow flexion	Elbow flexors	Concentric
2	Head stabilization	Head extensors	Isometric

❑ Raising the arm and rotating the head to breathe (from hand next to thigh to head out of water to breathe)

Rank	*Motion*	*Muscle Group Involved*	*Contraction Type*
1	Spinal rotation (toward surface)	Spine rotators	Concentric
1	Scapula upward rotation	Scapula upward rotators	Concentric
1	Arm hyperextension	Arm extensors	Concentric
1	Arm abduction	Arm abductors	Concentric
1	Arm external rotation	Arm external rotators	Concentric
1	Head rotation (to breathe)	Head rotators	Concentric
2	Elbow flexion	Elbow flexors	Concentric
2	Wrist flexion	Wrist flexors	Concentric
2	Finger flexion	Finger flexors	Concentric

❑ Arm reach (from out of water with head during breath, to entry in water just prior to the start of the arm pull)

Rank	*Motion*	*Muscle Group Involved*	*Contraction Type*
1	Spinal rotation (back toward water)	Spine rotators	Concentric
1	Scapula upward rotation	Scapula upward rotators	Concentric
1	Arm abduction	Arm abductors	Concentric
1	Arm external rotation	Arm external rotators	Concentric
1	Wrist extension	Wrist extensors	Concentric
1	Finger extension	Finger extensors	Concentric
1	Elbow extension	Elbow extensors	Concentric
1	Head rotation (back into water)	Head rotators	Concentric
1	Arm flexion	Arm flexors	Concentric

❑ Flutter kick—downward kicking movement

Rank	Motion	Muscle Group Involved	Contraction Type
1	Hip-leg flexion	Hip-leg flexors	Concentric
1	Ankle dorsiflexion	Ankle dorsiflexors	Concentric
1	Knee extension	Knee extensors	Concentric

❑ Flutter kick—upward kicking movement

Rank	Motion	Muscle Group Involved	Contraction Type
1	Hip-leg extension	Hip-leg extensors	Concentric
1	Ankle plantar flexion	Ankle plantar flexors	Concentric
1	Knee flexion	Knee flexors	Concentric

Serving in Tennis

The classic overhead tennis serve

❑ Gripping the racquet

Rank	Motion	Muscle Group Involved	Contraction Type
1	Thumb flexion	Thumb flexors	Isometric
1	Finger flexion	Finger flexors	Isometric

❑ Ball toss and backswing

Rank	Motion	Muscle Group Involved	Contraction Type
1	Hip-leg flexion	Hip-leg extensors	Eccentric
1	Knee flexion	Knee flexors	Eccentric
1	Ankle dorsiflexion	Ankle plantar flexors	Eccentric
1	Spinal rotation (backwards)	Spine rotators	Concentric
1	Arm abduction (toss arm)	Arm abductors	Concentric
1	Elbow flexion (toss arm)	Elbow flexors	Concentric
1	Wrist flexion (toss arm)	Wrist flexors	Concentric
1	Arm abduction (racquet arm)	Arm abductors	Concentric
1	Arm external rotation (racquet arm)	Arm external rotators	Concentric
1	Elbow flexion (racquet arm)	Elbow flexors	Concentric

❑ Extending upward and reaching for the ball with the tennis racquet

Rank	*Motion*	*Muscle Group Involved*	*Contraction Type*
1	Hip-leg extension	Hip-leg extensors	Concentric
1	Knee extension	Knee extensors	Concentric
1	Ankle plantar flexion	Ankle plantar flexors	Concentric
1	Spinal extension	Spine extensors	Concentric
1	Head extension	Head extensors	Concentric
1	Arm external rotation (racquet arm)	Arm external rotators	Concentric
1	Arm abduction (racquet arm)	Arm abductors	Concentric
1	Arm adduction (toss arm)	Arm adductors	Concentric
1	Wrist extension (racquet arm)	Wrist extensors	Concentric
1	Elbow extension (racquet arm)	Elbow extensors	Concentric
1	Spine rotation (forward)	Spine rotators	Concentric

❑ Power overhead swing (striking the ball)

Rank	*Motion*	*Muscle Group Involved*	*Contraction Type*
1	Arm extension (racquet arm)	Arm extensors	Concentric
1	Arm internal rotation (racquet arm)	Arm internal rotators	Concentric
1	Wrist flexion (racquet arm)	Wrist flexors	Concentric
1	Elbow extension (racquet arm)	Elbow extensors	Concentric
1	Arm adduction (racquet arm)	Arm adductors	Concentric
1	Spinal rotation (forward)	Spine rotators	Concentric
1	Arm internal rotation (toss arm)	Arm internal rotators	Concentric
1	Spinal flexion	Spine flexors	Concentric
1	Ankle plantar flexion	Ankle plantar flexors	Concentric
2	Hip-leg extension	Hip-leg extensors	Concentric
2	Knee extension	Knee extensors	Concentric

❑ Follow-through (slowing the racquet down)

Rank	*Motion*	*Muscle Group Involved*	*Contraction Type*
1	Arm extension (racquet arm)	Arm flexors	Eccentric
1	Arm internal rotation (racquet arm)	Arm external rotators	Eccentric
1	Wrist flexion (racquet arm)	Wrist extensors	Eccentric
1	Elbow extension (racquet arm)	Elbow flexors	Eccentric
1	Arm adduction (racquet arm)	Arm abductors	Eccentric
1	Spinal rotation (forward)	Spine rotators (antagonists)	Eccentric
1	Arm internal rotation (toss arm)	Arm external rotators	Eccentric
1	Spinal flexion	Spine extensors	Eccentric
1	Ankle dorsiflexion	Ankle plantar flexors	Eccentric
1	Hip-leg flexion	Hip-leg extensors	Eccentric
1	Knee flexion	Knee extensors	Eccentric

Appendix A • Quizzes

The quizzes in Appendix A can be printed out from the accompanying CD. They are all compatible with Scantron grading.

Quiz 1 - Planes and Actions

1. The frontal plane is always around what axis?
a. Frontal horizontal axis
b. Sagittal horizontal axis
c. Frontal axis
d. Vertical axis
e. Vertical horizontal axis

2. Hip flexion is in what plane?
a. Frontal plane
b. Sagittal plane
c. Sagittal horizontal plane
d. Transverse plane
e. Vertical plane

3. Hip adduction is in what plane?
a. Frontal plane
b. Sagittal plane
c. Sagittal horizontal plane
d. Transverse plane
e. Vertical plane

4. Pronation is in what plane?
a. Frontal plane
b. Sagittal plane
c. Sagittal horizontal plane
d. Horizontal plane
e. Vertical plane

5. In the drawing to the right, what is the action?
a. Arm abduction
b. Shoulder adduction
c. Arm extension
d. Arm flexion
e. Arm external rotation

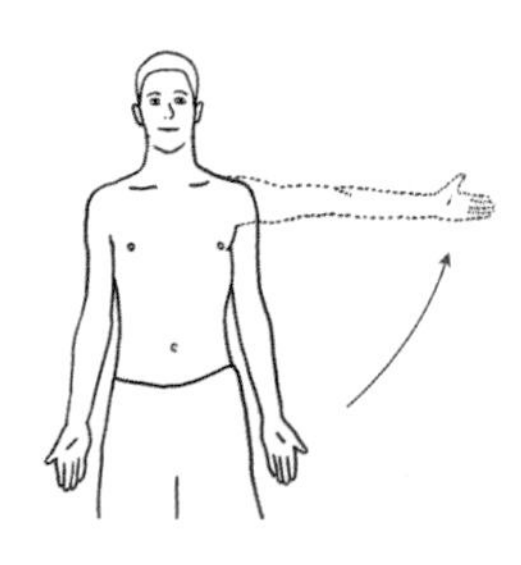

6. In the drawing to the right, what is the action?
a. Dorsiflexion
b. Plantar flexion
c. Ankle extension
d. Eversion
e. Foot rotation

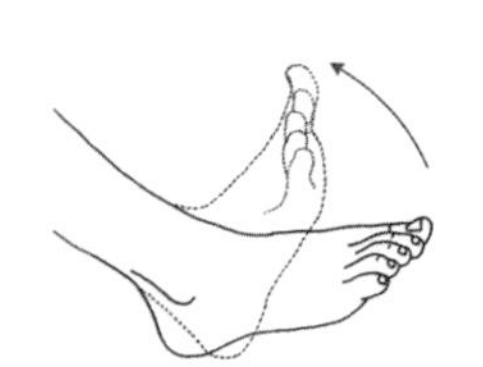

7. In the drawing to the right, what is the action?
a. Inversion
b. Eversion
c. Rotation
d. Retraction
e. Abduction

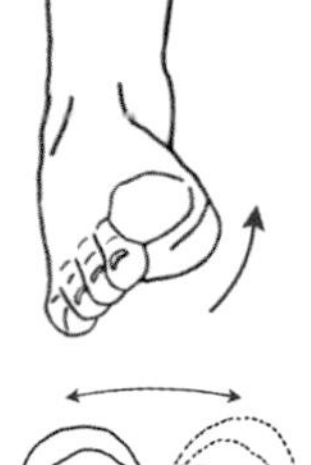

8. In the drawing to the right, what is the action?
a. Head flexion
b. Head extension
c. Neck extension
d. Neck flexion
e. Head lateral flexion

9. Arm internal rotation occurs around what axis?
a. Frontal horizontal axis
b. Sagittal horizontal axis
c. Frontal axis
d. Vertical axis
e. Sagittal axis

10. Spine lateral flexion occurs around what axis?
a. Frontal horizontal axis
b. Sagittal horizontal axis
c. Frontal axis
d. Vertical axis
e. Vertical horizontal axis

11. Dorsiflexion is in what plane?
a. Frontal plane
b. Sagittal plane
c. Sagittal horizontal plane
d. Horizontal plane
e. Vertical plane

12. Head rotation occurs around what axis?
a. Frontal horizontal axis
b. Sagittal horizontal axis
c. Frontal axis
d. Vertical axis
e. Sagittal axis

13. Ulna flexion involves what axis?
a. Frontal horizontal axis
b. Sagittal horizontal axis
c. Frontal axis
d. Vertical axis
e. Vertical horizontal axis

14. The region from the elbow to the hand is referred to as:
a. Brachial
b. Antibrachial
c. Popliteal
d. Axilla
e. Volar

15. The region from the elbow to the shoulder is referred to as:
a. Brachial
b. Antibrachial
c. Popliteal
d. Pectoral
e. Axilla

16. The region of the small of the back is referred to as:
a. Nuchae
b. Cervical
c. Lumbar
d. Iliac
e. Thoracic

17. What term refers to the head?
a. Hepato
b. Reno
c. Cephalo
d. Encephalon
e. Chondro

18. What term refers to the brain?
a. Hepato
b. Reno
c. Cephalo
d. Encephalon
e. Chondro

19. The characteristics of a diarthosis joint are:
a. Joint capsule, joint cavity, smooth joint surface
b. Joint capsule, free-moving, cartilaginous connection
c. Synovial fluid, synovial membrane, synarthrosis
d. Amphiarthrosis, joint surface smooth, joint capsule
e. Free-moving, fused bones, joint cavity

20. What is the function of bones?
a. Protection
b. Support
c. Locomotion
d. Manufacture of blood cells
e. All of the above

Quiz 2 - The Skull

Please refer to the following drawings concerning the questions in this quiz about the skull.

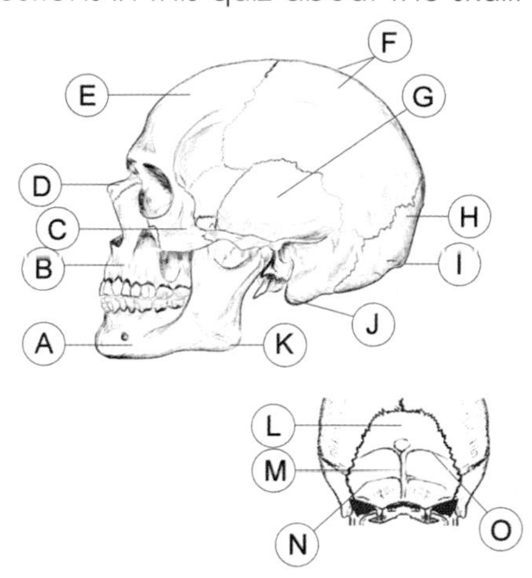

1. On the drawings of the skull, label E is the:
a. Frontal bone
b. Mastoid process
c. Parietal bone
d. Temporal bone
e. Occipital bone

2. On the drawings of the skull, label G is the:
a. Frontal bone
b. Mastoid process
c. Parietal bone
d. Temporal bone
e. Occipital bone

3. On the drawings of the skull, label F is the:
a. Frontal bone
b. Mastoid process
c. Parietal bone
d. Temporal bone
e. Occipital bone

4. On the drawings of the skull, label H is the:
a. Frontal bone
b. Mastoid process
c. Parietal bone
d. Temporal bone
e. Occipital bone

5. On the drawings of the skull, label I is the:
a. Temporal bone
b. Inferior nuchal line
c. Occipital protuberance
d. Nuchal ligament
e. Superior nuchal line

6. On the drawings of the skull, label O is the:
a. Temporal bone
b. Inferior nuchal line
c. Occipital bone
d. Nuchal ligament
e. Superior nuchal line

7. On the drawings of the skull, label M is the:
a. Temporal bone
b. Inferior nuchal line
c. Medial nuchal line
d. Nuchal ligament
e. Superior nuchal line

8. On the drawings of the skull, label N is the:
a. Temporal bone
b. Medial nuchal line
c. Occipital bone
d. Nuchal ligament
e. Inferior nuchal line

9. On the drawings of the skull, label J is the:
a. Temporal bone
b. Inferior nuchal line
c. Occipital bone
d. Mastoid process
e. Superior nuchal line

10. On the drawings of the skull, label A is the:
a. Mandible
b. Maxilla
c. Occipital bone
d. Mastoid process
e. Superior nuchal line

11. On the drawings of the skull, label B is the:
a. Mandible
b. Maxilla
c. Occipital bone
d. Mastoid process
e. Superior nuchal line

12. On the drawings of the skull, label C is the:
a. Zygomatic bone
b. Inferior nuchal line
c. Occipital bone
d. Mastoid process
e. Superior nuchal line

13. On the drawings of the skull, label D is the:
a. Zygomatic bone
b. Inferior nuchal line
c. Nasal bone
d. Maxilla
e. Superior nuchal line

14. On the drawings of the skull, label K is the:
a. Zygomatic bone
b. Inferior nuchal line
c. Nasal bone
d. Maxilla
e. Angle of the mandible

15. On the drawings of the skull, label L is the:
a. Temporal bone
b. Inferior nuchal line
c. Occipital bone
d. Mastoid process
e. Superior nuchal line

Quiz 3 - The Scapula

Please refer to the following drawings concerning the questions in this quiz about the scapula.

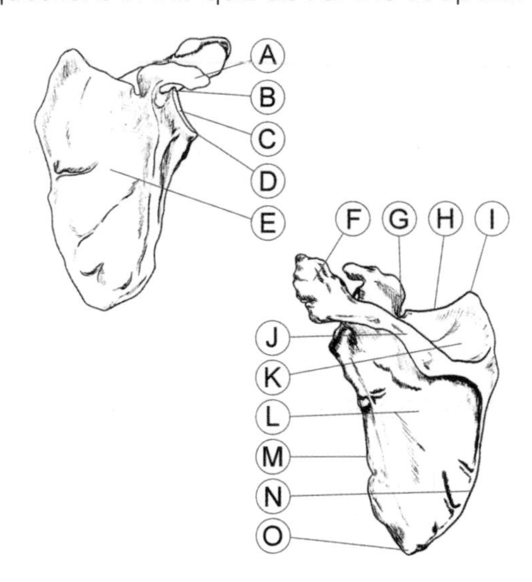

1. On the drawings of the scapula, label A is the:
a. Glenoid fossa
b. Spine
c. Supraglenoid tubercle
d. Superior angle
e. Coracoid process

2. On the drawings of the scapula, label B is the:
a. Glenoid fossa
b. Spine
c. Supraglenoid tubercle
d. Superior angle
e. Coracoid process

3. On the drawings of the scapula, label C is the:
a. Glenoid fossa
b. Spine
c. Supraglenoid tubercle
d. Superior angle
e. Coracoid process

4. On the drawings of the scapula, label D is the:
a. Glenoid fossa
b. Spine
c. Infraglenoid tubercle
d. Superior angle
e. Coracoid process

5. On the drawings of the scapula, label E is the:
a. Glenoid fossa
b. Subscapula fossa
c. Supraglenoid tubercle
d. Superior angle
e. Coracoid process

6. On the drawings of the scapula, label F is the:
a. Acromion process
b. Infraglenoid tubercle
c. Inferior angle
d. Infraspinatus fossa
e. Axillary border

7. On the drawings of the scapula, label G is the:
a. Acromion process
b. Infraglenoid tubercle
c. Scapula notch
d. Infraspinatus fossa
e. Axillary border

8. On the drawings of the scapula, label H is the:
a. Superior border
b. Infraglenoid tubercle
c. Inferior angle
d. Infraspinatus fossa
e. Axillary border

9. On the drawings of the scapula, label I is the:
a. Acromion process
b. Infraglenoid tubercle
c. Superior angle
d. Infraspinatus fossa
e. Axillary border

10. On the drawings of the scapula, label J is the:
a. Acromion process
b. Infraglenoid tubercle
c. Inferior angle
d. Infraspinatus fossa
e. Spine

11. On the drawings of the scapula, label K is the:
a. Supraspinatus fossa
b. Infraglenoid tubercle
c. Inferior angle
d. Infraspinatus fossa
e. Spine

12. On the drawings of the scapula, label L is the:
a. Acromion process
b. Infraglenoid tubercle
c. Inferior angle
d. Infraspinatus fossa
e. Spine

13. On the drawings of the scapula, label M is the:
a. Acromion process
b. Infraglenoid tubercle
c. Axillary border
d. Infraspinatus fossa
e. Spine

14. On the drawings of the scapula, label N is the:
a. Acromion process
b. Axillary border
c. Inferior angle
d. Vertebral border
e. Spine

15. On the drawings of the scapula, label O is the:
a. Acromion process
b. Infraglenoid tubercle
c. Inferior angle
d. Infraspinatus fossa
e. Spine

Quiz 4 - The Humerus

Please refer to the following drawings concerning the questions in this quiz about the humerus.

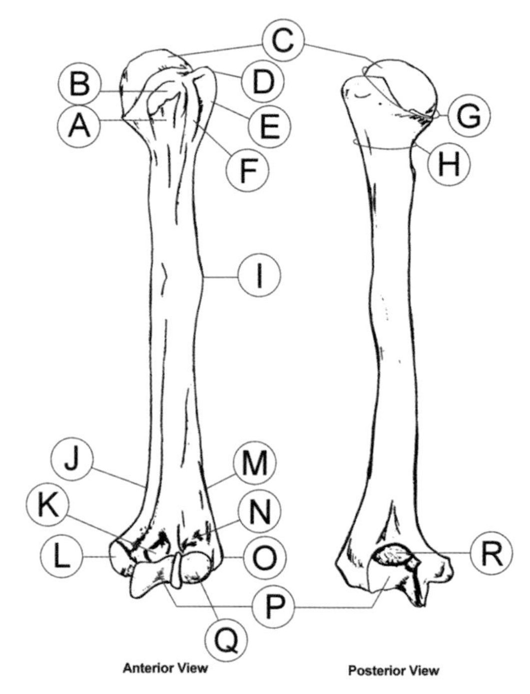

1. On the drawings of the humerus, label D is the:
a. Lesser tuberosity
b. Capitulum
c. Lateral epicondyle
d. Medial epicondyle
e. Greater tuberosity

2. On the drawings of the humerus, label B is the:
a. Lesser tuberosity
b. Capitulum
c. Lateral epicondyle
d. Medial epicondyle
e. Greater tuberosity

3. On the drawings of the humerus, label O is the:
a. Lesser tuberosity
b. Capitulum
c. Lateral epicondyle
d. Medial epicondyle
e. Greater tuberosity

4. On the drawings of the humerus, label Q is the:
a. Lesser tuberosity
b. Capitulum
c. Lateral epicondyle
d. Medial epicondyle
e. Greater tuberosity

5. On the drawings of the humerus, label L is the:
a. Lesser tuberosity
b. Capitulum
c. Lateral epicondyle
d. Medial epicondyle
e. Greater tuberosity

6. On the drawings of the humerus, label F is the:
a. Trochlear
b. Coronoid fossa
c. Olecranon fossa
d. Crest of the lesser tuberosity
e. Intertubular groove

7. On the drawings of the humerus, label K is the:
a. Trochlear
b. Coronoid fossa
c. Olecranon fossa
d. Crest of the lesser tuberosity
e. Intertubular groove

8. On the drawings of the humerus, label P is the:
a. Trochlear
b. Coronoid fossa
c. Olecranon fossa
d. Crest of the lesser tuberosity
e. Intertubular groove

9. On the drawings of the humerus, label R is the:
a. Trochlear
b. Coronoid fossa
c. Olecranon fossa
d. Crest of the lesser tuberosity
e. Intertubular groove

10. On the drawings of the humerus, label A is the:
a. Trochlear
b. Coronoid fossa
c. Olecranon fossa
d. Crest of the lesser tuberosity
e. Intertubular groove

11. On the drawings of the humerus, label C is the:
a. Deltoid tuberosity
b. Radial fossa
c. Crest of the greater tuberosity
d. Neck
e. Head

12. On the drawings of the humerus, label E is the:
a. Deltoid tuberosity
b. Radial fossa
c. Crest of the greater tuberosity
d. Neck
e. Head

13. On the drawings of the humerus, label H is the:
a. Deltoid tuberosity
b. Radial fossa
c. Crest of the greater tuberosity
d. Surgical neck
e. Head

14. On the drawings of the humerus, label I is the:

a. Deltoid tuberosity
b. Radial fossa
c. Crest of the greater tuberosity
d. Neck
e. Head

15. On the drawings of the humerus, label N is the:

a. Deltoid tuberosity
b. Radial fossa
c. Crest of the greater tuberosity
d. Neck
e. Head

16. On the drawings of the humerus, label J is the:

a. Lesser tuberosity
b. Capitulum
c. Lateral supracondylar ridge
d. Medial supracondylar ridge
e. Greater tuberosity

Quiz 5 - The Radius and Ulna

Please refer to the following drawings concerning the questions in this quiz about the radius and ulna.

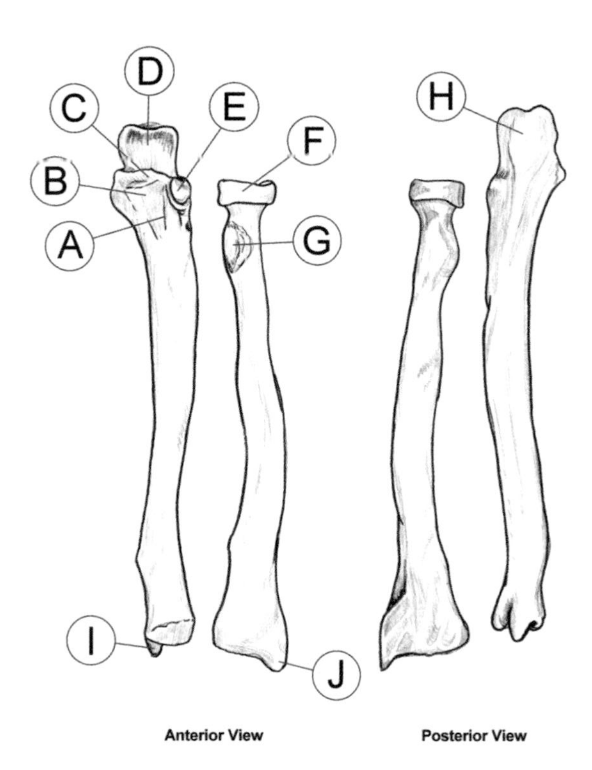

1. **On the drawings of the radius and ulna, label D is the:**
   a. Trochlear or semilunar notch
   b. Head of the radius
   c. Radial notch
   d. Olecranon process
   e. Coronoid process

2. **On the drawings of the radius and ulna, label F is the:**
   a. Trochlear or semilunar notch
   b. Head of the radius
   c. Radial notch
   d. Olecranon process
   e. Coronoid process

3. **On the drawings of the radius and ulna, label E is the:**
   a. Trochlear or semilunar notch
   b. Head of the radius
   c. Radial notch
   d. Olecranon process
   e. Coronoid process

4. **On the drawings of the radius and ulna, label H is the:**
   a. Trochlear or semilunar notch
   b. Head of the radius
   c. Radial notch
   d. Olecranon process
   e. Coronoid process

5. **On the drawings of the radius and ulna, label C is the:**
   a. Trochlear or semilunar notch
   b. Head of the radius
   c. Radial notch
   d. Olecranon process
   e. Coronoid process

6. **On the drawings of the radius and ulna, label I is the:**
   a. Styloid process of the radius
   b. Radial tuberosity
   c. Styloid process of the ulna
   d. Ulna tuberosity
   e. Supinator crest

7. **On the drawings of the radius and ulna, label J is the:**
   a. Styloid process of the radius
   b. Radial tuberosity
   c. Styloid process of the ulna
   d. Ulna tuberosity
   e. Supinator crest

8. **On the drawings of the radius and ulna, label G is the:**
   a. Styloid process of the radius
   b. Radial tuberosity
   c. Styloid process of the ulna
   d. Ulna tuberosity
   e. Supinator crest

9. **On the drawings of the radius and ulna, label B is the:**
   a. Styloid process of the radius
   b. Radial tuberosity
   c. Styloid process of the ulna
   d. Ulna tuberosity
   e. Supinator crest

10. **On the drawings of the radius and ulna, label A is the:**
    a. Styloid process of the radius
    b. Radial tuberosity
    c. Styloid process of the ulna
    d. Ulna tuberosity
    e. Supinator crest

Quiz 6 - The Hand

Please refer to the following drawings concerning the questions in this quiz about the hand.

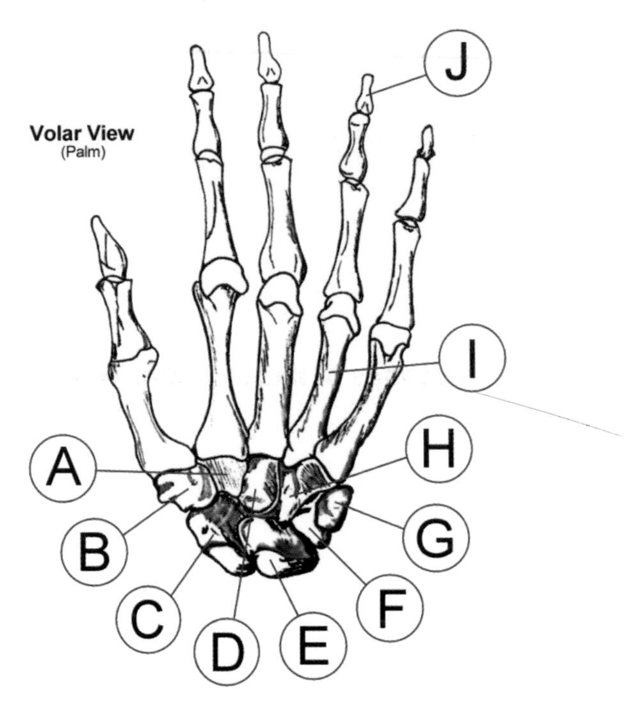

1. **On the drawing of the hand, label F is the:**
   a. Triangularis or triquetrum
   b. Capitate
   c. Navicular or scaphoid
   d. Fourth metacarpal
   e. Hamate

2. **On the drawing of the hand, label D is the:**
   a. Triangularis or triquetrum
   b. Capitate
   c. Navicular or scaphoid
   d. Fourth metacarpal
   e. Hamate

3. **On the drawing of the hand, label C is the:**
   a. Triangularis or triquetrum
   b. Capitate
   c. Navicular or scaphoid
   d. Fourth metacarpal
   e. Hamate

4. **On the drawing of the hand, label I is the:**
   a. Triangularis or triquetrum
   b. Capitate
   c. Navicular or scaphoid
   d. Fourth metacarpal
   e. Hamate

5. **On the drawing of the hand, label H is the:**
   a. Triangularis or triquetrum
   b. Capitate
   c. Navicular or scaphoid
   d. Fourth metacarpal
   e. Hamate

6. **On the drawing of the hand, label B is the:**
   a. Greater multangular or trapezium
   b. Lunate
   c. Trapezoid
   d. Second metacarpal
   e. Pisiform

7. **On the drawing of the hand, label A is the:**
   a. Greater multangular or trapezium
   b. Lunate
   c. Lesser multangular or trapezoid
   d. Second metacarpal
   e. Pisiform

8. **On the drawing of the hand, label E is the:**
   a. Greater multangular or trapezium
   b. Lunate
   c. Lesser multangular or trapezoid
   d. Second metacarpal
   e. Pisiform

9. **On the drawing of the hand, label J is the:**
   a. Greater multangular or trapezium
   b. Lunate
   c. Lesser multangular or trapezoid
   d. Third phalange
   e. Pisiform

10. **On the drawing of the hand, label G is the:**
    a. Greater multangular or trapezium
    b. Lunate
    c. Lesser multangular or trapezoid
    d. Second metacarpal
    e. Pisiform

Quiz 7 - The Vertebra

Please refer to the following drawings concerning the questions in this quiz about the vertebra.

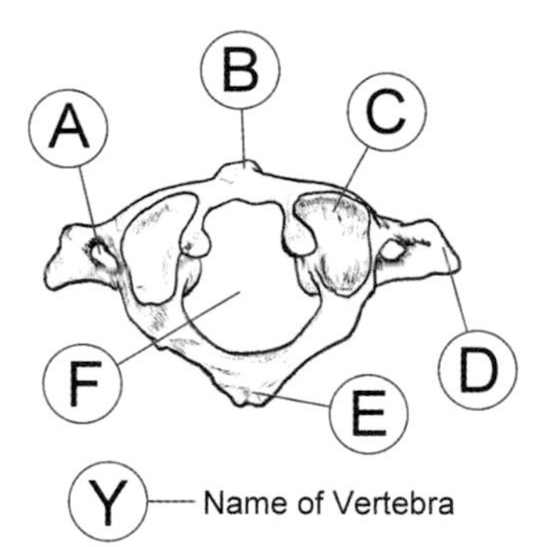

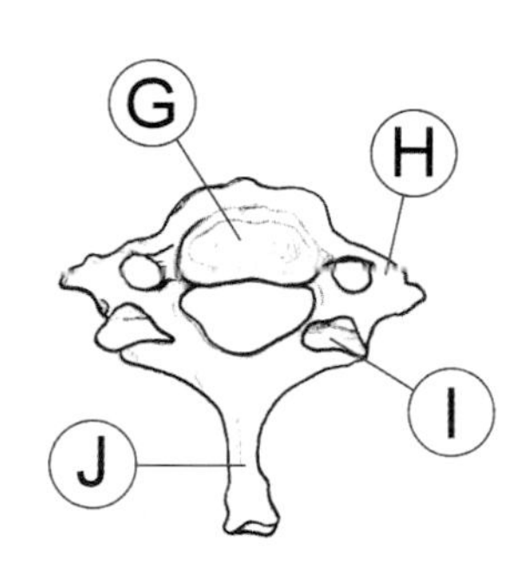

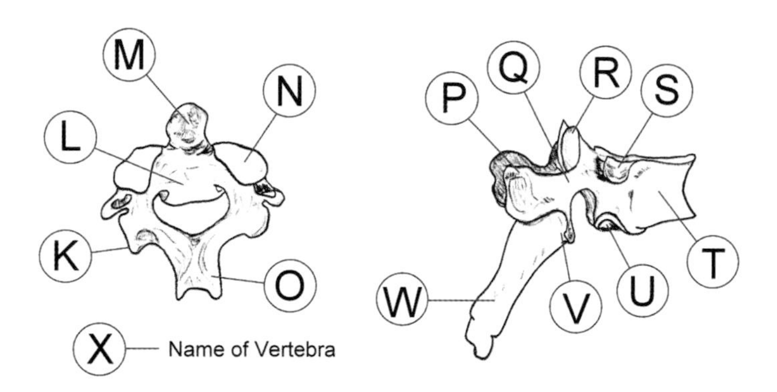

1. **On the drawings of the vertebra, label Y is the:**
   a. Superior articular process
   b. Vertebral foramen
   c. Atlas
   d. Transverse process
   e. Spiny process

2. **On the drawings of the vertebra, label C is the:**
   a. Superior articular process
   b. Vertebral foramen
   c. Atlas
   d. Transverse process
   e. Spiny process

3. **On the drawings of the vertebra, label D is the:**
   a. Superior articular process
   b. Vertebral foramen
   c. Atlas
   d. Transverse process
   e. Spiny process

4. **On the drawings of the vertebra, label E is the:**
   a. Superior articular process
   b. Vertebral foramen
   c. Atlas
   d. Transverse process
   e. Posterior tubercle

5. **On the drawings of the vertebra, label F is the:**
   a. Superior articular process
   b. Vertebral foramen
   c. Atlas
   d. Transverse process
   e. Spiny process

6. **On the drawings of the vertebra, label A is the:**
   a. Axis
   b. Dens
   c. Transverse foramen
   d. Anterior tubercle
   e. Superior articular process

7. **On the drawings of the vertebra, label B is the:**
   a. Axis
   b. Dens
   c. Transverse foramen
   d. Anterior tubercle
   e. Superior articular process

8. **On the drawings of the vertebra, label X is the:**
   a. Axis
   b. Dens
   c. Transverse foramen
   d. Anterior tubercle
   e. Superior articular process

9. **On the drawings of the vertebra, label M is the:**
   a. Axis
   b. Dens
   c. Transverse foramen
   d. Anterior tubercle
   e. Superior articular process

10. **On the drawings of the vertebra, label N is the:**
    a. Axis
    b. Dens
    c. Transverse foramen
    d. Anterior tubercle
    e. Superior articular process

11. **On the drawings of the vertebra, label L is the:**
    a. Superior articular process
    b. Spine
    c. Body
    d. Inferior articular process
    e. Inferior tubercle

12. **On the drawings of the vertebra, label K is the:**
    a. Superior articular process
    b. Spine
    c. Body
    d. Inferior articular process
    e. Inferior tubercle

13. On the drawings of the vertebra, label O is the:
a. Superior articular process
b. Spine
c. Body
d. Inferior articular process
e. Inferior tubercle

14. On the drawings of the vertebra, label G is the:
a. Superior articular process
b. Spine
c. Body
d. Inferior articular process
e. Inferior tubercle

15. On the drawings of the vertebra, label I is the:
a. Superior articular process
b. Spine
c. Body
d. Inferior articular process
e. Inferior tubercle

16. On the drawings of the vertebra, label J is the:
a. Superior articular process
b. Spine
c. Body
d. Inferior articular process
e. Inferior tubercle

17. On the drawings of the vertebra, label H is the:
a. Superior articular process
b. Spine
c. Body
d. Superior demi-facet
e. Transverse process

18. On the drawings of the vertebra, label R is the:
a. Superior articular process
b. Spine
c. Body
d. Inferior articular process
e. Inferior tubercle

19. On the drawings of the vertebra, label S is the:
a. Superior articular process
b. Spine
c. Body
d. Superior demi-facet
e. Transverse process

20. On the drawings of the vertebra, label T is the:
a. Superior articular process
b. Spine
c. Body
d. Superior demi-facet
e. Transverse process

21. On the drawings of the vertebra, label U is the:
a. Pedicle
b. Spine
c. Inferior articular process
d. Inferior demi-facet
e. Transverse process

22. On the drawings of the vertebra, label V is the:
a. Pedicle
b. Spine
c. Inferior articular process
d. Inferior demi-facet
e. Transverse process

23. On the drawings of the vertebra, label W is the:
a. Pedicle
b. Spine
c. Inferior articular process
d. Inferior demi-facet
e. Transverse process

24. On the drawings of the vertebra, label P is the:
a. Pedicle
b. Spine
c. Inferior articular process
d. Inferior demi-facet
e. Transverse process

25. On the drawings of the vertebra, label Q is the:
a. Pedicle
b. Spine
c. Inferior articular process
d. Inferior demi-facet
e. Transverse process

Quiz 8 - The Innominate Bone

Please refer to the following drawings concerning the questions in this quiz about the innominate bone.

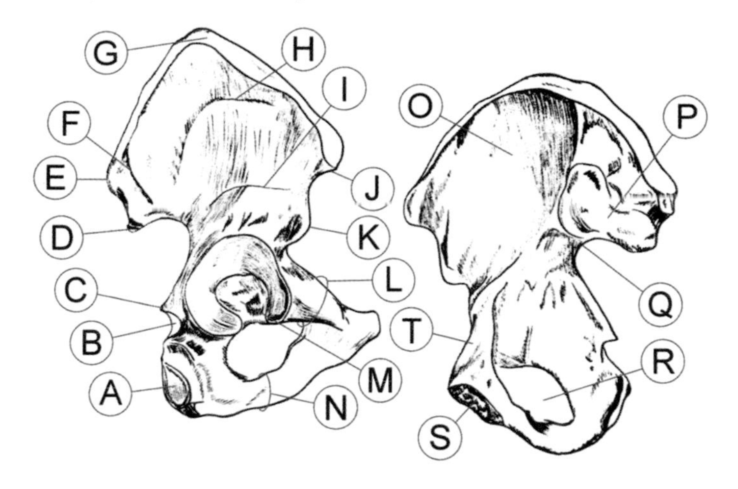

1. **On the drawings of the innominate bone, label E is the:**
   a. Inferior posterior spine
   b. Spine of the ischium
   c. Superior posterior spine
   d. Superior anterior spine
   e. Inferior anterior spine

2. **On the drawings of the innominate bone, label J is the:**
   a. Inferior posterior spine
   b. Spine of the ischium
   c. Superior posterior spine
   d. Superior anterior spine
   e. Inferior anterior spine

3. **On the drawings of the innominate bone, label C is the:**
   a. Inferior posterior spine
   b. Spine of the ischium
   c. Superior posterior spine
   d. Superior anterior spine
   e. Inferior anterior spine

4. **On the drawings of the innominate bone, label K is the:**
   a. Inferior posterior spine
   b. Spine of the ischium
   c. Superior posterior spine
   d. Superior anterior spine
   e. Inferior anterior spine

5. **On the drawings of the innominate bone, label D is the:**
   a. Inferior posterior spine
   b. Spine of the ischium
   c. Superior posterior spine
   d. Superior anterior spine
   e. Inferior anterior spine

6. **On the drawings of the innominate bone, label G is the:**
   a. Tuberosity of the ilium
   b. Anterior gluteal line
   c. Iliac crest
   d. Posterior gluteal line
   e. Inferior gluteal line

7. **On the drawings of the innominate bone, label H is the:**
   a. Tuberosity of the ilium
   b. Anterior gluteal line
   c. Iliac drest
   d. Posterior gluteal line
   e. Inferior gluteal line

8. **On the drawings of the innominate bone, label F is the:**
   a. Tuberosity of the ilium
   b. Anterior gluteal line
   c. Iliac crest
   d. Posterior gluteal line
   e. Inferior gluteal line

9. **On the drawings of the innominate bone, label I is the:**
   a. Tuberosity of the ilium
   b. Anterior gluteal line
   c. Iliac crest
   d. Posterior gluteal line
   e. Inferior gluteal line

10. **On the drawings of the innominate bone, label P is the:**
    a. Tuberosity of the ilium
    b. Anterior gluteal line
    c. Iliac crest
    d. Posterior gluteal line
    e. Inferior gluteal line

11. **On the drawings of the innominate bone, label A is the:**
    a. Iliac fossa
    b. Ischial tuberosity
    c. Obturator foramen
    d. Crest of the pubis
    e. Symphysis

12. **On the drawings of the innominate bone, label S is the:**
    a. Iliac fossa
    b. Ischial tuberosity
    c. Obturator foramen
    d. Crest of the pubis
    e. Symphysis

13. On the drawings of the innominate bone, label R is the:
a. Iliac fossa
b. Ischial tuberosity
c. Obturator foramen
d. Crest of the pubis
e. Symphysis

14. On the drawings of the innominate bone, label T is the:
a. Iliac fossa
b. Ischial tuberosity
c. Obturator foramen
d. Crest of the pubis
e. Symphysis

15. On the drawings of the innominate bone, label O is the:
a. Iliac fossa
b. Ischial tuberosity
c. Obturator foramen
d. Crest of the pubis
e. Symphysis

16. On the drawings of the innominate bone, label M is the:
a. Superior ramus
b. Lesser sciatic notch
c. Acetabulum
d. Greater sciatic notch
e. Inferior ramus of the ischium

17. On the drawings of the innominate bone, label B is the:
a. Superior ramus
b. Lesser sciatic notch
c. Acetabulum
d. Greater sciatic notch
e. Inferior ramus of the ischium

18. On the drawings of the innominate bone, label Q is the:
a. Superior ramus
b. Lesser sciatic notch
c. Acetabulum
d. Greater sciatic notch
e. Inferior ramus of the ischium

19. On the drawings of the innominate bone, label L is the:
a. Superior ramus of the pubis
b. Lesser sciatic notch
c. Acetabulum
d. Greater sciatic notch
e. Inferior ramus of the ischium

20. On the drawings of the innominate bone, label N is the:
a. Superior ramus
b. Lesser sciatic notch
c. Acetabulum
d. Greater sciatic notch
e. Inferior ramus of the ischium

Quiz 9 - The Femur

Please refer to the following drawings concerning the questions in this quiz about the femur.

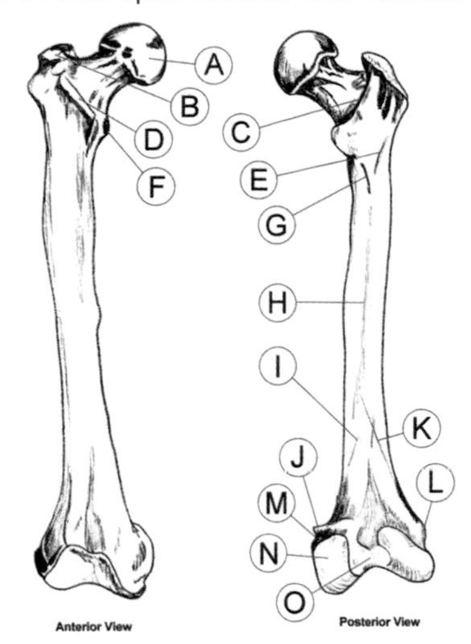

1. **On the drawings of the femur, label A is the:**
   a. Greater trochanter
   b. Intertrochanteric line
   c. Head
   d. Intertrochanteric crest
   e. Lesser trochanter

2. **On the drawings of the femur, label B is the:**
   a. Greater trochanter
   b. Intertrochanteric line
   c. Head
   d. Intertrochanteric crest
   e. Lesser trochanter

3. **On the drawings of the femur, label C is the:**
   a. Greater trochanter
   b. Intertrochanteric line
   c. Head
   d. Intertrochanteric crest
   e. Lesser trochanter

4. **On the drawings of the femur, label F is the:**
   a. Greater trochanter
   b. Intertrochanteric line
   c. Head
   d. Intertrochanteric crest
   e. Lesser trochanter

5. **On the drawings of the femur, label D is the:**
   a. Greater trochanter
   b. Intertrochanteric line
   c. Head
   d. Intertrochanteric crest
   e. Lesser trochanter

6. **On the drawings of the femur, label O is the:**
   a. Medial epicondyle
   b. Adductor tubercle
   c. Medial condyle
   d. Intercondylar fossa
   e. Lateral epicondyle

7. **On the drawings of the femur, label J is the:**
   a. Medial epicondyle
   b. Adductor tubercle
   c. Medial condyle
   d. Intercondylar fossa
   e. Lateral epicondyle

8. **On the drawings of the femur, label M is the:**
   a. Medial epicondyle
   b. Adductor tubercle
   c. Medial condyle
   d. Intercondylar fossa
   e. Lateral epicondyle

9. **On the drawings of the femur, label L is the:**
   a. Medial epicondyle
   b. Adductor tubercle
   c. Medial condyle
   d. Intercondylar fossa
   e. Lateral epicondyle

10. **On the drawings of the femur, label N is the:**
    a. Medial epicondyle
    b. Adductor tubercle
    c. Medial condyle
    d. Intercondylar fossa
    e. Lateral epicondyle

11. **On the drawings of the femur, label H is the:**
    a. Linea aspera
    b. Pectineal line
    c. Medial epicondylar ridge
    d. Gluteal line
    e. Lateral epicondylar ridge

12. **On the drawings of the femur, label G is the:**
    a. Linea aspera
    b. Pectineal line
    c. Medial epicondylar ridge
    d. Gluteal line
    e. Lateral epicondylar ridge

13. **On the drawings of the femur, label I is the:**
    a. Linea aspera
    b. Pectineal line
    c. Medial epicondylar ridge
    d. Gluteal line
    e. Lateral epicondylar ridge

14. **On the drawings of the femur, label E is the:**
    a. Linea aspera
    b. Pectineal line
    c. Medial epicondylar ridge
    d. Gluteal line
    e. Lateral epicondylar ridge

15. **On the drawings of the femur, label K is the:**
    a. Linea aspera
    b. Pectineal line
    c. Medial epicondylar ridge
    d. Gluteal line
    e. Lateral epicondylar ridge

Quiz 10 - The Tibia and Fibula

Please refer to the following drawings concerning the questions in this quiz about the tibia and fibula.

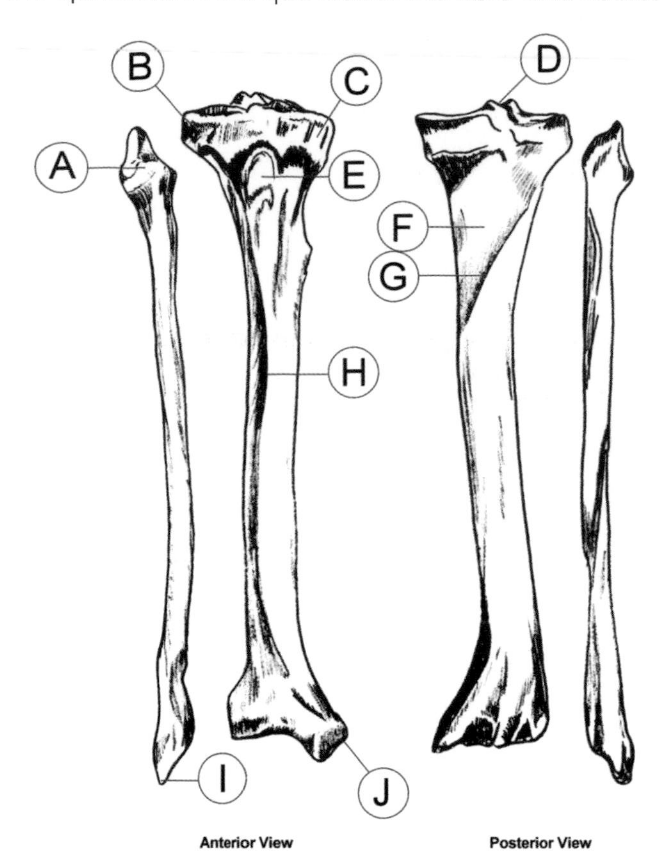

1. **On the drawings of the tibia and fibula, label H is the:**
   a. Popliteal line
   b. Tibia (anterior) tuberosity
   c. Medial condyle
   d. Anterior crest
   e. Head

2. **On the drawings of the tibia and fibula, label A is the:**
   a. Popliteal line
   b. Tibia (anterior) tuberosity
   c. Medial condyle
   d. Anterior crest
   e. Head of the fibula

3. **On the drawings of the tibia and fibula, label E is the:**
   a. Popliteal line
   b. Tibia (anterior) tuberosity
   c. Medial condyle
   d. Anterior crest
   e. Head

4. **On the drawings of the tibia and fibula, label C is the:**
   a. Popliteal line
   b. Tibia (anterior) tuberosity
   c. Medial condyle
   d. Anterior crest
   e. Head

5. **On the drawings of the tibia and fibula, label G is the:**
   a. Popliteal line
   b. Tibia (anterior) tuberosity
   c. Medial condyle
   d. Anterior crest
   e. Head

6. **On the drawings of the tibia and fibula, label B is the:**
   a. Lateral malleolus
   b. Shaft
   c. Lateral condyle
   d. Intercondylar eminence
   e. Medial malleolus

7. **On the drawings of the tibia and fibula, label I is the:**
   a. Lateral malleolus
   b. Shaft
   c. Lateral condyle
   d. Intercondylar eminence
   e. Medial malleolus

8. **On the drawings of the tibia and fibula, label F is the:**
   a. Lateral malleolus
   b. Shaft of fibula
   c. Lateral condyle
   d. Popliteal surface
   e. Medial malleolus

9. **On the drawings of the tibia and fibula, label J is the:**
   a. Lateral malleolus
   b. Shaft
   c. Lateral condyle
   d. Intercondylar eminence
   e. Medial malleolus

10. **On the drawings of the tibia and fibula, label D is the:**
    a. Lateral malleolus
    b. Shaft
    c. Lateral condyle
    d. Intercondylar eminence
    e. Medial malleolus

Quiz 11 - The Foot

Please refer to the following drawings concerning the questions in this quiz about the foot.

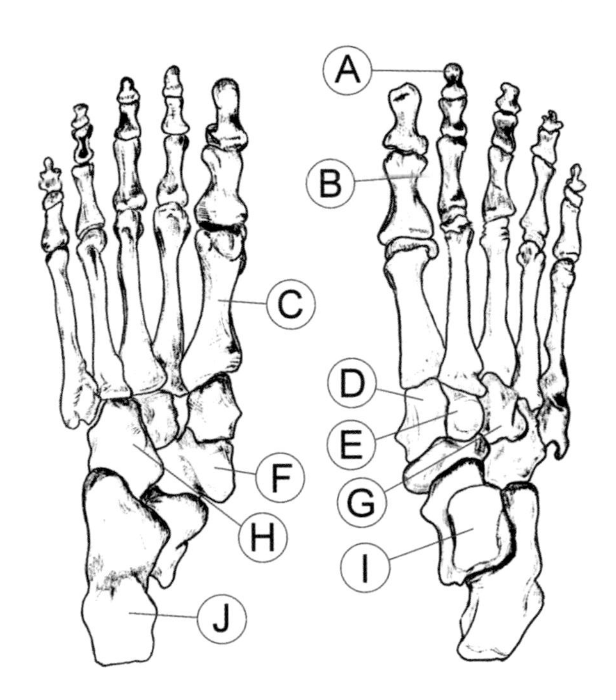

1. **On the drawings of the foot, label I is the:**
   a. Navicular
   b. Third cuneiform
   c. Talus
   d. First cuneiform
   e. Second cuneiform

2. **On the drawings of the foot, label F is the:**
   a. Navicular
   b. Third cuneiform
   c. Talus
   d. First cuneiform
   e. Second cuneiform

3. **On the drawings of the foot, label D is the:**
   a. Navicular
   b. Third cuneiform
   c. Talus
   d. First cuneiform
   e. Second cuneiform

4. **On the drawings of the foot, label G is the:**
   a. Navicular
   b. Third cuneiform
   c. Talus
   d. First cuneiform
   e. Second cuneiform

5. **On the drawings of the foot, label E is the:**
   a. Navicular
   b. Third cuneiform
   c. Talus
   d. First cuneiform
   e. Second cuneiform

6. **On the drawings of the foot, label C is the:**
   a. Third phalange
   b. Cuboid
   c. First phalange
   d. Calcaneus
   e. First metatarsal

7. **On the drawings of the foot, label J is the:**
   a. Third phalange
   b. Cuboid
   c. First phalange
   d. Calcaneus
   e. First metatarsal

8. **On the drawings of the foot, label B is the:**
   a. Third phalange
   b. Cuboid
   c. First phalange
   d. Calcaneus
   e. First metatarsal

9. **On the drawings of the foot, label A is the:**
   a. Third phalange
   b. Cuboid
   c. First phalange
   d. Calcaneus
   e. First metatarsal

10. **On the drawings of the foot, label H is the:**
    a. Third phalange
    b. Cuboid
    c. First phalange
    d. Calcaneus
    e. First metatarsal

Quiz 12 - Muscles and Movers

1. **The four properties of muscle cells are:**
   a. Contractility, extensibility, flexibility, tonicity
   b. Contractility, compressibility, flexibility, tonicity
   c. Contractility, extensibility, isokinetic, tonicity
   d. Contractility, extensibility, flexibility, adaptability
   e. Contractility, extensibility, elasticity, tonicity

2. **The five classifications of muscle contractions by anatomists and kinesiologists are:**
   a. Isometric, isotonic, concentric, eccentric, static
   b. Isometric, isotonic, isokinetic, eccentric, static
   c. Isometric, isotonic, concentric, eccentric, dynamic
   d. Isometric, isokinetic, concentric, eccentric, static
   e. Isometric, isotonic, isostatic, eccentric, static

3. **To what muscle classifications do the biceps belong?**
   a. Skeletal, visceral, smooth
   b. Skeletal, voluntary, smooth
   c. Skeletal, striated, voluntary
   d. Skeletal, involuntary, smooth
   e. Smooth, visceral, involuntary

4. **To what muscle classification does the stomach muscle belong?**
   a. Skeletal, visceral, smooth
   b. Skeletal, voluntary, smooth
   c. Skeletal, striated, voluntary
   d. Skeletal, involuntary, smooth
   e. Smooth, visceral, involuntary

5. **To what classification does the heart belong?**
   a. Syncytium
   b. Striated
   c. Smooth
   d. Visceral
   e. Voluntary/involuntary

6. **Prime Movers are the muscles that do the major movement.**
   a. True
   b. False

7. **During push-ups, the stabilizers are:**
   a. Hip extensors and spine extensors
   b. Hip flexors and spine flexors
   c. Hip extensors and spine flexors
   d. Hip flexors and spine extensors
   e. Hip flexors and knee flexors

8. **During push-ups, what are the prime movers?**
   a. Elbow flexors and arm flexors
   b. Elbow flexors and arm extensors
   c. Elbow extensors and arm extensors
   d. Elbow extensors and arm flexors
   e. Elbow extensors and forearm supinators

9. **During pull-ups, what are the prime movers?**
   a. Elbow flexors and arm flexors
   b. Elbow flexors and arm extensors
   c. Elbow extensors and arm extensors
   d. Elbow extensors and arm flexors
   e. Elbow extensors and forearm supinators

10. **During a sit-up, the prime movers are:**
    a. Hip flexors and spine flexors
    b. Hip flexors and knee flexors
    c. Hip flexors and spine extensors
    d. Spine extensors and hip extensors
    e. Leg flexors and spine extensors

Quiz 13 - Insertions, Origins, and Actions

Choose the most correct answer for the following questions.

1. The origin of the sternocleidomastoid is:
a. Manubrium of the sternum
b. Lateral third of the clavicle
c. The anterior 3rd, 4th, & 5th ribs
d. The medial third of the clavicle and tho anterior surface of the manubrium
e. Coracoid process

2. The origin of the splenius capitis is:
a. Ligamentum nuchae
b. Spines of the 1-4 thoracic vertebra
c. Seventh cervical spiny process
d. The lower ligamentum nuchae, the seventh cervical vertebra and the upper four thoracic vertebra
e. Mastoid process and adjacent occipital bone

3. The origin of the pectoralis minor is:
a. Manubrium of the sternum
b. Lateral third of the clavicle
c. The anterior 3rd, 4th , & 5th ribs
d. Coracoid process
e. Upper eight ribs

4. The insertion of the sternocleidomastoid is:
a. Mastoid process
b. Lateral third of the clavicle
c. The anterior 3rd, 4th, & 5th ribs
d. Mastoid process and the adjacent occipital bone
e. Coracoid process

5. The insertion of the splenius capitis is:
a. Mastoid process
b. Lateral third of the clavicle
c. The anterior 3rd, 4th, & 5th ribs
d. Mastoid process and the adjacent occipital bone
e. Coracoid process

6. The actions of the sternocleidomastoid are:
a. Head flexion and lateral flexion
b. Head extension and lateral flexion
c. Head rotation and flexion
d. Head lateral flexion, rotation, and flexion
e. Head lateral flexion, rotation, and extension

7. The actions of the splenius capitis are:
a. Head flexion and lateral flexion
b. Head extension and lateral flexion
c. Head rotation and flexion
d. Head lateral flexion, rotation, and flexion
e. Head lateral flexion, rotation, and extension

8. The origin of the serratus anterior is:
a. Manubrium of the sternum
b. Upper 8-9 ribs
c. The anterior 3rd, 4th, & 5th ribs
d. Vertebral border of the scapula
e. Costal surface of the scapula

9. The insertion of the serratus anterior is:
a. Manubrium of the sternum
b. Upper 8-9 ribs
c. The anterior 3rd, 4th, & 5th ribs
d. The costal surface of the vertebral border of the scapula
e. Costal surface of the scapula

10. The origin of the trapezius is:
a. Occipital protuberance
b. Superior nuchal line
c. The ligamentum nuchae
d. The seventh cervical and all thoracic vertebra
e. All of the above

11. The insertion of the trapezius is:
a. Occipital protuberance
b. Lateral third of the clavicle
c. The spine of the scapula
d. The lateral third of the clavicle, the acromion process and spine of the scapula
e. Vertebral border of the scapula

12. The origins of the rhomboids are:
a. The vertebral border of the scapula
b. Superior nuchal line
c. The ligamentum nuchae
d. The seventh cervical and all thoracic vertebra
e. Lower ligamentum nuchae, seventh cervical and first five thoracic vertebra

13. The insertion of the rhomboids is:
a. The vertebral border of the scapula below the spine of the scapula
b. Superior nuchal line
c. The ligamentum nuchae
d. The seventh cervical and all thoracic vertebra
e. The lower ligamentum nuchae, seventh cervical and first five thoracic vertebra

14. The action of the rhomboids is:
a. Scapula elevation, downward rotation, adduction
b. Scapula elevation, upward rotation, adduction
c. Scapula elevation, downward rotation, abduction
d. Scapula elevation, depression, adduction
e. Scapula elevation, abduction

15. The action of the trapezius is:
a. Scapula elevation, downward rotation, abduction, depression
b. Scapula depression, downward rotation, adduction
c. Scapula elevation, downward rotation, abduction
d. Scapula elevation, depression, adduction, upward rotation
e. Scapula elevation, depression, adduction, downward rotation

16. The origin of the levator scapula is:
a. Ligamentum nuchae, seventh cervical
b. Lower ligamentum nuchae, seventh cervical, 1-4 thoracic vertebra
c. 1-4 cervical vertebra
d. 4-5-6 cervical vertebra
e. Ligamentum nuchae, seventh cervical, all thoracic vertebra

17. The insertion of the levator scapula is:
a. Vertebral border of the scapula below the spine of the scapula
b. Vertebral border of the scapula
c. Spine and acromion process of the scapula
d. Superior border of the scapula
e. Vertebral border of the scapula above the spine of the scapula

18. The action of the levator scapula is:
a. Scapula elevation, downward rotation, adduction
b. Scapula elevation, upward rotation, adduction
c. Scapula elevation, upward rotation
d. Scapula elevation, depression, downward rotation
e. Scapula elevation, abduction, downward rotation

19. The origin of the coracobrachialis is:
a. Acromion process
b. Acromion process and coracoid process
c. Lateral third of the clavicle
d. Coracoid process
e. Anterior 3,4,5 ribs

20. The insertion of the coracobrachialis is:
a. Anterior 3, 4, 5 ribs
b. Medial middle third of the humerus
c. Brachial tuberosity
d. Lower third of the humerus
e. Anterior humerus

21. The action of the coracobrachialis is:
a. Arm flexion, adduction, external rotation
b. Arm extension and external rotation
c. Arm flexion and abduction
d. Arm flexion and internal rotation
e. Arm flexion, adduction, and internal rotation

22. The action of the lower fibers of the anterior deltoid is:
a. Arm flexion, adduction, internal rotation
b. Arm flexion, adduction, external rotation
c. Arm extension, adduction, external rotation
d. Arm extension, adduction, internal rotation
e. Arm abduction

23. The action of the lower fibers of the posterior deltoid is:
a. Arm flexion, adduction, internal rotation
b. Arm flexion, adduction, external rotation
c. Arm extension, adduction, external rotation
d. Arm extension, adduction, internal rotation
e. Arm abduction

24. The action of the middle deltoid is:
a. Arm flexion, adduction, internal rotation
b. Arm flexion, adduction, external rotation
c. Arm extension, adduction, external rotation
d. Arm extension, adduction, internal rotation
e. Arm abduction

25. The origin of the deltoid is:
a. Medial two-thirds clavicle, acromion process, coracoid process
b. Clavicle and spine of the scapula
c. Lateral one-thirds clavicle, acromion process, and spine of the scapula
d. Lateral one-thirds clavicle, superior border, and spine of scapula
e. Lateral one-thirds clavicle, acromion process, and axillary border

26. Give the action of the biceps.
a. Elbow flexion
b. Elbow extension and arm extension
c. Arm flexion, elbow flexion, and supination
d. Elbow flexion and supination
e. Elbow flexion and pronation

27. Give the action of the brachialis.
a. Elbow flexion
b. Elbow extension and arm extension
c. Arm flexion, elbow flexion, supination
d. Elbow flexion and supination
e. Elbow flexion and pronation

28. Give the action of the triceps.
a. Elbow flexion
b. Elbow extension and arm extension & adduction
c. Arm flexion, elbow flexion, supination
d. Elbow flexion and supination
e. Elbow flexion and pronation

29. Give the action of the latissimus dorsi.
a. Arm flexion, arm adduction, arm internal rotation
b. Arm extension, arm adduction, arm internal rotation
c. Arm extension, elbow extension, arm internal rotation
d. Arm internal rotation, arm flexion, arm adduction
e. Arm extension, arm adduction, external rotation

30. Give the action of the teres major.
a. Arm flexion, arm adduction, arm internal rotation
b. Arm extension, arm adduction, arm internal rotation
c. Arm extension, elbow extension, arm internal rotation
d. Arm internal rotation, arm flexion, arm adduction
e. Arm extension, arm adduction, external rotation

31. Give the action of the supraspinatus.
a. Arm abduction, arm flexion, arm internal rotation
b. Arm adduction, arm flexion, arm internal rotation
c. Arm abduction
d. Arm abduction, arm internal rotation
e. Arm adduction

32. Give the action of the pronator teres.
a. Elbow flexion
b. Elbow extension and arm extension
c. Arm flexion, elbow flexion, supination
d. Elbow flexion and supination
e. Elbow flexion and pronation

33. Give the action of the brachioradialis.
a. Elbow flexion
b. Elbow extension and arm extension
c. Arm flexion, elbow flexion, supination
d. Elbow flexion and supination
e. Elbow flexion and pronation

34. Give the action of the rectus abdominis.
a. Spine flexion, rotation to the opposite side, and compression of the abdomen
b. Extension of the spine
c. Compression of the abdomen
d. Flexion of the spine and compression of the abdomen
e. Spine flexion and rotation to the same side; compression of the abdomen

35. Give the action of the external abdominal oblique.
a. Spine flexion and rotation to the opposite side; compression of the abdomen
b. Extension of the spine
c. Compression of the abdomen
d. Flexion of the spine and compression of the abdomen
e. Spine flexion and rotation to the same side; compression of the abdomen

36. Give the action of the transverse abdominis.
a. Spine flexion and rotation to the opposite side; compression of the abdomen
b. Extension of the spine
c. Compression of the abdomen
d. Flexion of the spine and compression of the abdomen
e. Spine flexion and rotation to the same side; compression of the abdomen

37. Give the action of the internal abdominal oblique.
a. Spine flexion and rotation to the opposite side; compression of the abdomen
b. Extension of the spine
c. Compression of the abdomen
d. Flexion of the spine and compression of the abdomen
e. Spine flexion and rotation to the same side; compression of the abdomen

38. The action of the iliacus is:
a. Hip flexion, abduction, external rotation
b. Hip flexion, adduction, external rotation
c. Hip extension, external rotation, adduction
d. Hip flexion, adduction, internal rotation
e. Leg flexion, abduction, external rotation

39. The action of the gluteus maximus is:
a. Hip flexion, abduction, external rotation
b. Hip flexion, adduction, (possible external rotation)
c. Hip extension, external rotation, abduction & adduction
d. Hip flexion, adduction, internal rotation
e. Leg flexion, abduction, external rotation

40. The action of the sartorius is:
a. Hip flexion, abduction, external rotation, knee flexion
b. Hip flexion, adduction, (possible external rotation)
c. Hip extension, external rotation, adduction
d. Hip flexion, adduction, internal rotation, knee flexion
e. Hip flexion, abduction, external rotation, knee extension

41. The action of the gracilis is:
a. Hip flexion, abduction, external rotation
b. Hip flexion, adduction, (possible external rotation)
c. Hip extension, external rotation, adduction
d. Hip flexion, abduction, internal rotation
e. Hip flexion, abduction, external rotation

42. The action of the adductor longus is:
a. Hip flexion, abduction, external rotation
b. Hip flexion, adduction, external rotation
c. Hip extension, external rotation, adduction
d. Hip flexion, adduction, internal rotation
e. Leg flexion, abduction, external rotation

43. The action of the pectineus is:
a. Hip flexion, abduction, external rotation
b. Hip flexion, adduction, external rotation
c. Hip extension, external rotation, adduction
d. Hip flexion, adduction, internal rotation
e. Leg flexion, abduction, external rotation

44. The action of the tensor fascia lata is:
a. Hip flexion, abduction, external rotation
b. Hip flexion, adduction, external rotation
c. Hip extension, external rotation, adduction
d. Hip flexion, adduction, internal rotation
e. Hip flexion, abduction, internal rotation

45. The action of the psoas is:
a. Hip flexion, abduction, external rotation
b. Hip flexion, adduction, external rotation
c. Hip extension, external rotation, adduction
d. Hip flexion, adduction, internal rotation
e. Hip flexion, abduction, internal rotation

46. The origin of the gluteus maximus is the:
a. Pectineal line and upper one-third of the linea aspera of the femur
b. Post. gluteal line, post. crest of ilium, post/lat. sacrum & coccyx
c. Middle one-third of the linea aspera
d. Pectineal line of femur
e. Linea aspera, medial epicondylar ridge, adductor tubercle of femur

47. The origin of the psoas is the:
a. Posterior crest of the ilium
b. The symphysis of the pubic bone
c. Inferior ramus of the pubic bone
d. Iliac fossa of the ilium and adjacent sacrum
e. Transverse processes and bodies of the five lumbar vertebra

48. The action of the gastrocnemius is:
a. Plantar flexion
b. Knee flexion, plantar flexion
c. Knee extension, plantar flexion
d. Plantar flexion, hip extension
e. Dorsiflexion

49. Identify the ankle evertor.
a. Tibialis anterior
b. Tibialis posterior
c. Extensor hallicus longus
d. Flexor hallicus longus
e. Peroneus brevis

50. Identify the ankle dorsiflexors.
a. Extensor hallicus longus and extensor digitorum longus
b. Tibialis posterior and flexor hallicus longus
c. Soleus
d. Peroneus longus
e. Peroneus brevis and tertius

51. Identify the ankle plantar flexors.
a. Flexor digitorum longus
b. Extensor hallicus longus
c. Extensor digitorum longus
d. Tibialis anterior
e. Flexor digitorum

52. Identify the ankle invertor.
a. Extensor digitorum longus
b. Flexor digitorum longus
c. Peroneus tertius
d. Peroneus longus
e. Peroneus brevis

53. The action of the tibialis posterior is:
a. Dorsiflexion, eversion
b. Plantar flexion, inversion
c. Plantar flexion, toe flexion
d. Plantar flexion, eversion
e. Toe flexion, eversion

54. The action of the extensor digitorum longus is:
a. Dorsiflexion, eversion
b. Plantar flexion, inversion
c. Plantar flexion, toe flexion
d. Plantar flexion, eversion
e. Toe flexion, eversion

55. The action of the flexor digitorum longus is:
a. Dorsiflexion, eversion, toe flexion
b. Plantar flexion, inversion, toe flexion
c. Plantar flexion, toe flexion
d. Plantar flexion, eversion, toe flexion
e. Toe flexion, eversion

56. The action of the peroneus longus is:
a. Dorsiflexion, eversion
b. Plantar flexion, inversion
c. Plantar flexion, toe flexion
d. Plantar flexion, eversion
e. Toe flexion, eversion

57. The insertion of the peroneus brevis & tertius is:
a. Junction of the 2nd & 3rd metatarsal, & 2nd & 3rd cuneiform
b. Base of the first metatarsal, first cuneiform
c. Base of the first metatarsal
d. Navicular, cuboid
e. Base of the fifth metatarsal

58. The origin of the longissimus dorsi is the:
a. Lower two thoracic vertebra and upper two lumbar vertebra
b. Lumbar dorsal fascia
c. Transverse processes of upper six thoracic vertebra
d. Lower one-fourth nuchal ligament, seventh cervical, first and second thoracic vertebra
e. Spines of the lower four cervical vertebra and upper four thoracic vertebra

59. The origin of the spinalis dorsi is the:
a. 11th & 12 thoracic vertebra and 1st & 2nd lumbar vertebra
b. Lumbodorsal fascia
c. Transverse processes of upper six thoracic vertebra
d. Lower one-fourth nuchal ligament, seventh cervical, first and second thoracic vertebra
e. Spines of the lower four cervical vertebra and upper four thoracic vertebra

60. The origin of the longissimus cervicus is the:
a. Lower two thoracic vertebra and upper two lumbar vertebra
b. Lumbar dorsal fascia
c. Transverse processes of upper 4-5 thoracic vertebra
d. Lower one-fourth nuchal ligament, seventh cervical, first and second thoracic vertebra
e. Spines of the lower four cervical vertebra and upper four thoracic vertebra

61. The origin of the spinalis cervicis is the:
a. Lower two thoracic vertebra and upper two lumbar vertebra
b. Lumbar dorsal fascia
c. Transverse processes of upper six thoracic vertebra
d. Lower one-fourth nuchal ligament, seventh cervical, first and second thoracic vertebra
e. Spines of the lower four cervical vertebra & upper four thoracic vertebra

62. The origin of the spinalis capitus is the:
a. Lower two thoracic vertebra and upper two lumbar vertebra
b. Lumbar dorsal fascia
c. Transverse processes of upper six thoracic vertebra
d. Lower one-fourth nuchal ligament, seventh cervical, first and second thoracic vertebra
e. Transverse processes of the lower four cervical and upper six or seven thoracic vertebra

63. The insertion of the flexor pollicis brevis is the:
a. The flexor retinaculum and trapezium
b. Pisiform
c. The lateral and medial side of the base of the first phalange of the thumb
d. Ulna side of the first metacarpal
e. The flexor retinaculum and hamate

64. The insertion of the abductor digiti minimi (hand) is the:
a. The flexor retinaculum and trapezium
b. Pisiform
c. The lateral and medial side of the base of the first phalange of the thumb
d. Ulna side of the first phalange of fifth finger
e. The flexor retinaculum and hamate

65. The insertion of the abductor pollicis brevis is the:
a. The flexor retinaculum and trapezium
b. Base of the first phalange of thumb, lateral side
c. Base of the first phalange of thumb, medial side
d. All of the radial side of the first metacarpal
e. The lateral and medial side of the base of the first phalange of the thumb

66. The insertion of the abductor digiti minimi (foot):
a. Tendons of the flexor digitorum longus
b. Base of the first phalange of the little toe, lateral side
c. The lateral and medial side of the base of the first phalange of the little toe
d. Calcaneum
e. Base of the first phalange of the great toe, medial side

67. The insertion of the flexor hallicus brevis:
a. Tendons of the flexor digitorum longus
b. Base of the first phalange of the little toe, lateral side
c. The lateral and medial side of the base of the first phalange of the great toe
d. Calcaneum
e. Base of the first phalange of the great toe, medial side

Quiz 14 - Scapula Attachments

Refer to the following drawings concerning the following questions in this quiz about the scapula attachments.

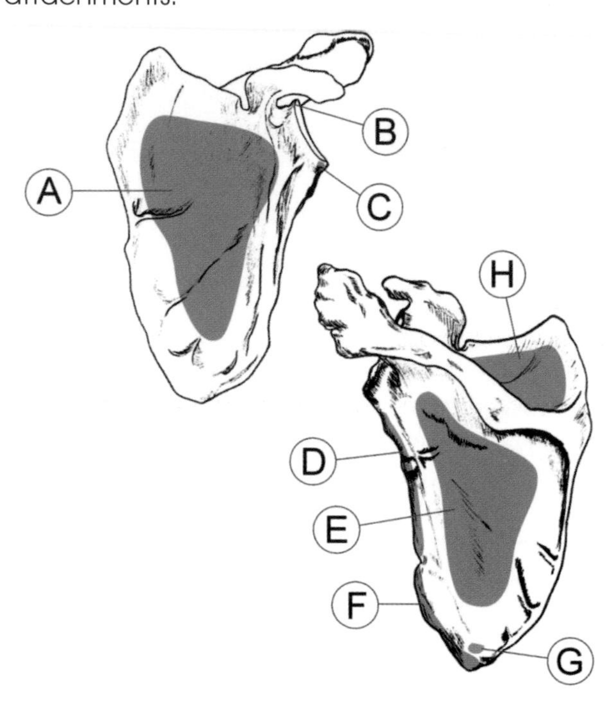

1. On the drawings of the scapula, label A is the:
a. Teres major
b. Latissimus dorsi
c. Infraspinatus
d. SubScapularis
e. Teres minor

2. On the drawings of the scapula, label D is the:
a. Teres major
b. Latissimus dorsi
c. Infraspinatus
d. SubScapularis
e. Teres minor

3. On the drawings of the scapula, label G is the:
a. Teres major
b. Latissimus dorsi
c. Infraspinatus
d. SubScapularis
e. Teres minor

4. On the drawings of the scapula, label F is the:
a. Teres major
b. Latissimus dorsi
c. Infraspinatus
d. SubScapularis
e. Teres minor

5. On the drawings of the scapula, label E is the:
a. Teres major
b. Latissimus dorsi
c. Infraspinatus
d. SubScapularis
e. Teres minor

6. On the drawings of the scapula, label H is the:
a. Supraspinatus
b. Biceps brachii
c. Triceps
d. Latissimus dorsi
e. Teres minor

7. On the drawings of the scapula, label C is the:
a. Supraspinatus
b. Biceps brachii
c. Triceps
d. Latissimus dorsi
e. Teres minor

8. On the drawings of the scapula, label B is the:
a. Supraspinatus
b. Biceps brachii
c. Triceps
d. Latissimus dorsi
e. Teres minor

Quiz 15 - Humerus Attachments

Refer to the following drawings concerning the questions in this quiz about humerus attachments.

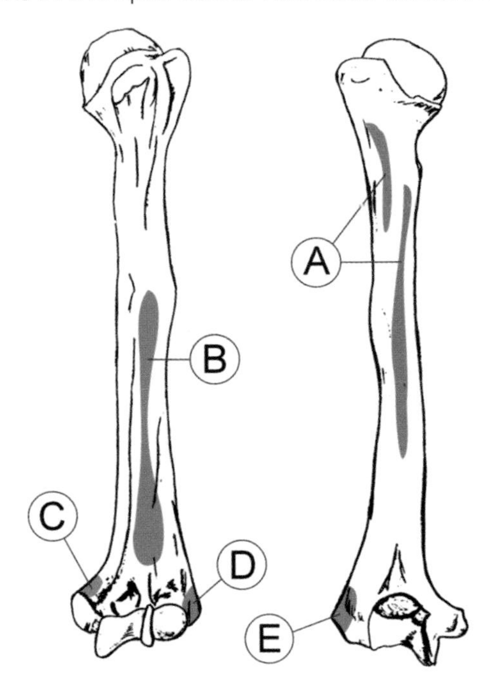

1. On the drawings of the humerus, what is label A?
a. Deltoid
b. Brachialis
c. Anconeus
d. Triceps
e. Coraobrachialis

2. On the drawings of the humerus, what is label E?
a. Anconeus
b. Brachialis
c. Deltoid
d. Triceps
e. Coracobrachialis

3. On the drawings of the humerus, what is label B?
a. Anconeus
b. Brachialis
c. Deltoid
d. Triceps
e. Coracobrachialis

4. On the drawings of the humerus, what is label D?
a. Flexor digitorum superficialis
b. Brachialis
c. Flexor digitorum
d. Triceps
e. Supinator

5. On the drawings of the humerus, what is label C?
a. Flexor digitorum superficialis
b. Brachialis
c. Flexor digitorum
d. Triceps
e. Supinator

Quiz 16 - Forearm Attachments

Refer to the following drawings concerning questions #1-12 in this quiz about forearm attachments.

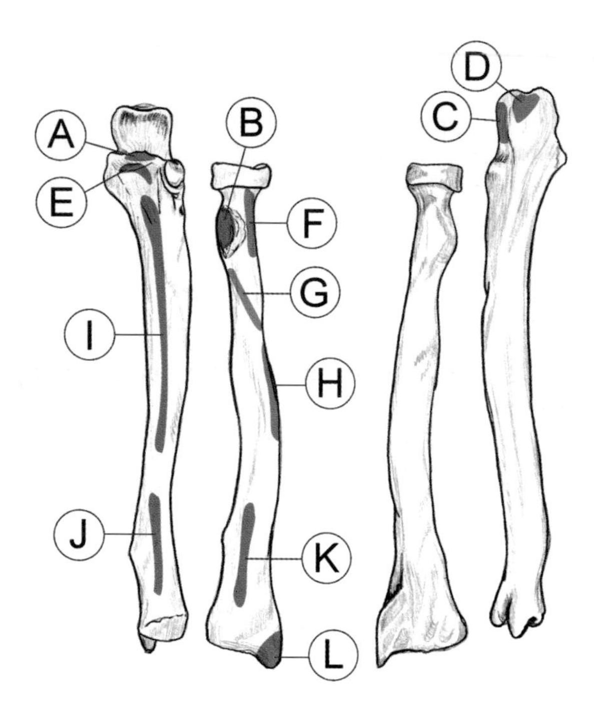

1. **On the drawings of the radius and ulna, what is label H?**
   a. Biceps
   b. Supinator
   c. Brachialis
   d. Pronator teres

2. **On the drawings of the radius and ulna, what is label F?**
   a. Biceps
   b. Supinator
   c. Brachialis
   d. Pronator teres

3. **On the drawings of the radius and ulna, what is label E?**
   a. Flexor digitorum profundus
   b. Flexor digitorum superficialis
   c. Pronator quadratus
   d. Flexor digitorum

4. **On the drawings of the radius and ulna, what is label I?**
   a. Flexor digitorum profundus
   b. Flexor digitorum superficialis
   c. Pronator quadratus
   d. Pronator teres

5. **On the drawings of the radius and ulna, what is label G?**
   a. Flexor digitorum profundus
   b. Flexor digitorum superficialis
   c. Pronator quadratus
   d. Pronator teres

6. **On the drawings of the radius and ulna, what is label J?**
   a. Flexor digitorum profundus
   b. Flexor digitorum superficialis
   c. Pronator quadratus
   d. Pronator teres

7. **On the drawings of the radius and ulna, what is label K?**
   a. Triceps
   b. Brachioradialis
   c. Anconeus
   d. Pronator quadratus

8. **On the drawings of the radius and ulna, what is label L?**
   a. Triceps
   b. Brachioradialis
   c. Anconeus
   d. Pronator quadratus

9. **On the drawings of the radius and ulna, what is label A?**
   a. Flexor digitorum profundus
   b. Flexor digitorum superficialis
   c. Brachialis
   d. Pronator teres

10. **On the drawings of the radius and ulna, what is label B?**
    a. Biceps
    b. Brachialis
    c. Pronator quadratus
    d. Pronator teres

11. **On the drawings of the radius and ulna, what is label C?**
    a. Triceps
    b. Brachioradialis
    c. Anconeus
    d. Pronator quadratus

12. **On the drawings of the radius and ulna, what is label D?**
    a. Triceps
    b. Brachioradialis
    c. Anconeus
    d. Pronator quadratus

Refer to the following drawings concerning questions #13-17 in this quiz about forearm attachments.

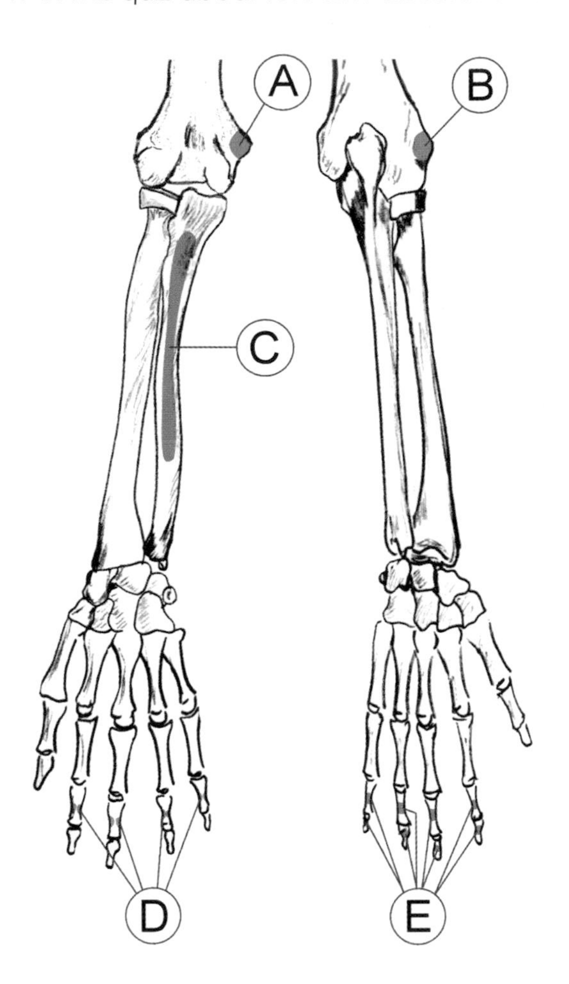

13. On the drawings of the forearm and hand, what is label B?

a. Flexor digitorum profundus
b. Flexor digitorum superficialis
c. Extensor digitorum
d. Pronator quadratus
e. Flexor digitorum

14. On the drawings of the forearm and hand, what is label A?

a. Flexor digitorum profundus
b. Flexor digitorum superficialis
c. Extensor digitorum
d. Pronator quadratus
e. Flexor digitorum

15. On the drawings of the forearm and hand, what is label C?

a. Flexor digitorum profundus
b. Flexor digitorum superficialis
c. Extensor digitorum
d. Pronator quadratus
e. Flexor digitorum

16. On the drawings of the forearm and hand, what is label D?

a. Flexor digitorum profundus
b. Flexor digitorum superficialis
c. Extensor digitorum
d. Pronator quadratus
e. Flexor digitorum

17. On the drawings of the forearm and hand, what is label E?

a. Flexor digitorum profundus
b. Flexor digitorum superficialis
c. Extensor digitorum
d. Pronator quadratus
e. Flexor digitorum

Quiz 17 - Innominate Attachments

Refer to the following drawings (#1 and #2) concerning the questions in this quiz about innominate attachments.

Innominate Bone - #1

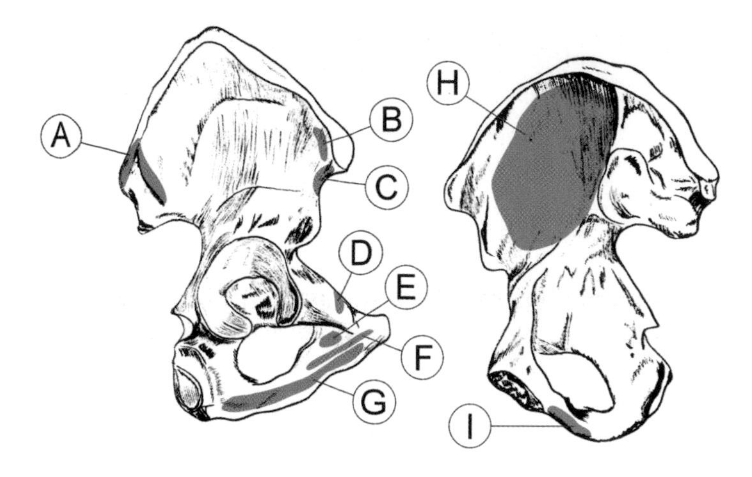

1. **On drawing #1 of the innominate bone, what muscle has its origin or insertion at label I?**
   a. Sartorius
   b. Gluteus maximus
   c. Gracilis
   d. Adductor magnus
   e. Adductor longus

2. **On drawing #1 of the innominate bone, what muscle has its origin or insertion at label E?**
   a. Sartorius
   b. Gluteus maximus
   c. Gracilis
   d. Adductor magnus
   e. Adductor longus

3. **On drawing #1 of the innominate bone, what muscle has its origin or insertion at label C?**
   a. Sartorius
   b. Gluteus maximus
   c. Gracilis
   d. Adductor magnus
   e. Adductor longus

4. **On drawing #1 of the innominate bone, what muscle has its origin or insertion at label A?**
   a. Sartorius
   b. Gluteus maximus
   c. Gracilis
   d. Adductor magnus
   e. Adductor longus

5. **On drawing #1 of the innominate bone, what muscle has its origin or insertion at label B?**
   a. Tensor fascia lata
   b. Pectineus
   c. Adductor longus
   d. Iliacus
   e. Adductor brevis

6. **On drawing #1 of the innominate bone, what muscle has its origin or insertion at label H?**
   a. Tensor fascia lata
   b. Pectineus
   c. Adductor longus
   d. Iliacus
   e. Adductor brevis

7. **On drawing #1 of the innominate bone, what muscle has its origin or insertion at label F?**
   a. Tensor fascia lata
   b. Pectineus
   c. Adductor longus
   d. Iliacus
   e. Adductor brevis

8. **On drawing #1 of the innominate bone, what muscle has its origin or insertion at label D?**
   a. Tensor fascia lata
   b. Pectineus
   c. Adductor longus
   d. Iliacus
   e. Adductor brevis

9. **On drawing #1 of the innominate bone, what muscle has its origin or insertion at label G?**
   a. Tensor fascia lata
   b. Pectineus
   c. Adductor longus
   d. Iliacus
   e. Adductor magnus

Innominate Bone - #2

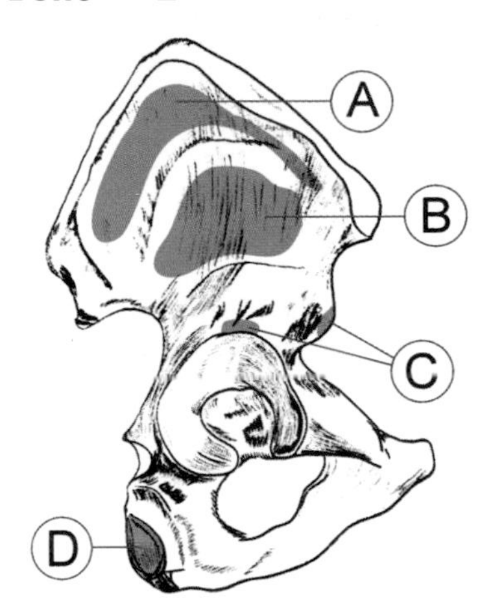

10. On drawing #2 of the innominate bone, what muscle(s) has its origin or insertion at label D?
a. Biceps femoris
b. Semitendinosis
c. Semimembranosis
d. Hamstrings
e. All of the above

11. On drawing #2 of the innominate bone, what muscle has its origin or insertion at label C?
a. Biceps femoris
b. Semitendinosis
c. Semimembranosis
d. Rectus femoris
e. Vastus intermedius

12. On drawing #2 of the innominate bone, what muscle has its origin or insertion at label A?
a. Biceps femoris
b. Semitendinosis
c. Semimembranosis
d. Rectus femoris
e. Gluteus medius

13. On drawing #2 of the innominate bone, what muscle has its origin or insertion at label B?
a. Biceps femoris
b. Gluteus minimus
c. Semimembranosis
d. Rectus femoris
e. Vastus intermedius

Quiz 18 - Femur Attachments

Refer to the following drawings (#1 and #2) concerning the questions in this quiz about femur attachments.

Femur - #1

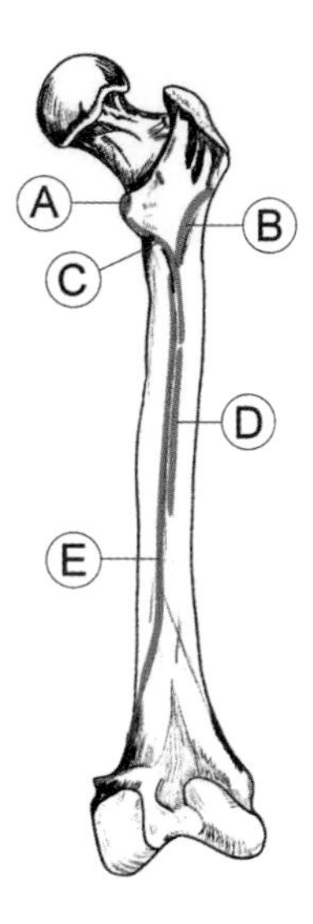

1. **On drawing #1 of the femur, what muscle has its origin or insertion at label D?**
   a. Adductor magnus
   b. Adductor brevis
   c. Iliacus
   d. Gluteus maximus
   e. Adductor longus

2. **On drawing #1 of the femur, what is at label A?**
   a. Adductor magnus
   b. Adductor brevis
   c. Iliacus
   d. Gluteus maximus
   e. Adductor longus

3. **On drawing #1 of the femur, what is at label C?**
   a. Adductor magnus
   b. Adductor brevis
   c. Iliacus
   d. Gluteus maximus
   e. Adductor longus

4. **On drawing #1 of the femur, what is at label B?**
   a. Adductor magnus
   b. Adductor brevis
   c. Pectineus
   d. Gluteus maximus
   e. Adductor longus

5. **On drawing #1 of the femur, what muscle is at label E?**
   a. Adductor magnus
   b. Adductor brevis
   c. Iliacus
   d. Gluteus maximus
   e. Adductor longus

Femur - #2

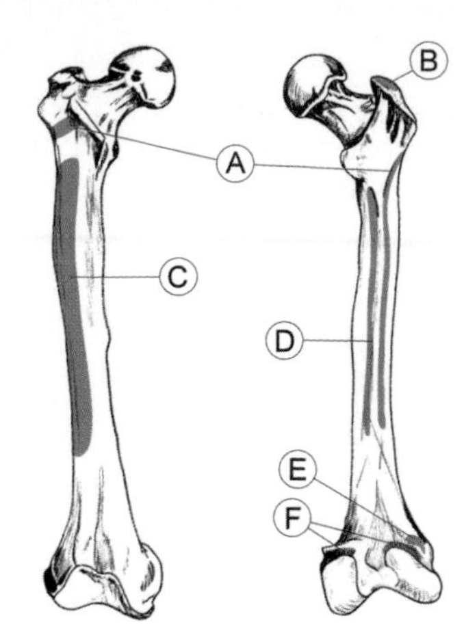

6. **On drawing #2 of the femur, what muscle has its origin or insertion at label A?**
   a. Vastus lateralis
   b. Vastus medialis
   c. Vastus intermedius
   d. Rectus femoris
   e. Gluteus medius

7. **On drawing #2 of the femur, what muscle has its origin or insertion at label D?**
   a. Vastus lateralis
   b. Vastus medialis
   c. Vastus intermedius
   d. Rectus femoris
   e. Gluteus medius

8. **On drawing #2 of the femur, what muscle has its origin or insertion at label C?**
   a. Vastus lateralis
   b. Vastus medialis
   c. Vastus intermedius
   d. Rectus femoris
   e. Gluteus medius

9. **On drawing #2 of the femur, what muscle has its origin or insertion at label B?**
   a. Vastus lateralis
   b. Vastus medialis
   c. Vastus intermedius
   d. Rectus femoris
   e. Gluteus medius

10. **On drawing #2 of the femur, what muscle has its origin or insertion at label E?**
    a. Vastus lateralis
    b. Vastus medialis
    c. Plantaris
    d. Rectus femoris
    e. Gluteus medius

11. **On drawing #2 of the femur, what muscle has its origin or insertion at label F?**
    a. Vastus lateralis
    b. Gastrocnemius
    c. Plantaris
    d. Rectus femoris
    e. Gluteus medius

Quiz 19 - Tibia and Fibula Attachments

Refer to the following drawings (#1 and #2) concerning the questions in this quiz about the tibia and fibula attachments.

Tibia and Fibula - #1

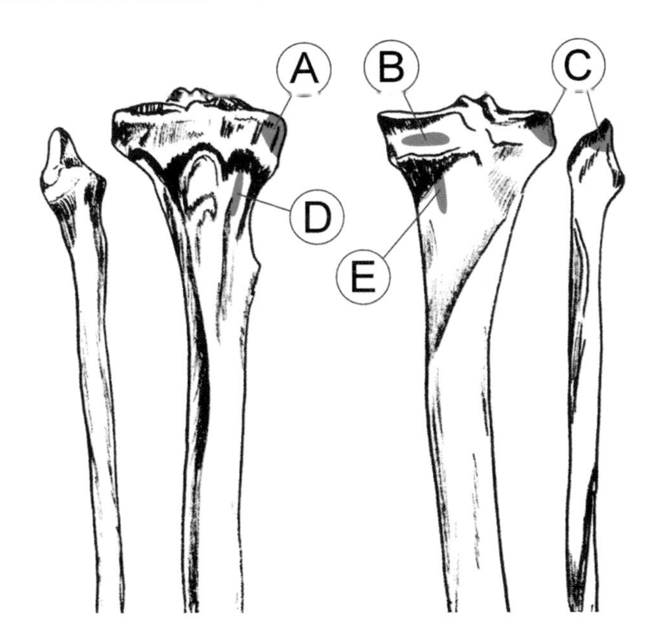

Tibia and Fibula - #2

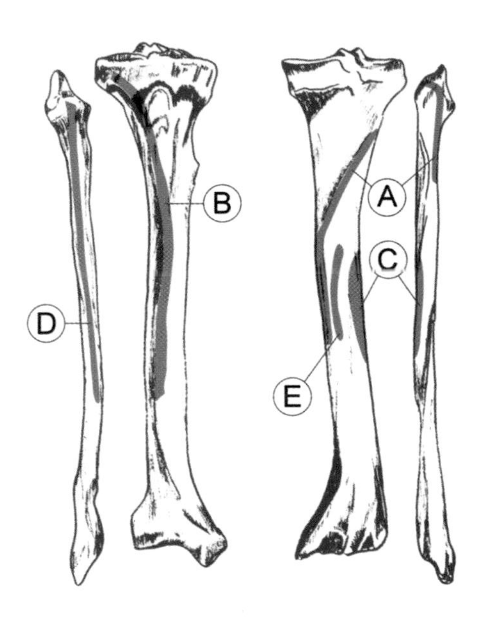

1. **On drawing #1 of the tibia and fibula, what muscle has its origin or insertion at label D?**
   a. Biceps femoris
   b. Semitendinosis
   c. Semimembranosis
   d. Rectus femoris
   e. Adductor magnus

2. **On drawing #1 of the tibia and fibula, what muscle has its origin or insertion at label B?**
   a. Biceps femoris
   b. Semitendinosis
   c. Semimembranosis
   d. Rectus femoris
   e. Adductor magnus

3. **On drawing #1 of the tibia and fibula, what muscle has its origin or insertion at label C?**
   a. Biceps femoris
   b. Semitendinosis
   c. Semimembranosis
   d. Rectus femoris
   e. Adductor magnus

4. **On drawing #2 of the tibia and fibula, what muscle has its origin or insertion at label D?**
   a. Tibialis anterior
   b. Tibialis posterior
   c. Extensor digitorum longus
   d. Flexor digitorum longus
   e. Soleus

5. **On drawing #2 of the tibia and fibula, what muscle has its origin or insertion at label B?**
   a. Tibialis anterior
   b. Tibialis posterior
   c. Extensor digitorum longus
   d. Flexor digitorum longus
   e. Soleus

6. **On drawing #2 of the tibia and fibula, what muscle has its origin or insertion at label C?**
   a. Tibialis anterior
   b. Tibialis posterior
   c. Extensor digitorum longus
   d. Flexor digitorum longus
   e. Soleus

7. **On drawing #2 of the tibia and fibula, what muscle has its origin or insertion at label E?**
   a. Tibialis anterior
   b. Tibialis posterior
   c. Extensor digitorum longus
   d. Flexor digitorum longus
   e. Soleus

8. **On drawing #2 of the tibia and fibula, what muscle has its origin or insertion at label A?**
   a. Tibialis anterior
   b. Tibialis posterior
   c. Extensor digitorum longus
   d. Flexor digitorum longus
   e. Soleus

Quiz 20 - Exercise and Activity Analysis

For the following questions, select the most correct answer from those listed.

1. Name the prime movers in Figure 20-1.
a. Knee extensors, hip flexors
b. Knee flexors, hip Flexors
c. Knee extensors, hip extensors
d. Spine extensors, hip extensors
e. Hip flexors, knee extensors, dorsiflexors

Figure 20-1

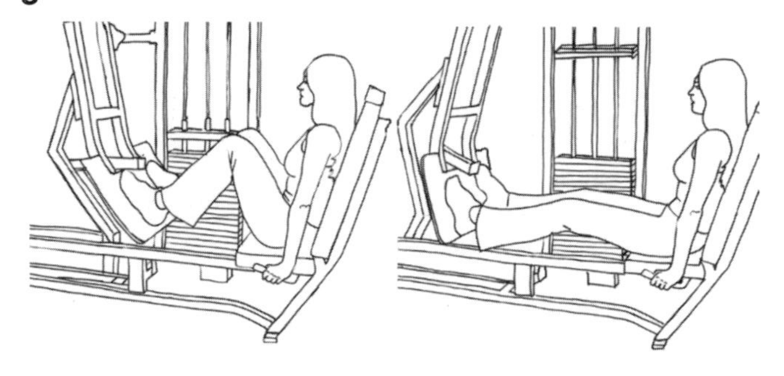

2. Name the prime movers in Figure 20-2.
a. Shoulder abductors, wrist flexors
b. Arm abductors
c. Shoulder abductors, elbow flexors
d. Shoulder abductors, hip flexors
e. Arm abductors, shoulder outward rotators

Figure 20-2

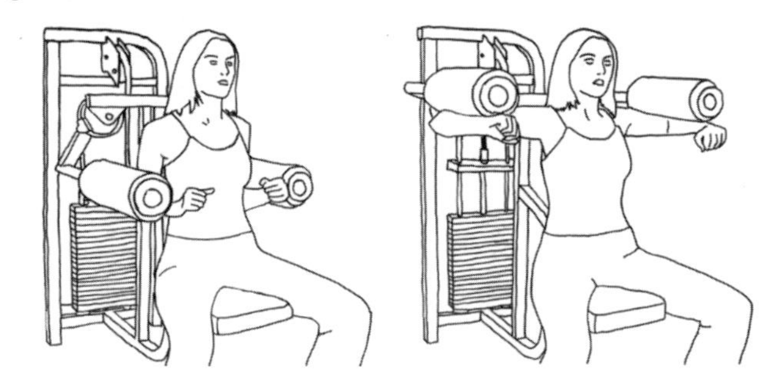

3. Name the prime movers in Figure 20-3.
a. Knee flexors
b. Plantar flexors
c. Elbow extensors
d. Hip flexors
e. Knee extensors

Figure 20-3

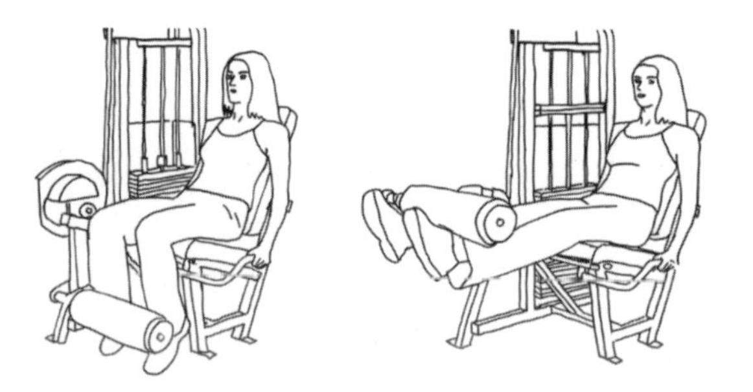

4. Name the prime movers in Figure 20-4.
a. Elbow extensors, shoulder abductors
b. Shoulder abductors
c. Elbow flexors, arm abductors/flexors
d. Spine extensors, arm extensors
e. Arm adductors/extensors, elbow flexors

Figure 20-4

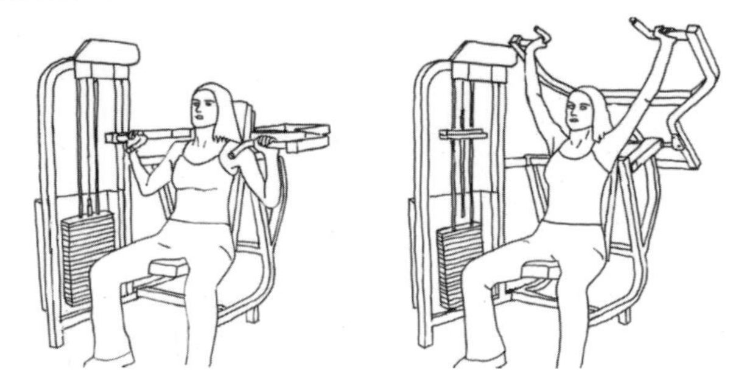

5. Name the Prime Movers in Figure 20-5.
a. Elbow extensors
b. Shoulder flexors
c. Elbow extensors, arm abductors/flexors
d. Spine extensors, elbow extensors
e. Arm adductors/extensors, elbow flexors

Figure 20-5

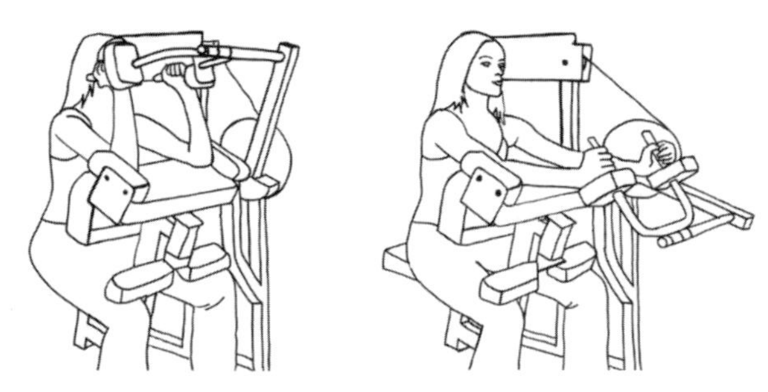

6. Name the prime movers in Figure 20-6.
a. Knee flexors
b. Knee flexors, hip extensors
c. Spine extensors
d. Hip extensors
e. Hip flexors

Figure 20-6

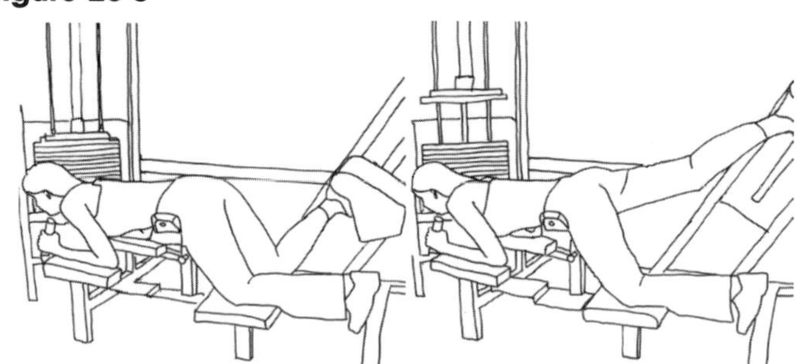

7. Name the prime movers in Figure 20-7.
a. Elbow extensors, shoulder flexors
b. Elbow extensors, shoulder extensors
c. Elbow flexors, shoulder flexors
d. Elbow flexors, scapula adductors
e. Elbow flexors, shoulder extensors

Figure 20-7

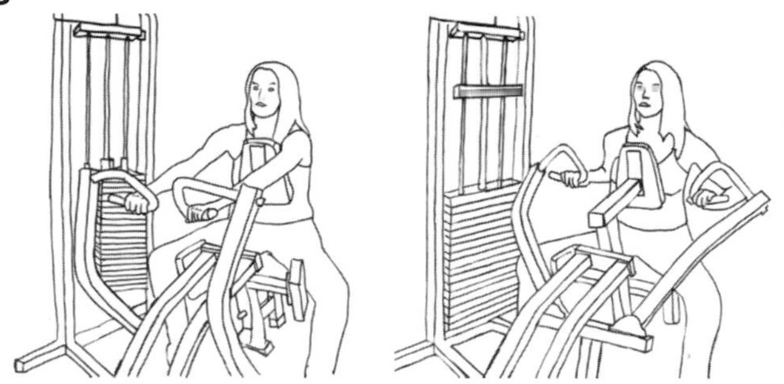

8. Name the prime movers in Figure 20-8.
a. Elbow extensors
b. Elbow flexors
c. Shoulder flexors
d. Wrist flexors
e. Finger, wrist flexors

Figure 20-8

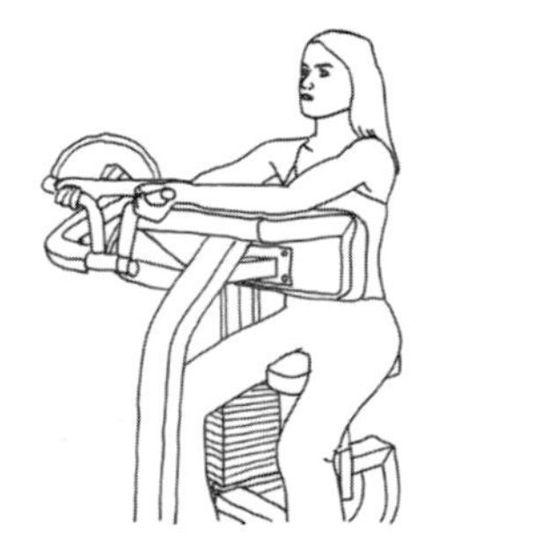

9. Name the prime movers in Figure 20-9.
a. Spine extensors
b. Trunk flexors
c. Hip extensors
d. Hip flexors
e. Spine and neck extensors

Figure 20-9

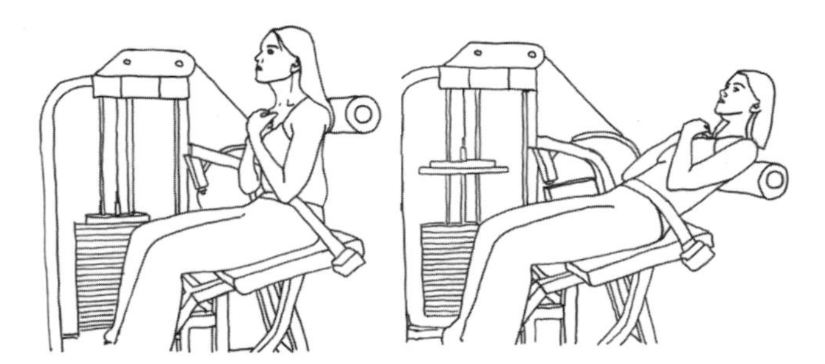

10. Name the prime movers in Figure 20-10.
a. Arm horizontal abductors
b. Shoulder adductors
c. Elbow flexors
d. Wrist extensors
e. Arm external rotators

Figure 20-10

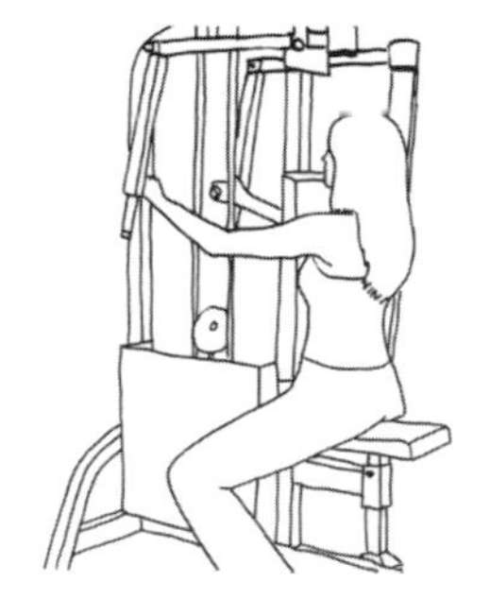

11. Name the prime movers in Figure 20-11.
a. Leg abductors, knee extensors
b. Hip adductors. knee flexors
c. Hip abductors, and leg adductors
d. Hip abductors
e. Spine extensors

Figure 20-11

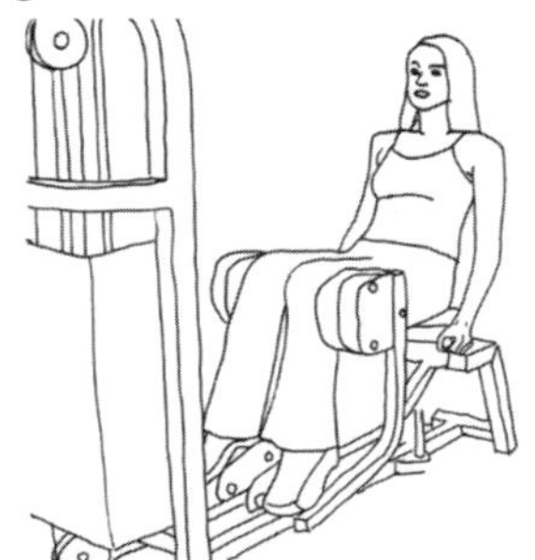

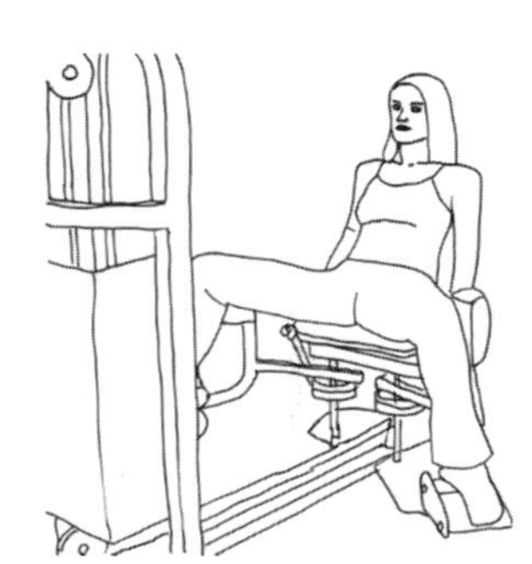

12. Name the prime movers in Figure 20-12 (pull-up).
a. Elbow extensors, shoulder flexors
b. Elbow extensors, shoulder extensors
c. Elbow flexors, shoulder flexors
d. Elbow flexors, scapula adductors
e. Elbow flexors, shoulder extensors

Figure 20-12

13. Name the prime movers in Figure 20-13 for the upward movement.
a. Elbow extensors, shoulder flexors
b. Elbow extensors, shoulder extensors
c. Elbow flexors, shoulder flexors
d. Elbow flexors, scapula adductors
e. Elbow flexors, shoulder extensors

Figure 20-13

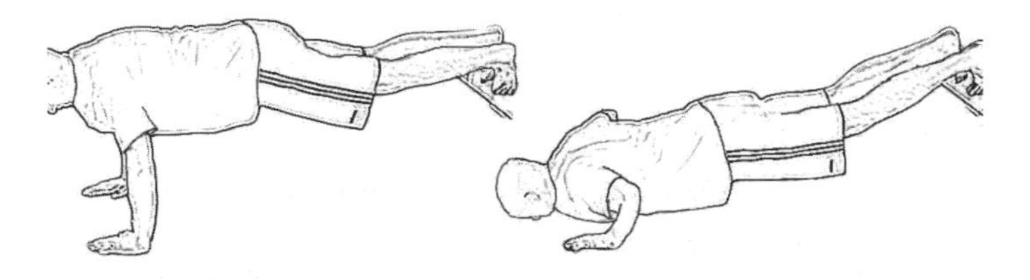

14. Name the prime movers in Figure 20-13 for the downward movement.
a. Elbow extensors, shoulder flexors
b. Elbow extensors, shoulder extensors
c. Elbow flexors, shoulder flexors
d. Elbow flexors, scapula adductors
e. Elbow flexors, shoulder extensors

15. Name the prime movers in Figure 20-14.
a. Hip adductors
b. Leg abductors
c. Knee extensors
d. Trunk lateral flexors
e. Hip flexors

Figure 20-14

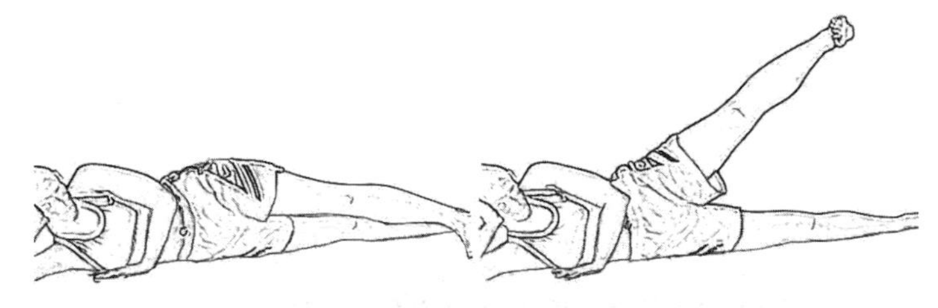

16. Name the prime movers in Figure 20-15 for the sitting-up movement.
a. Hip flexors
b. Hip extensors
c. Hip flexors, spine flexors
d. Hip flexors, knee flexors
e. Spine flexors, neck flexors

Figure 20-15

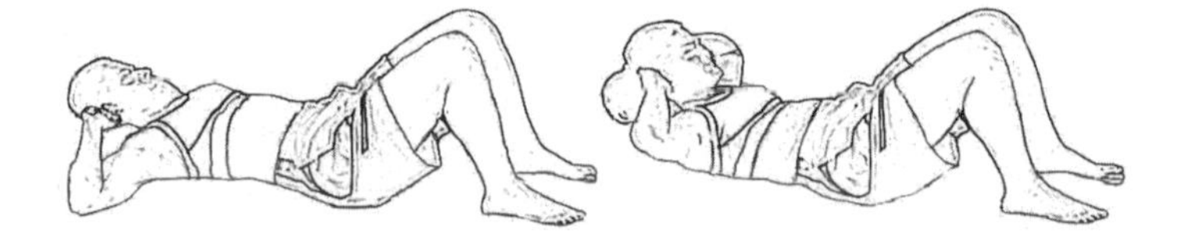

17. Name the prime movers in Figure 20-15 for the curling-down movement.
a. Hip flexors
b. Hip extensor
c. Hip flexors, spine flexors
d. Hip flexors, knee flexors
e. Spine flexors, neck flexors

18. Name the prime movers in Figure 20-16 for the up movement.
a. Elbow extensors, shoulder flexors
b. Elbow extensors, shoulder extensors
c. Elbow flexors, shoulder flexors
d. Elbow flexors, scapula adductors
e. Elbow flexors, shoulder extensors

Figure 20-16

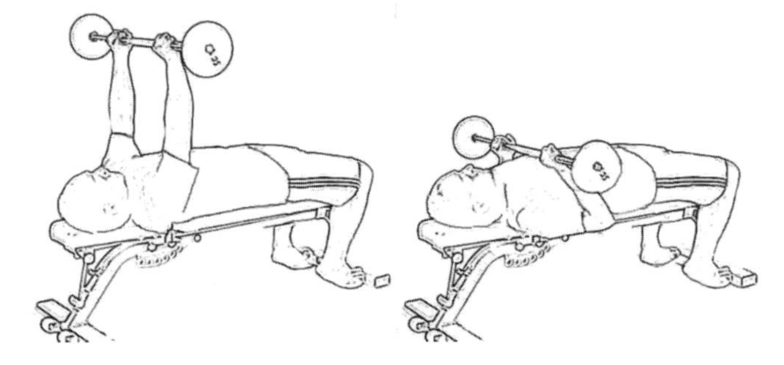

19. Name the prime movers in Figure 20-16 for the downward movement.
a. Elbow extensors, shoulder flexors
b. Elbow extensors, shoulder extensors
c. Elbow flexors, shoulder flexors
d. Elbow flexors, scapula adductors
e. Elbow flexors, shoulder extensors

20. Name the prime movers in Figure 20-17 for the outward movement.
a. Knee extensors, hip extensors
b. Knee flexors, hip flexors
c. Knee extensors, hip flexors
d. Knee flexors, hip extensors
e. Knee extensors, plantar flexors

Figure 20-17

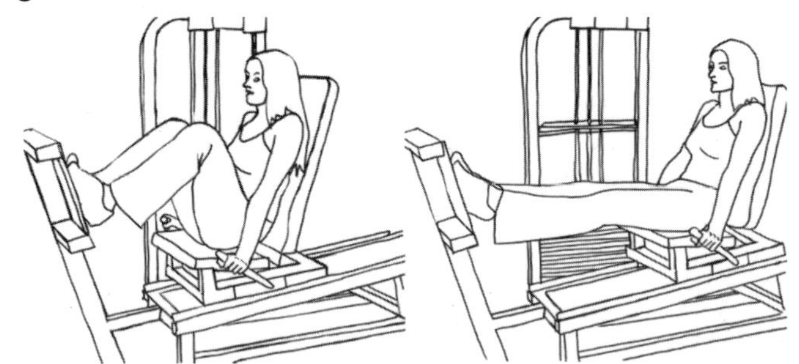

21. Name the prime movers in Figure 20-17 for the inward movement.
a. Knee extensors, hip extensors
b. Knee flexors, hip flexors
c. Knee extensors, hip flexors
d. Knee flexors, hip extensors
e. Knee extensors, plantar flexors

22. Name the prime movers in Figure 20-18 for the downward movement.
a. Knee extensors, hip extensors
b. Knee flexors, hip flexors
c. Knee extensors, hip flexors
d. Knee flexors, hip extensors
e. Knee extensors, plantar flexors

Figure 20-18

23. Name the prime movers in Figure 20-19.
a. Knee extensors
b. Knee flexors
c. Hip flexors
d. Hip extensors
e. Dorsiflexors

Figure 20-19

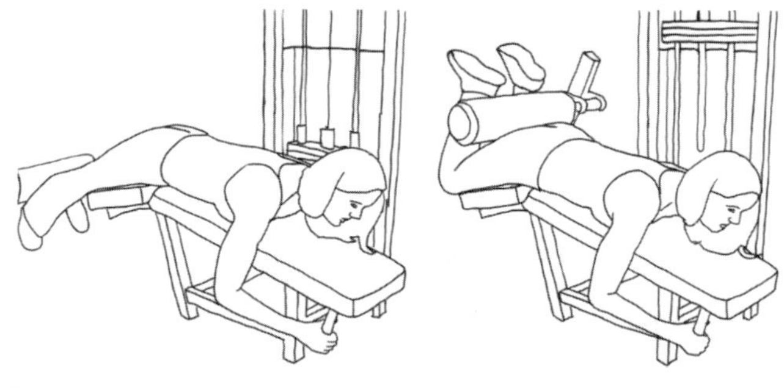

24. Name the prime movers used during the upward step when standing and stepping up onto a bench.
a. Knee extensors, hip extensors
b. Knee flexors, hip flexors
c. Knee extensors, hip flexors
d. Knee flexors, hip extensors
e. Knee extensors, plantar flexors

25. Name the prime movers used during the downward step when standing on a bench and stepping down backwards.
a. Knee extensors, hip extensors
b. Knee flexors, hip flexors
c. Knee extensors, hip flexors
d. Knee flexors, hip extensors
e. Knee extensors, plantar flexors

26. Name the prime movers in Figure 20-20 for the swimmers racing dive.
a. Knee extensors, hip extensors, plantar flexors, shoulder flexors
b. Knee flexors, hip flexors, dorsiflexors, shoulder flexion
c. Knee extensors, hip flexors, plantar flexors, shoulder flexors
d. Knee flexors, hip extensors, plantar flexors, shoulder flexors
e. Knee extensors, plantar flexors, shoulder extensors, hip extensors

Figure 20-20

27. In recreational cycling, the prime movers for the leg down-stroke are:
a. Knee extensors, hip extensors
b. Knee flexors, hip flexors
c. Knee extensors, hip flexors
d. Knee flexors, hip extensors
e. Knee extensors, plantar flexors

28. In racing cycling (using toe clips), the prime movers for the leg up-stroke are:
a. Knee extensors, hip extensors
b. Knee flexors, hip flexors
c. Knee extensors, hip flexors
d. Knee flexors, hip extensors
e. Knee extensors, plantar flexors

29. Name the prime movers in Figure 20-21 for the lower extremity.
a. Knee extensors, hip extensors
b. Knee flexors, hip flexors
c. Knee extensors, hip flexors
d. Knee flexors, hip extensors
e. Knee extensors, plantar flexors

Figure 20-21

30. Name the prime movers in Figure 20-22 for the butterfly pull of the upper extremity.

a. Shoulder extensors
b. Shoulder flexors
c. Arm flexors
d. Elbow flexors
e. Head extensors

Figure 20-22

31. Name the prime movers in Figure 20-23 for the volleyball serve of the upper extremity.

a. Shoulder extensors and elbow extensors
b. Shoulder flexors and elbow flexors
c. Arm flexors
d. Elbow flexors
e. Head extensors

Figure 20-23

32. Name the prime movers in Figure 20-24 for the kayak paddle for the left arm.

a. Shoulder extensors and elbow extensors
b. Shoulder flexors and elbow flexors
c. Arm flexors and elbow flexors
d. Elbow flexors and shoulder extensors
e. Elbow extensors and shoulder flexors

Figure 20-24

33. Name the prime movers in Figure 20-25 for the underhand softball pitch of the pitching arm.

a. Shoulder extensors
b. Shoulder flexors
c. Arm extensors
d. Elbow flexors
e. Elbow extensors

Figure 20-25

34. Name the prime movers in Figure 20-26 for the forward pass in football of the throwing arm.

a. Shoulder extensors and elbow extensors
b. Shoulder flexors and elbow flexors
c. Arm flexors and elbow flexors
d. Elbow flexors and shoulder extensors
e. Elbow extensors and shoulder flexors

Figure 20-26

35. Name the prime movers in Figure 20-27 for the basketball free throw of the arms.

a. shoulder extensors, elbow extensors, and wrist flexors
b. shoulder flexors, elbow flexors, and wrist flexors
c. Arm flexors, elbow flexors, and wrist extensors
d. Elbow flexors, shoulder extensors, and wrist flexors
e. Elbow extensors, shoulder flexors, and wrist flexors

Figure 20-27

36. Name the prime movers in Figure 20-28.

a. Hip extensors, spine flexors, and neck extensors
b. Hip flexors, spine flexors, and neck flexors
c. Hip extensors, spine extensors, and neck extensors
d. Hip extensors, spine extensors, and neck flexors
e. Hip flexors, spine extensors, and neck extensors

Figure 20-28

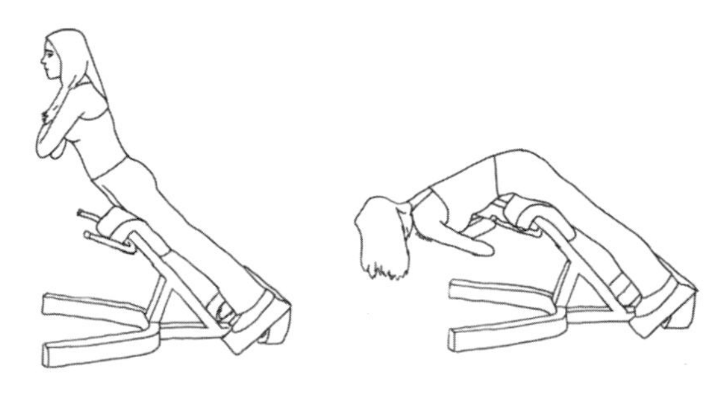

37. In the track sprinting start (from blocks), the prime movers are:

a. Hip extensors, spine extensors and neck extensors
b. Hip flexors, knee flexors and dorsiflexors
c. Hip extensors, knee extensors and plantar flexors
d. Hip extensors, knee extensors and dorsiflexors
e. Hip flexors, knee flexors and plantar flexors

38. In downhill skiing, the prime movers in the lower extremity are:

a. Eccentric and concentric contraction of the hip and knee extensors
b. Concentric contraction of the hip and knee flexors
c. Static contraction of the hip and knee extensors and the dorsiflexors
d. Eccentric contraction of the hip and knee flexors and the plantar flexors
e. Static contraction of the hip and knee extensors and the plantar and dorsiflexors

39. In the football punt kick, the prime movers of the kicking leg are:

a. Hip extensors, knee extensors, and dorsiflexors
b. Hip flexors, knee flexors, and dorsiflexors
c. Hip extensors, knee extensors, and plantar flexors
d. Hip flexors, knee extensors, and plantar flexors
e. Hip flexors, knee flexors, and plantar flexors

40. In the basketball double-hand chest pass, the prime movers of the upper extremity are:

a. Shoulder extensors, elbow extensors, and wrist flexors
b. Shoulder flexors, elbow extensors, and wrist flexors
c. Arm flexors, elbow flexors, and wrist extensors
d. Elbow flexors, shoulder extensors, and wrist flexors
e. Elbow extensors, shoulder extensors, and wrist extensors

41. Name the prime movers in Figure 20-29.

a. Supinators
b. Wrist flexors
c. Pronators
d. Wrist extensors
e. Wrist hyperextensors

Figure 20-29

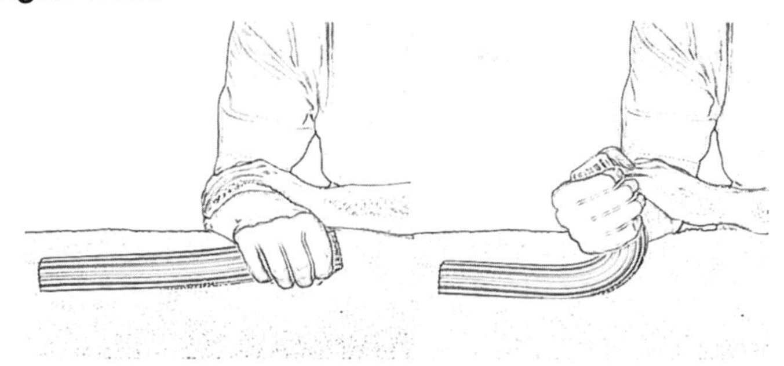

42. Name the prime movers in Figure 20-30.

a. Leg adductors
b. Hip lateral flexors
c. Knee abductors
d. Hip adductors
e. Hip abductors

Figure 20-30

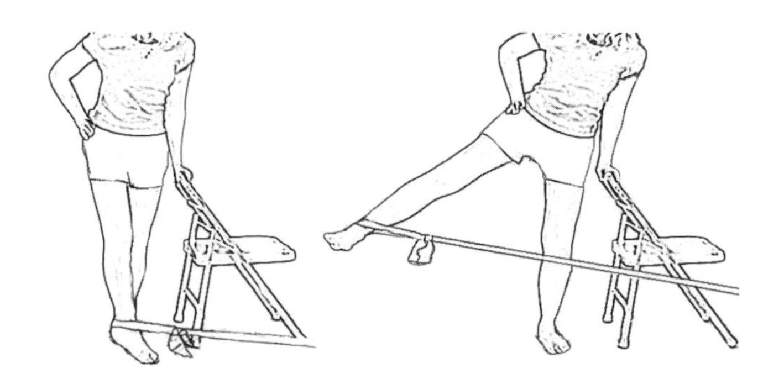

43. Name the prime movers in Figure 20-31.
a. Shoulder flexors
b. Arm extensors
c. Elbow extensors
d. Spine extensors
e. Wrist flexors

Figure 20-31

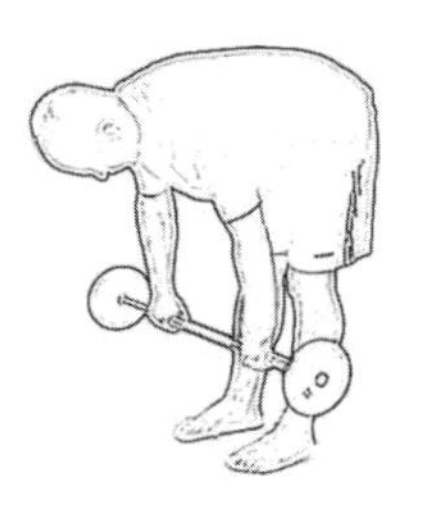
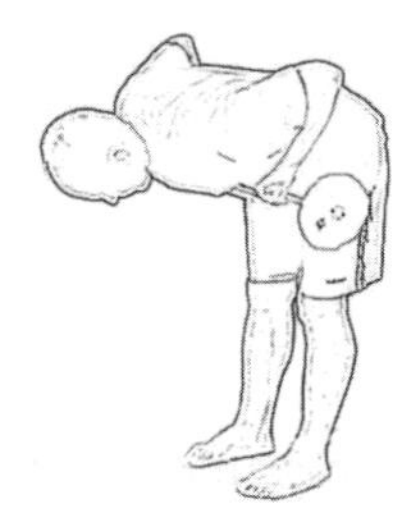

44. Name the prime movers in Figure 20-32.
a. Wrist flexors
b. Supinators
c. Radial flexors
d. Pronators
e. Ulna flexors

Figure 20-32

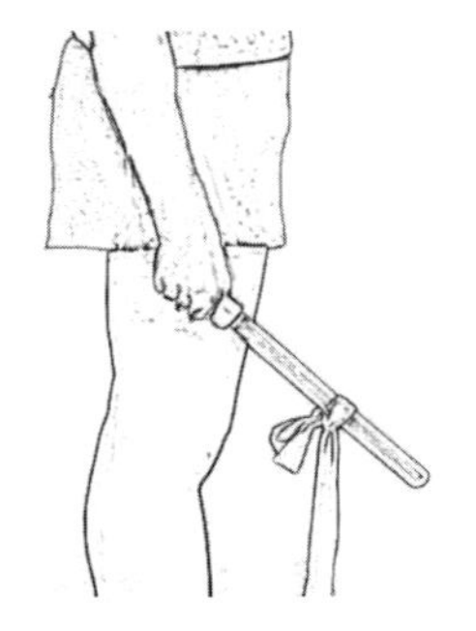

45. Name the Prime Movers in Figure 20-33.
a. Neck flexors
b. Head lateral flexors
c. Neck extensors
d. Head protraction
e. Head hyperextensors

Figure 20-33

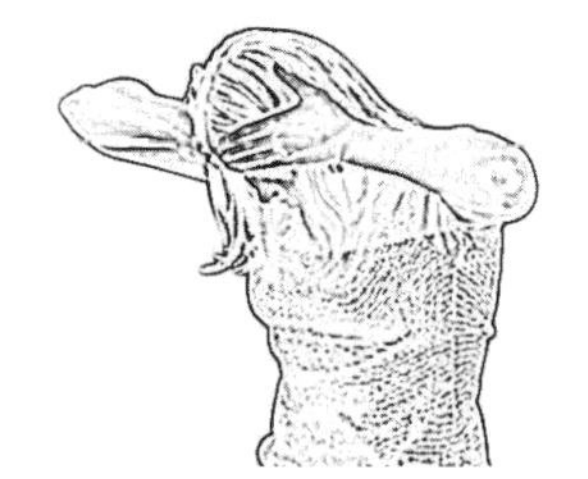

46. Name the prime movers in Figure 20-34.
a. Shoulder flexors
b. Arm extensors
c. Elbow extensors
d. Shoulder abductors
e. Wrist flexors

Figure 20-34

47. Name the prime movers in Figure 20-35 of the javelin-throwing arm.
a. Elbow flexors and shoulder extensors
b. Elbow extensors and shoulder flexors
c. Elbow extensors and shoulder extensors
d. Shoulder extensors and elbow flexors
e. Elbow extensors and arm flexors

Figure 20-35

48. Name the prime movers in Figure 20-36 of the shot putting arm.
a. Elbow flexors and trunk rotators
b. Elbow extensors, shoulder flexors, and spine flexors
c. Elbow flexors, shoulder extensors, and trunk rotators
d. Elbow flexors, arm extensors, and trunk rotators
e. Elbow extensors and trunk rotators

Figure 20-36

Quiz Answers

Quiz 1 - Planes and Actions
1b, 2b, 3a, 4d, 5a, 6a, 7a, 8e, 9d, 10b, 11b, 12d, 13b, 14b, 15a, 16c, 17c, 18d, 19a, 20e

Quiz 2 - The Skull
1a, 2d, 3c, 4e, 5c, 6e, 7c, 8e, 9d, 10a, 11b, 12a, 13c, 14e, 15c

Quiz 3 - The Scapula
1e, 2c, 3a, 4c, 5b, 6a, 7c, 8a, 9c, 10e, 11a, 12d, 13c, 14d, 15c

Quiz 4 - The Humerus
1e, 2a, 3c, 4b, 5d, 6e, 7b, 8a, 9c, 10d, 11e, 12c, 13d, 14a, 15b, 16d

Quiz 5 - The Radius and Ulna
1a, 2b, 3c, 4d, 5e, 6c, 7a, 8b, 9d, 10e

Quiz 6 - The Hand
1a, 2b, 3c, 4d, 5e, 6a, 7c, 8b, 9d, 10e

Quiz 7 - The Vertebra
1c, 2a, 3d, 4e, 5b, 6c, 7d, 8a, 9b, 10e, 11c, 12d, 13b, 14c, 15a, 16b, 17e, 18a, 19d, 20c, 21d, 22c, 23b, 24e, 25a

Quiz 8 - The Innominate Bone
1c, 2d, 3b, 4e, 5a, 6c, 7b, 8d, 9e, 10a, 11b, 12e, 13c, 14d, 15a, 16c, 17b, 18d, 19a, 20e

Quiz 9 - The Femur
1c, 2a, 3d, 4e, 5b, 6d, 7b, 8a, 9e, 10c, 11a, 12b, 13c, 14d, 15e

Quiz 10 - The Tibia and Fibula
1d, 2e, 3b, 4c, 5a, 6c, 7a, 8d, 9e, 10d

Quiz 11 - The Foot
1c, 2a, 3d, 4b, 5e, 6e, 7d, 8c, 9a, 10b

Quiz 12 - Muscles and Movers
1e, 2a, 3c, 4e, 5a, 6a, 7b, 8d, 9b, 10a

Quiz 13 - Insertions, Origins, and Actions
1d, 2d, 3c, 4d, 5d, 6d, 7e, 8b, 9d, 10e, 11d, 12e, 13a, 14a, 15d, 16c, 17e, 18a, 19d, 20b, 21a, 22a, 23c, 24e, 25c, 26c, 27a, 28b, 29b, 30b, 31c, 32e, 33d, 34d, 35a, 36c, 37e, 38b, 39c, 40a, 41b, 42b, 43b, 44e, 45b, 46b, 47e, 48b, 49e, 50a, 51a, 52b, 53b, 54a, 55b, 56d, 57e, 58b, 59a, 60c, 61d, 62e, 63c, 64d, 65b, 66b, 67e

Quiz 14 - Scapula Attachments
1d, 2e, 3b, 4a, 5c, 6a, 7c, 8b

Quiz 15 - Humerus Attachments
1d, 2a, 3b, 4e, 5a

Quiz 16 - Forearm Attachments
1d, 2b, 3b, 4a, 5d, 6c, 7d, 8b, 9c, 10a, 11c, 12a, 13c, 14b, 15a, 16a, 17c

Quiz 17 - Innominate Attachments
1c, 2e, 3a, 4b, 5a, 6d, 7e, 8b, 9e, 10e, 11d, 12e, 13b

Quiz 18 - Femur Attachments
1e, 2c, 3b, 4d, 5a, 6a, 7b, 8c, 9e, 10c, 11b

Quiz 19 - Tibia and Fibula Attachments
1b, 2c, 3a, 4c, 5a, 6b, 7d, 8e

Quiz 20 - Exercise and Activity Analysis
1c, 2b, 3e, 4a, 5a, 6d, 7e, 8b, 9c, 10a, 11d, 12e, 13a, 14a, 15b, 16e, 17e, 18a, 19a, 20a, 21a, 22a, 23b, 24a, 25a, 26a, 27a, 28b, 29a, 30a, 31a, 32a, 33b, 34a, 35a, 36c, 37c, 38a, 39d, 40b, 41a, 42e, 43b, 44c, 45a, 46d, 47c, 48e

Appendix B • Blank Drawings

This appendix includes blank skeletal sketches so that muscles and bony markings can be drawn on them. On the skeleton, the muscle can be drawn clearly showing the origin and insertion. Such a drawing could be done using a reference color, e.g., green origin (O) and blue insertion (I), and then the muscle could be outlined in red. The reason several skeletal drawings are included is so that a drawing of one muscle will not overlap other muscles, thereby hiding the origin or insertion of another muscle. Therefore, the total muscle can be seen and reviewed at a glance. While most muscle charts that appear in anatomy books show the surface muscles, the origin and insertion of these muscles is usually hidden by other muscles. By drawing the muscles, a unique reference for all the muscles is provided that can be seen by just looking at the drawing. In other words, these drawings should be thought of as a sort of "muscle coloring book."

The individual bone drawings can be used to identify where a muscle attaches to that bone (either origin or insertion). Again, these reference points could be color-coded. As such, these drawings will tell the user exactly where each muscle attaches to the bone. The belly of the muscle is not drawn, just where it attaches. This approach provides a useful method of identifying whether the individual doing the drawings knows the exact origin or insertion of a particular muscle.

The individual bone drawings can also be used to show the location of the various bony markings. Remember that the bony marking are usually the origin or insertion of a muscle. Therefore, by knowing the bony markings, you are also learning the origins and insertions.

The drawings in this appendix can also be printed from the companion CD software that accompanies this book. Keep in mind that one of the easiest ways of learning the muscles is to draw them on a blank drawing.

Another learning benefit that can be gained by effectively utilizing the blank drawings presented in this appendix involves muscle actions. Instead of rote memorization of the action of each specific muscle, you should be able to figure out a particular muscle's action from where it starts and where it goes. Imagine a muscle being a stretched rubber band, stretching from the origin to the insertion. If the origin was held stable and the rubber band was allowed to shorten, what would happen to the insertion? That would be the action of that muscle.

Anterior Upper Skeleton

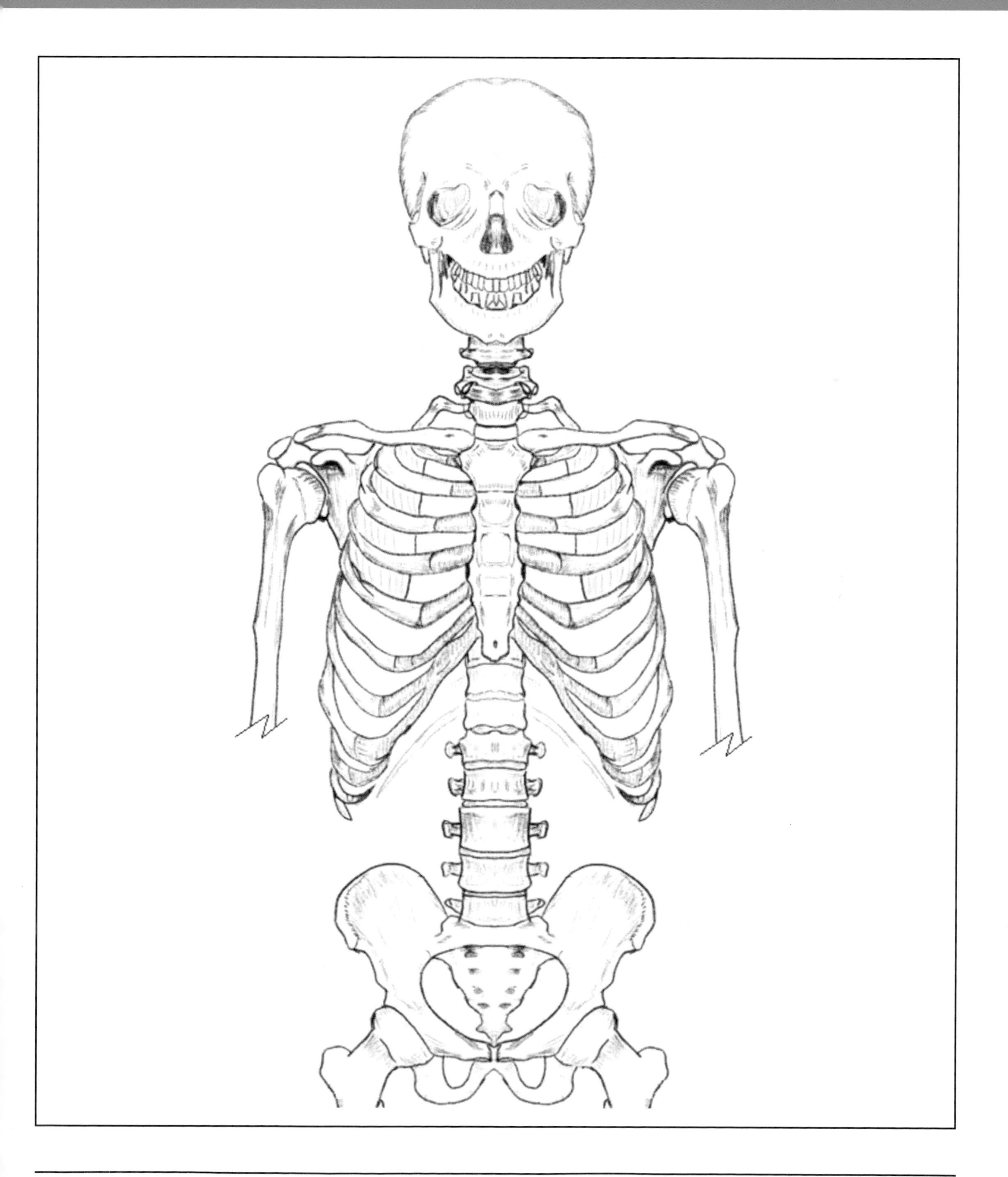

Posterior Upper Skeleton

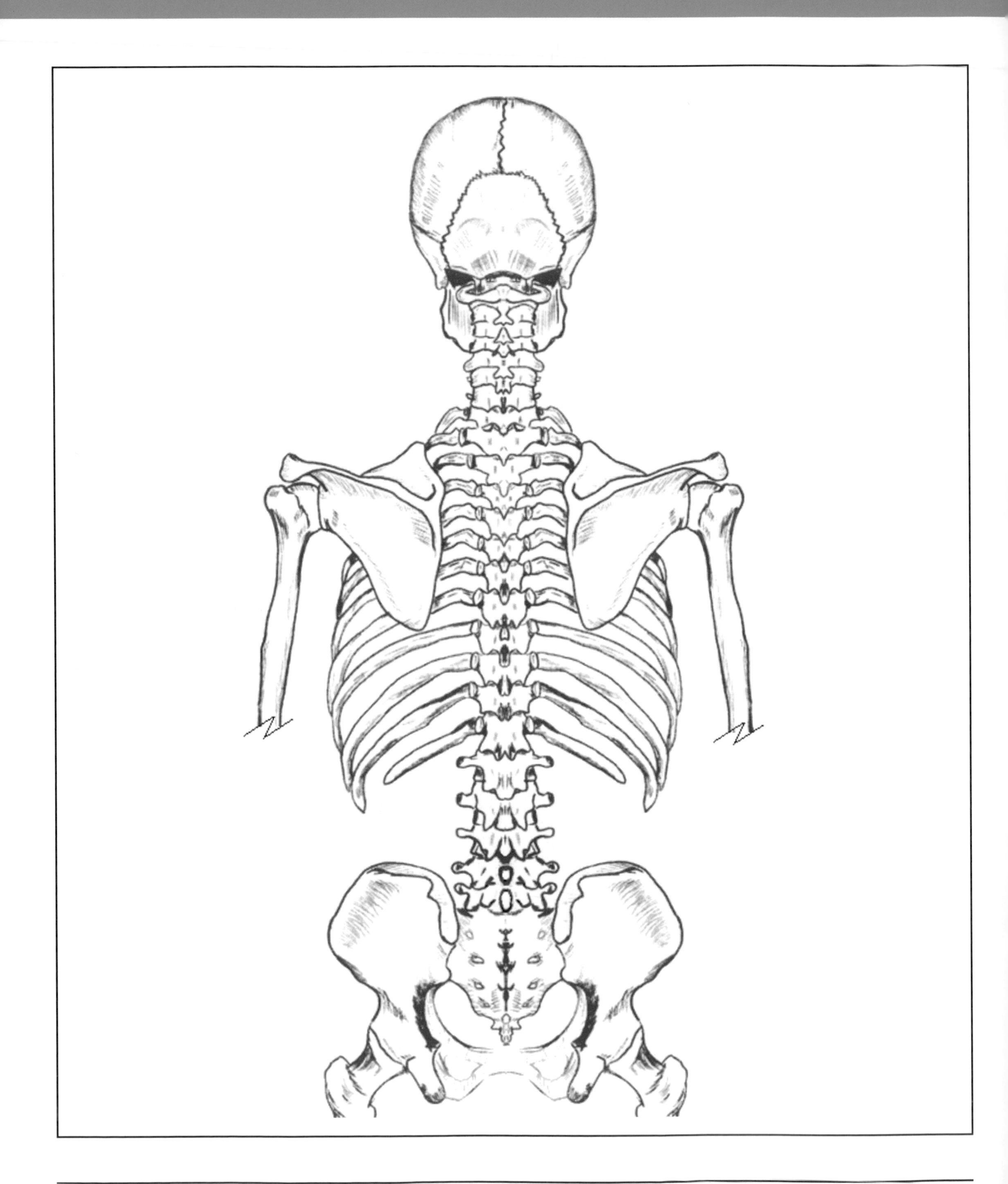

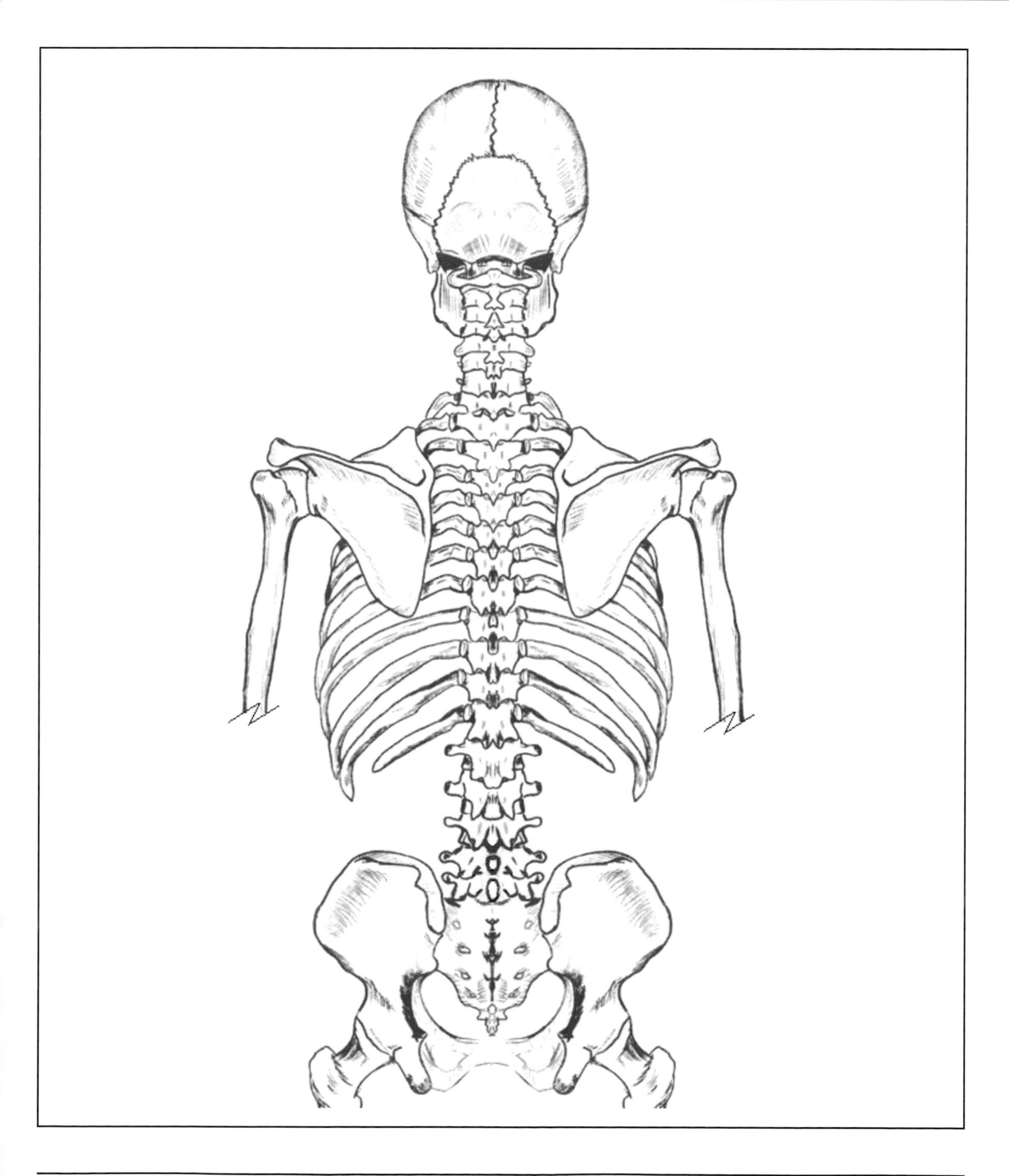

Posterior Upper Skeleton (No Scapula)

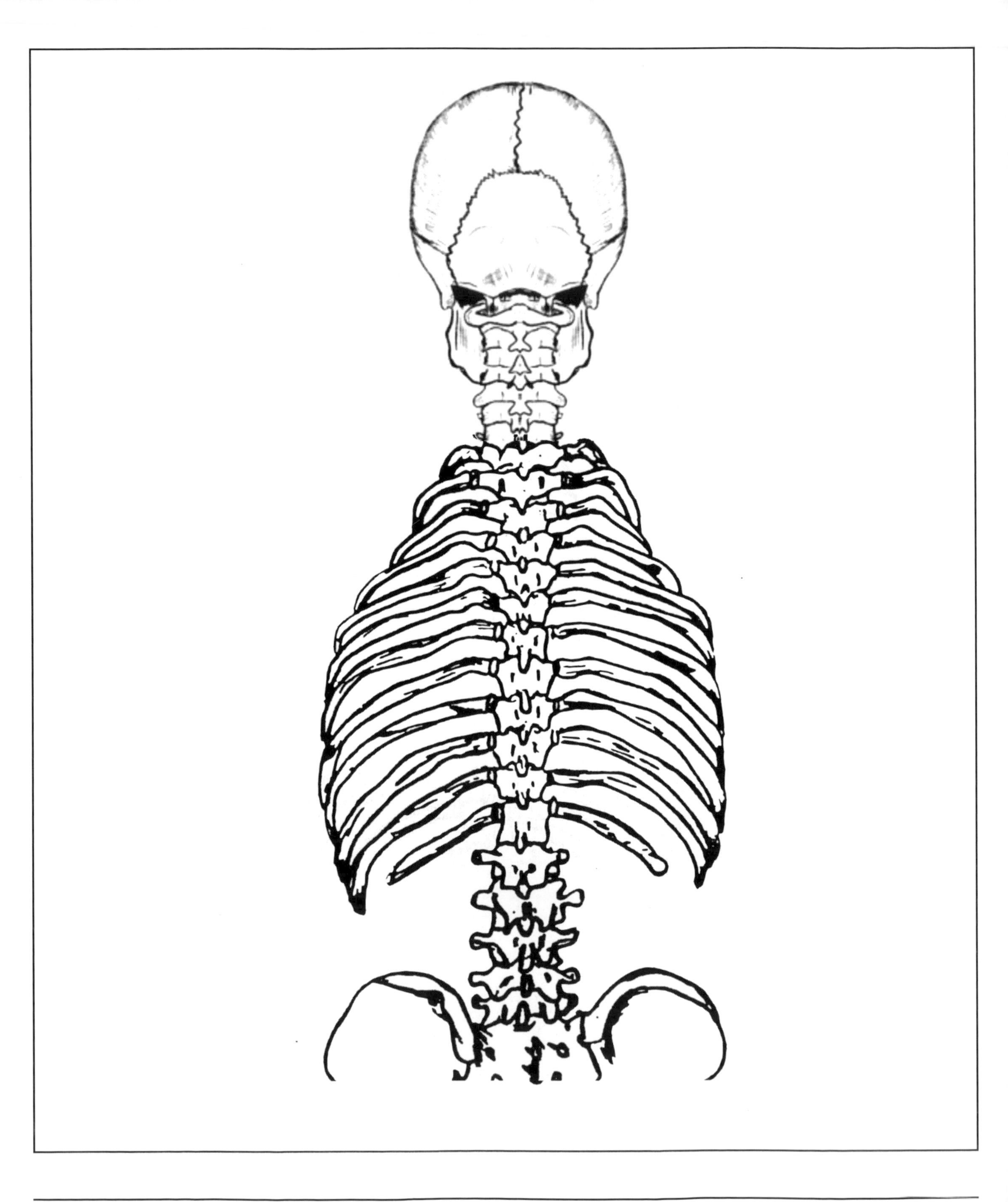

Posterior Upper Skeleton (No Scapula)

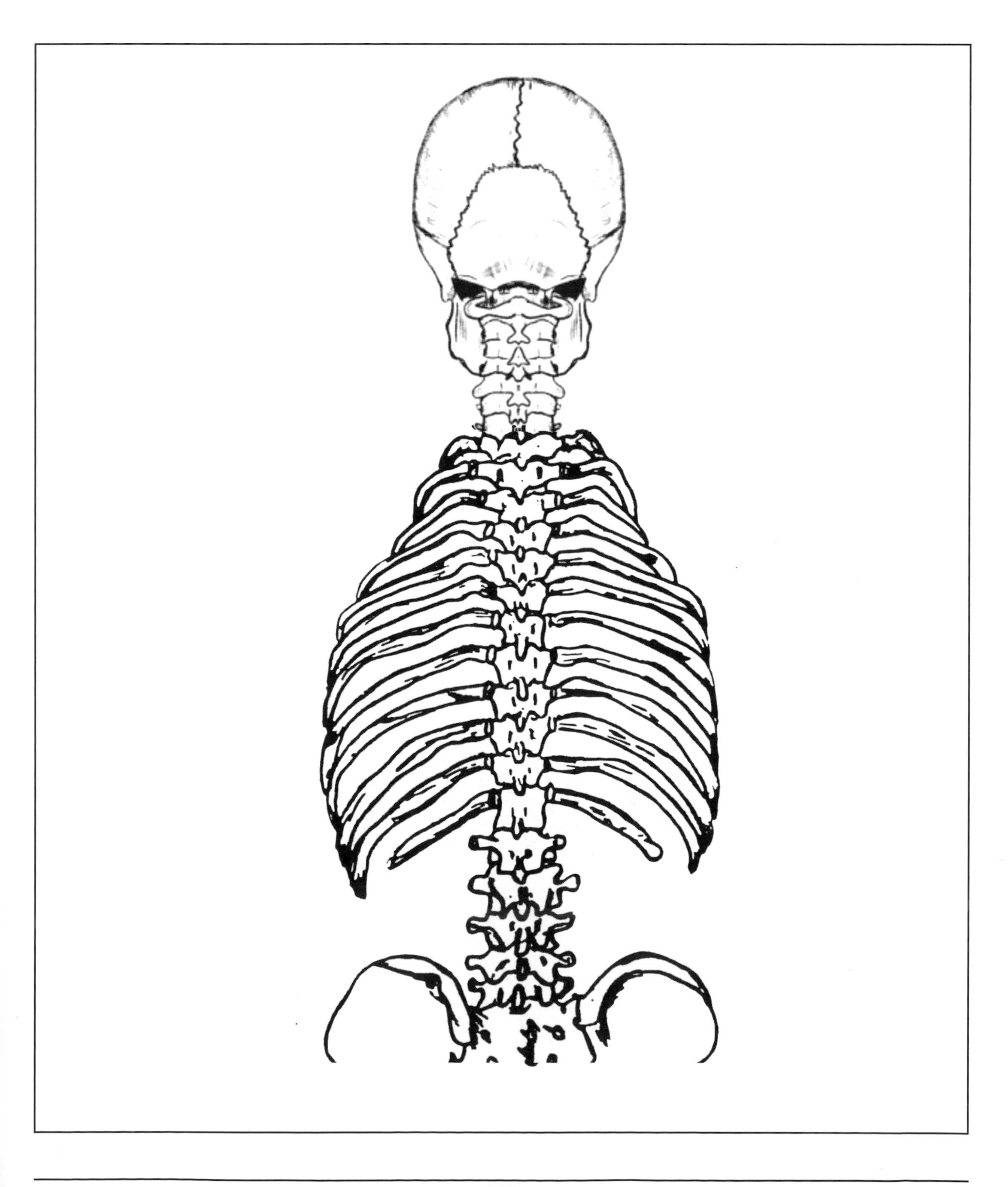

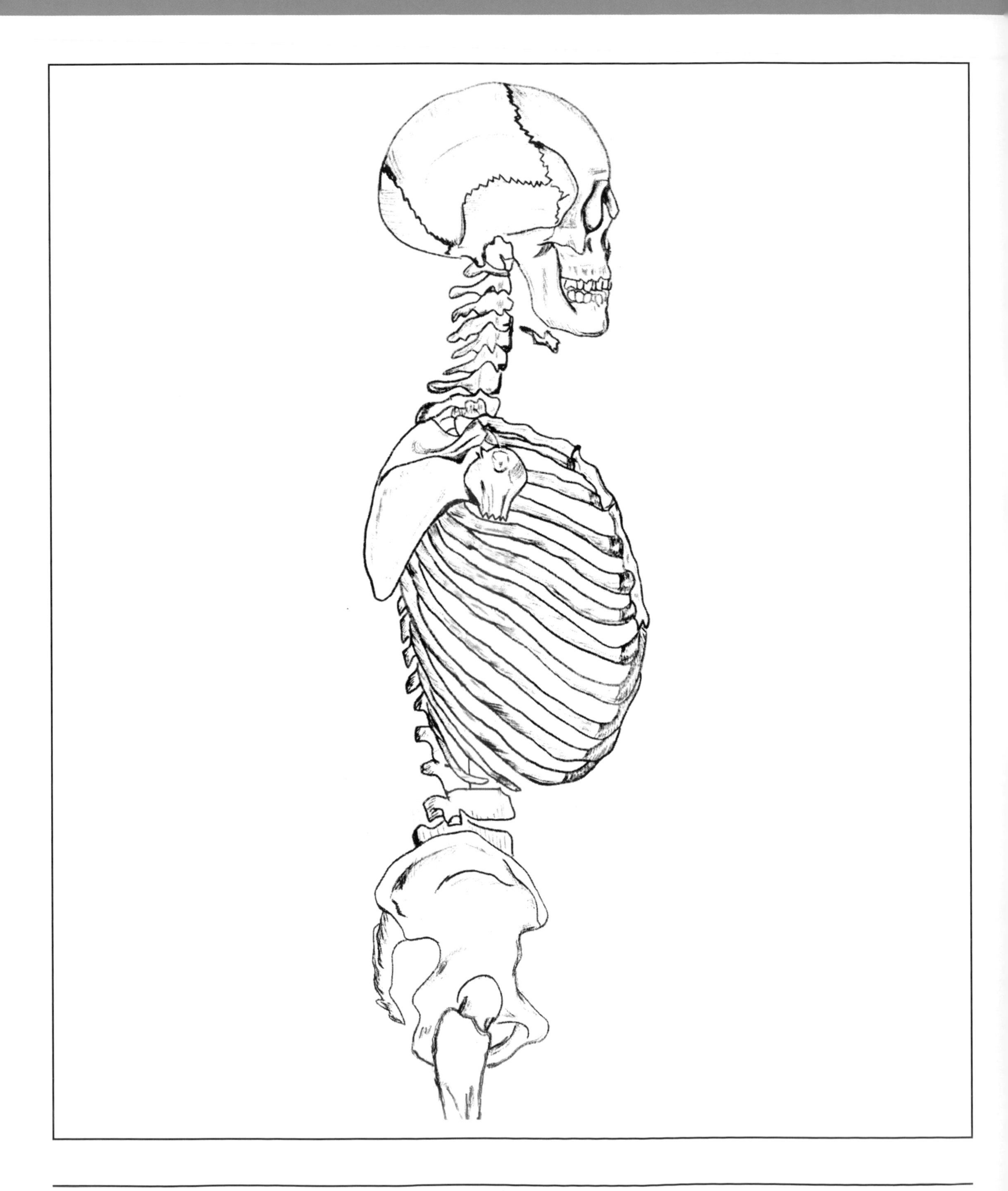

Anterior Lower Skeleton

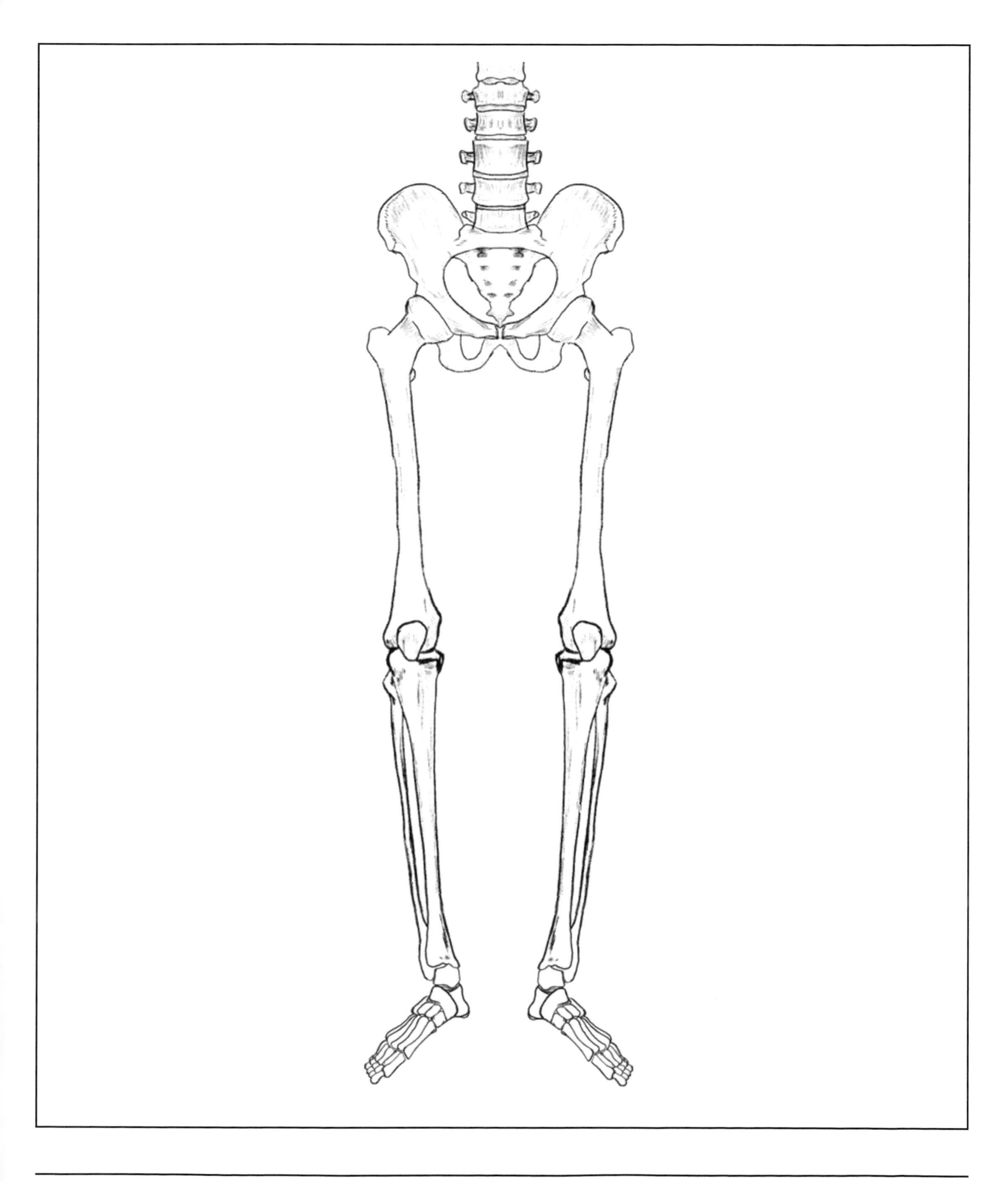

Posterior Lower Skeleton

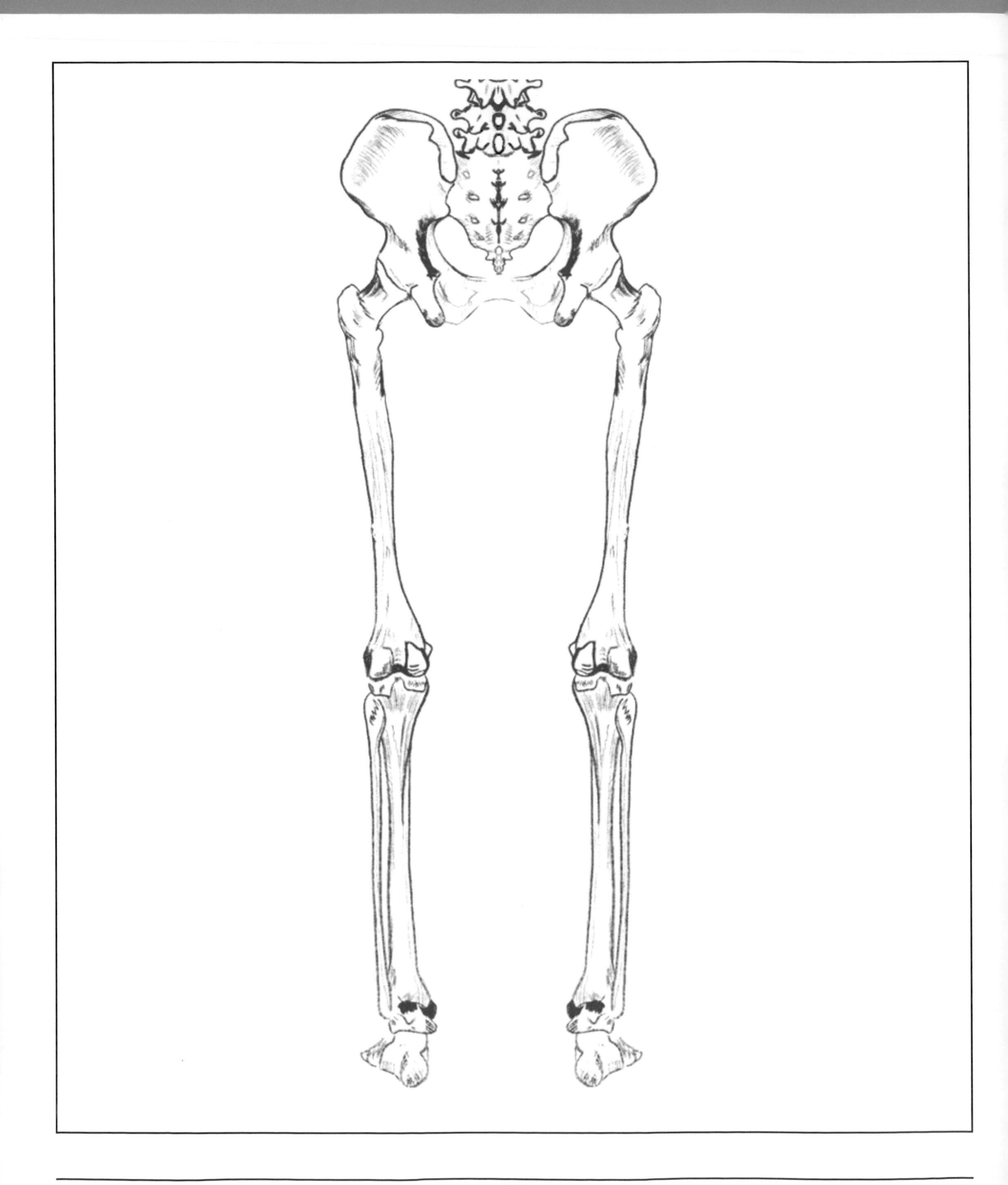

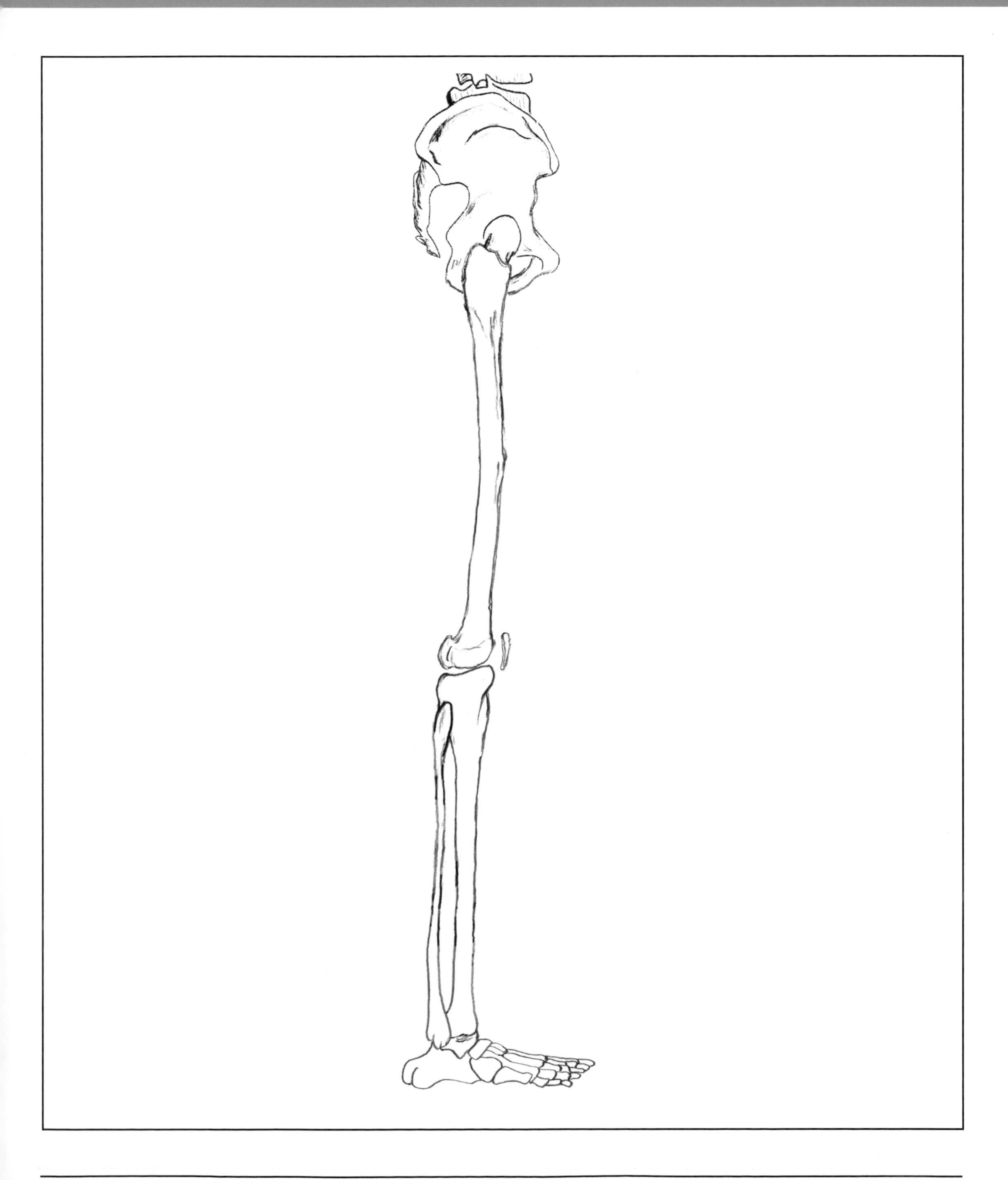

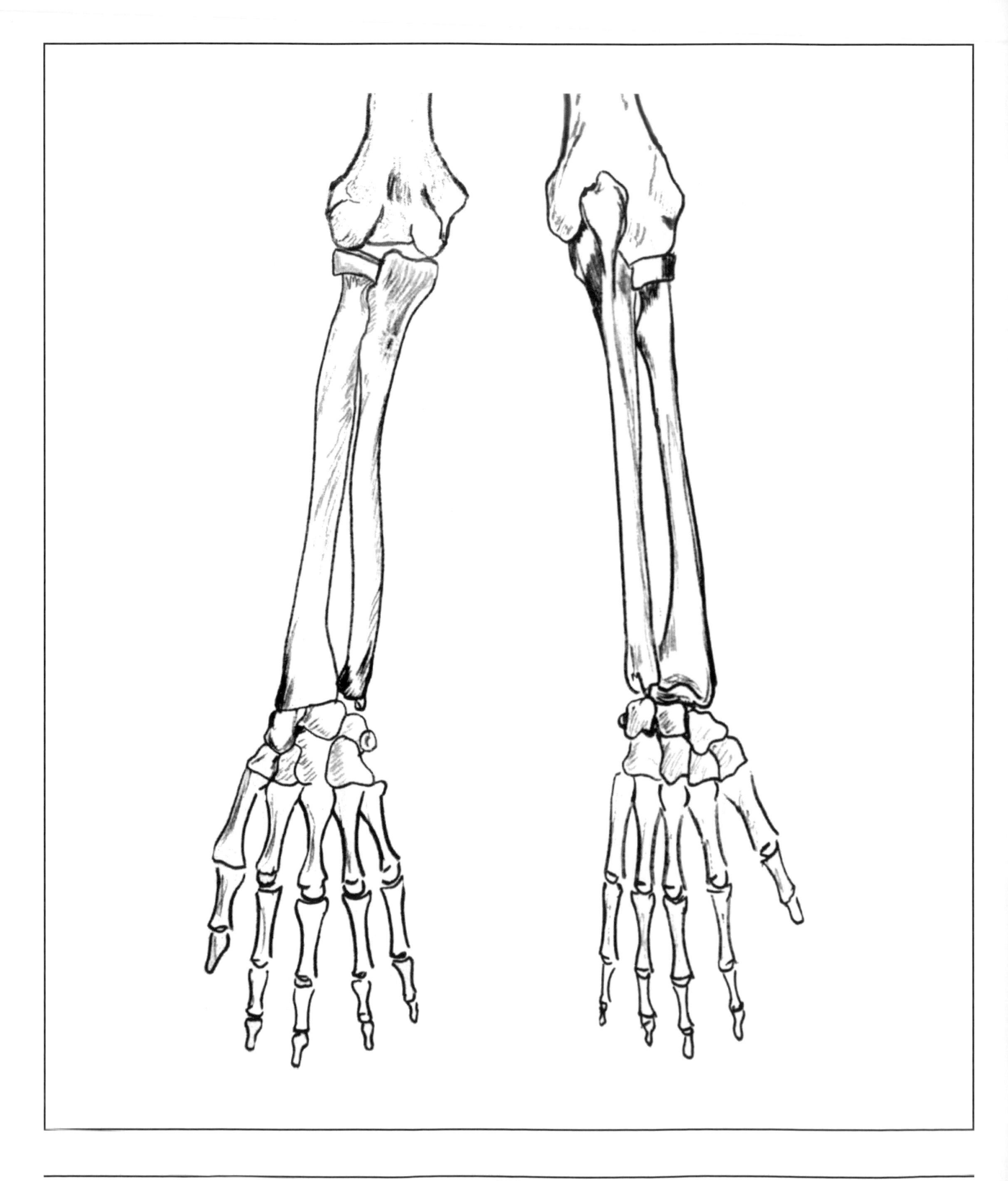

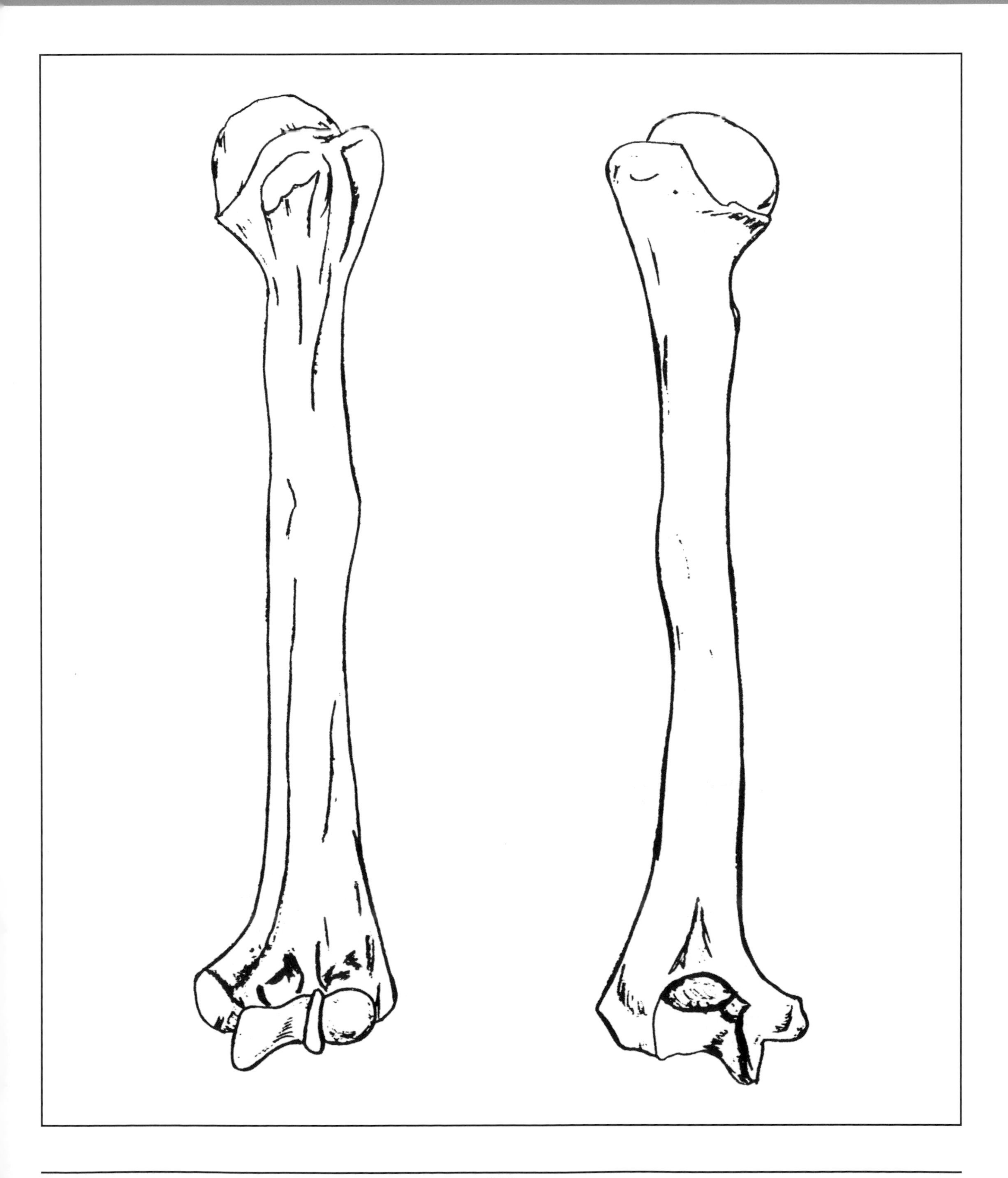

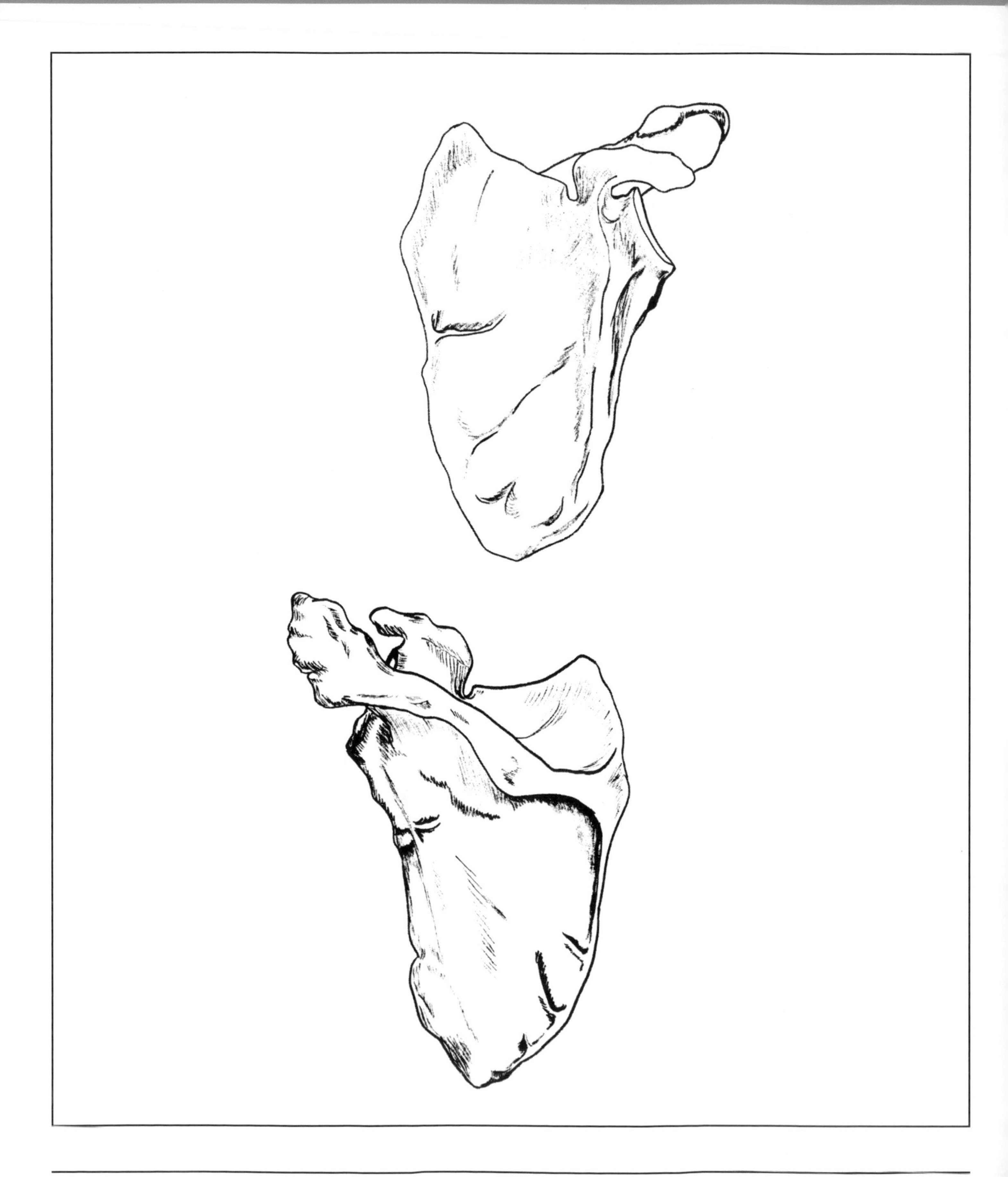

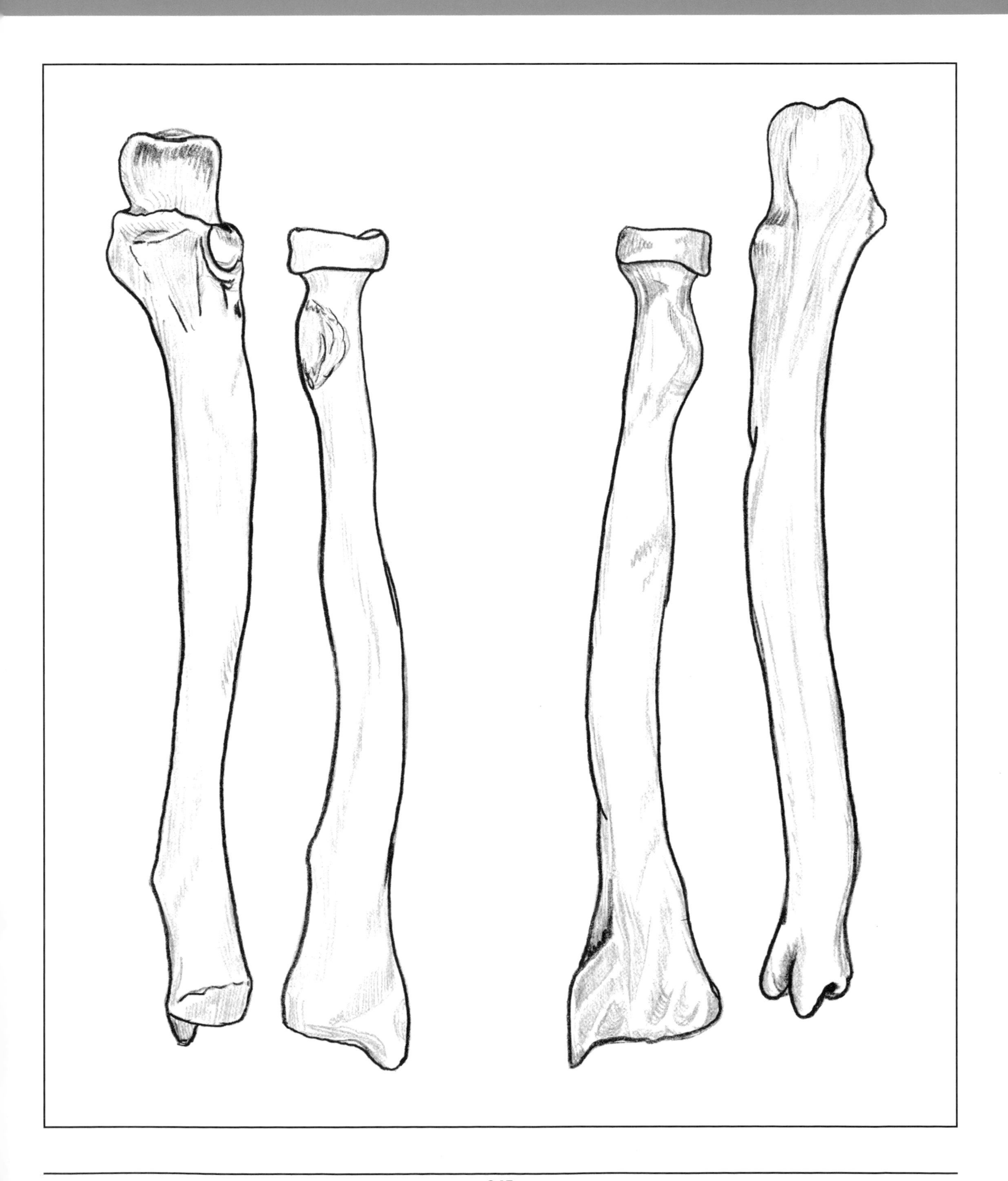

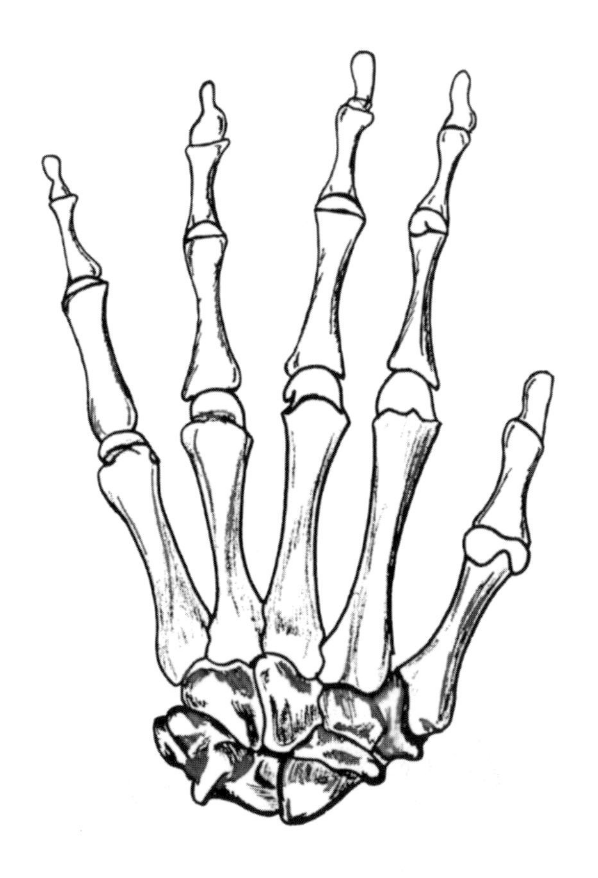

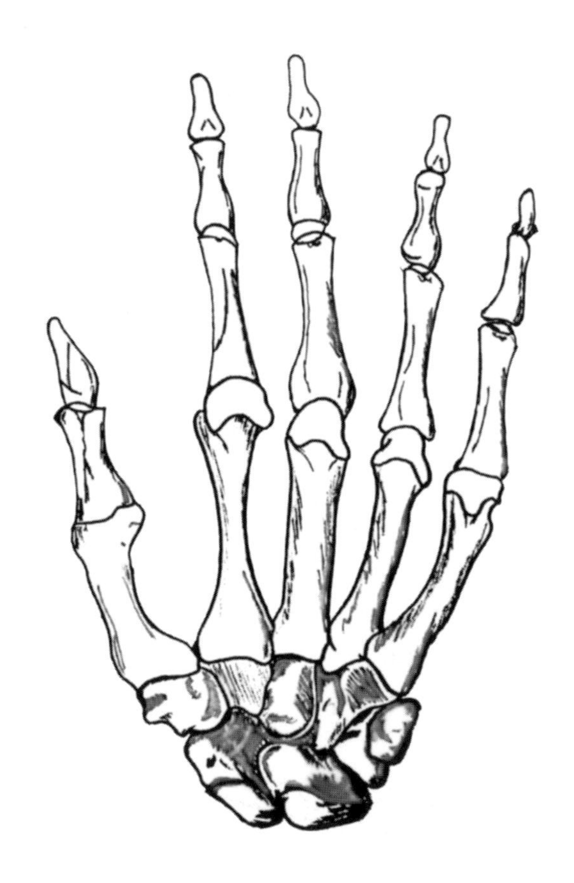

INNOMINATE

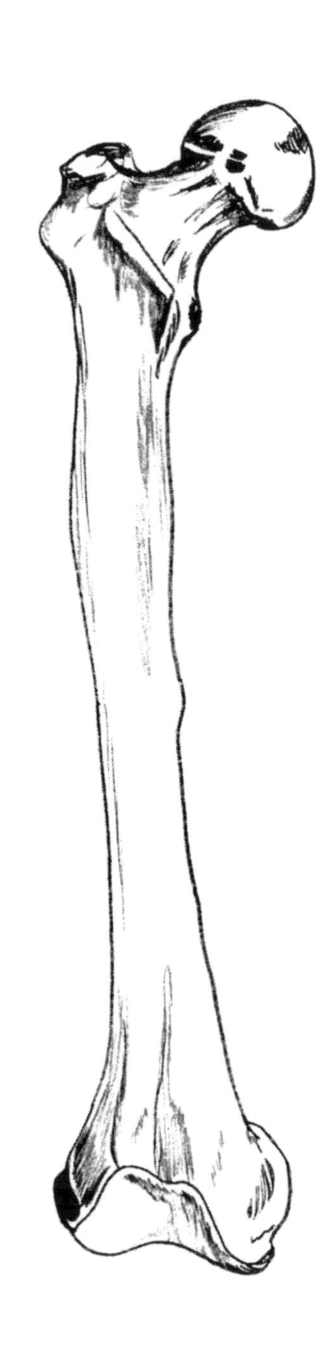

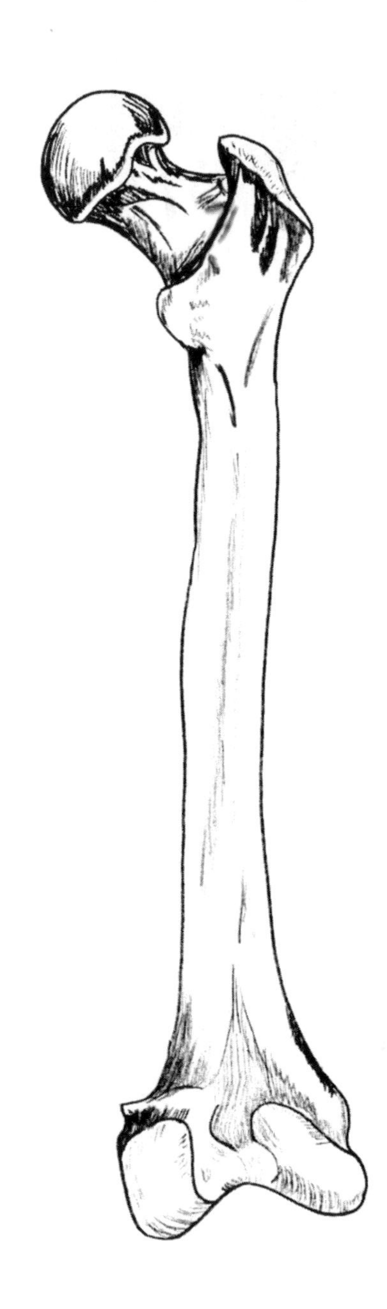

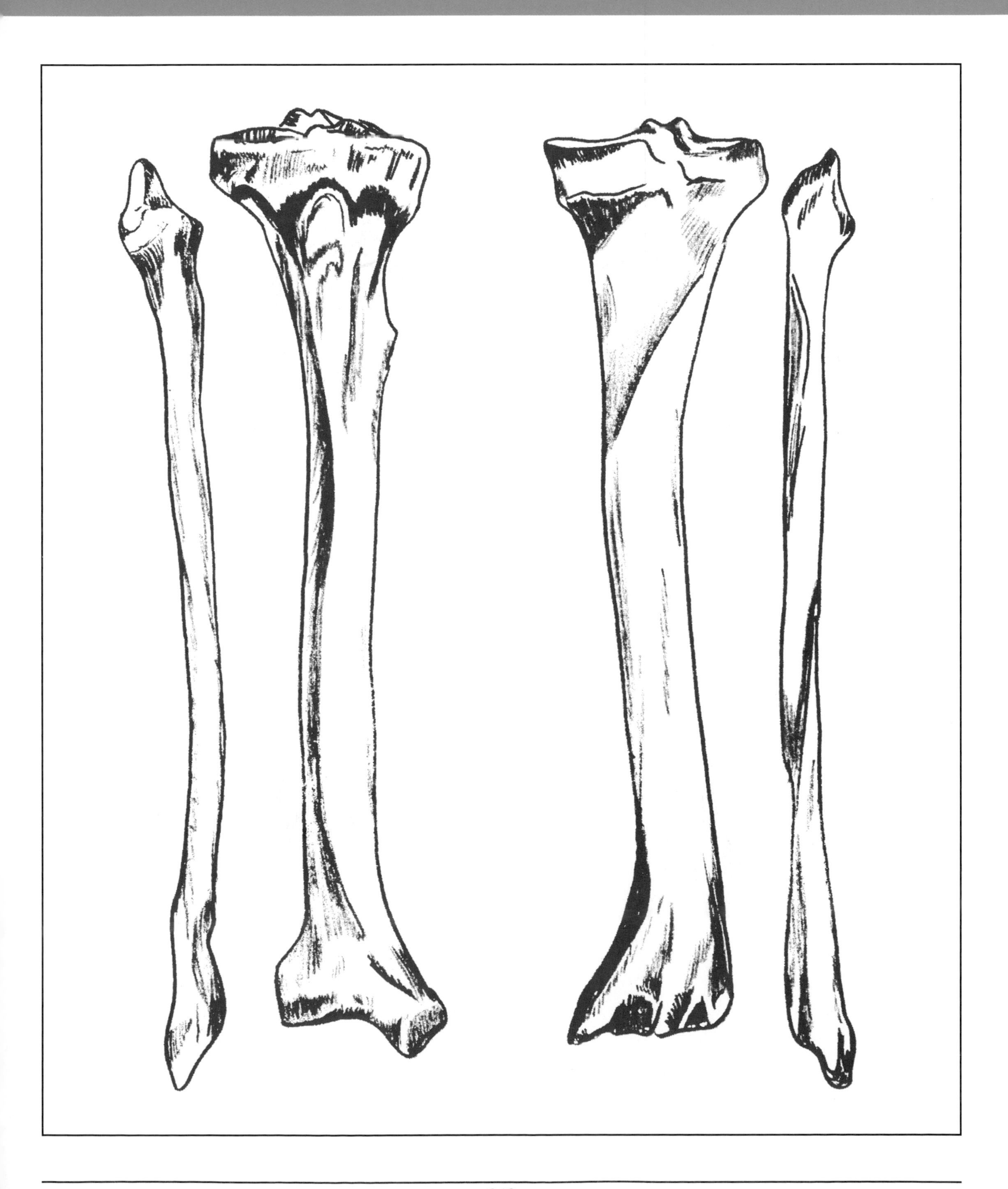

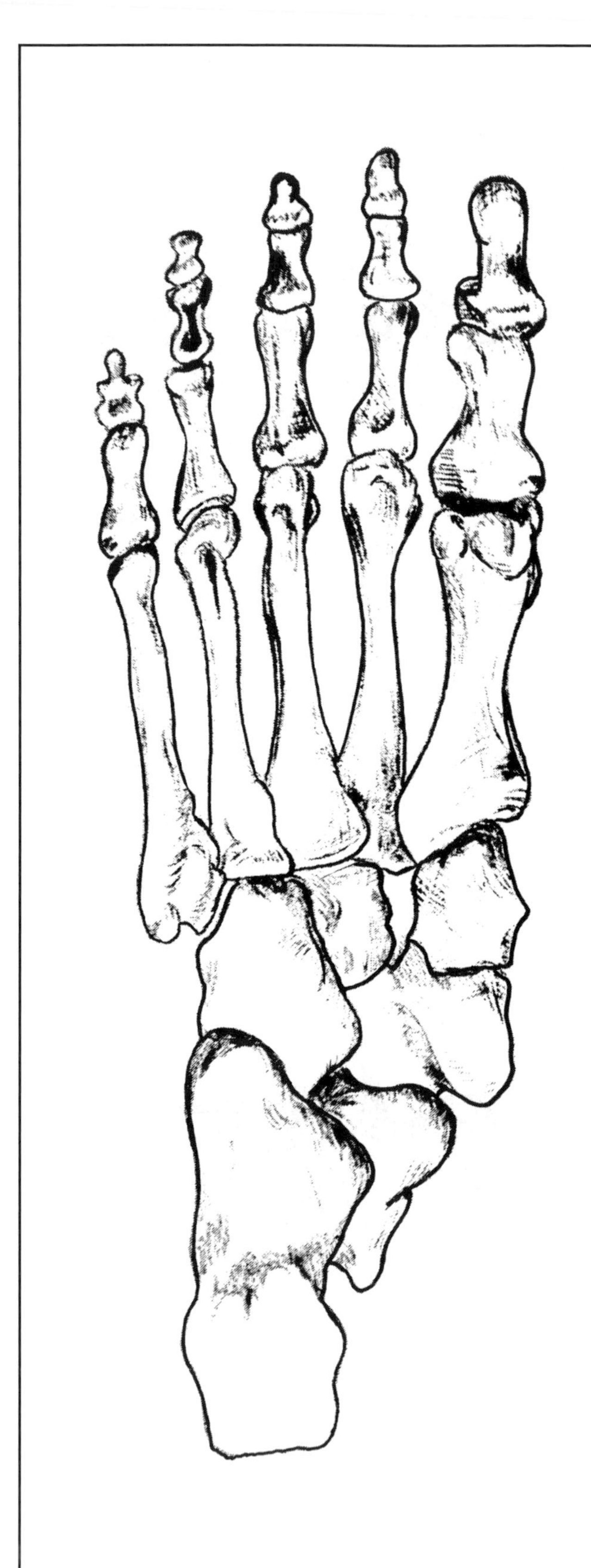

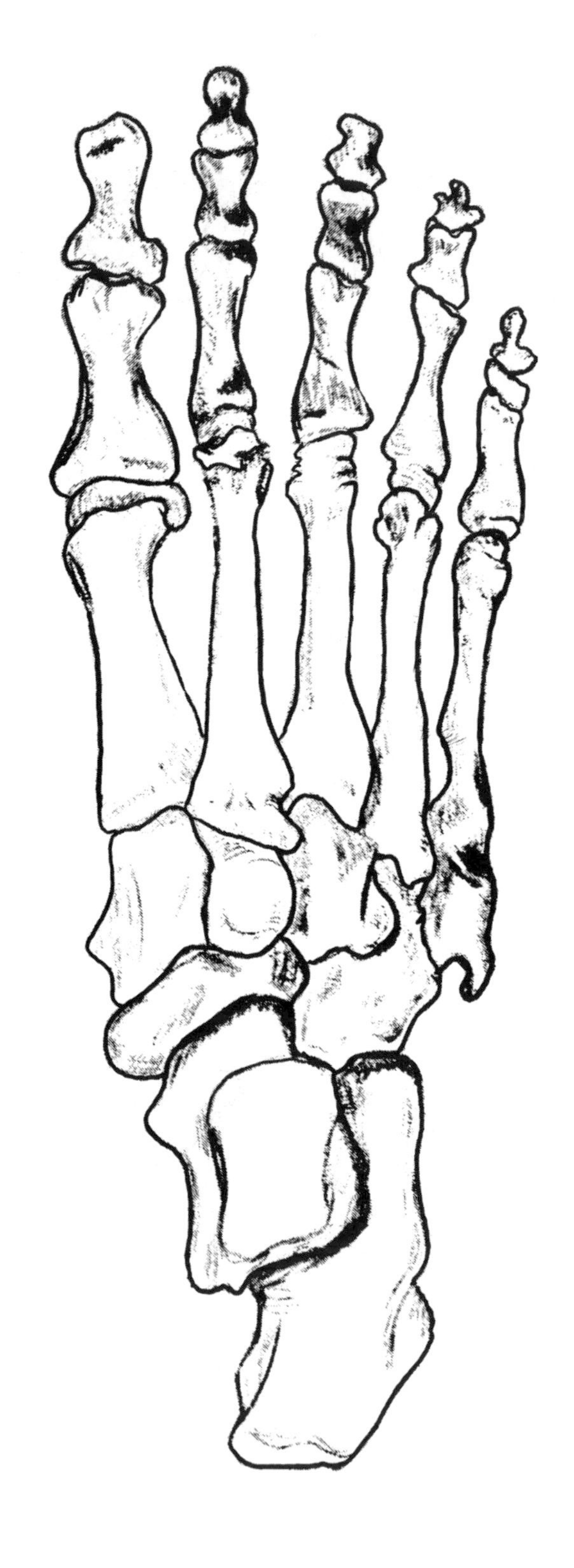

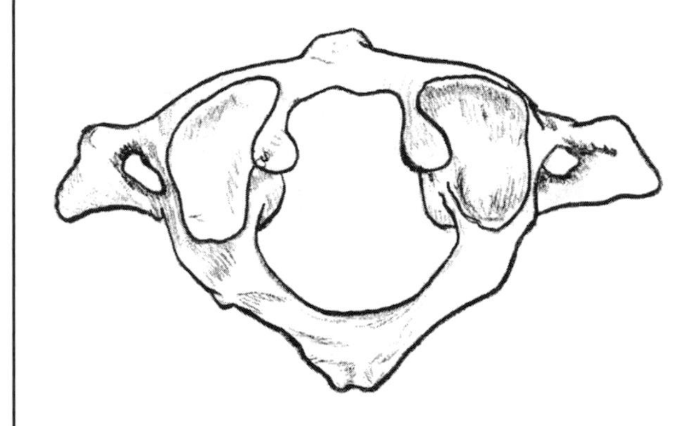

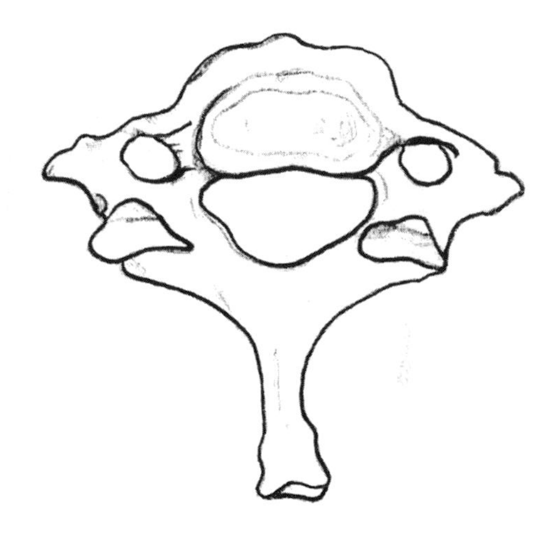

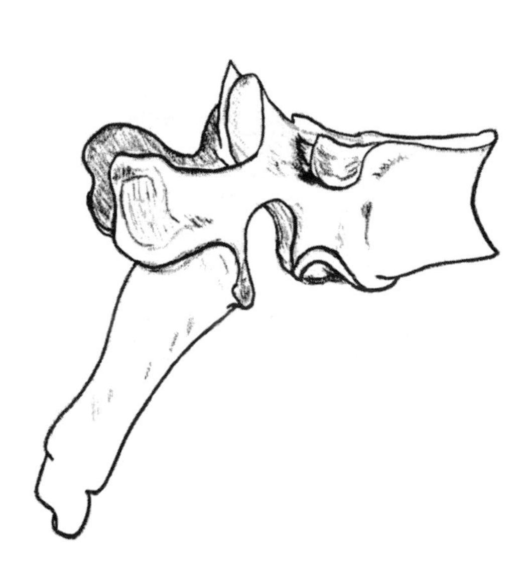

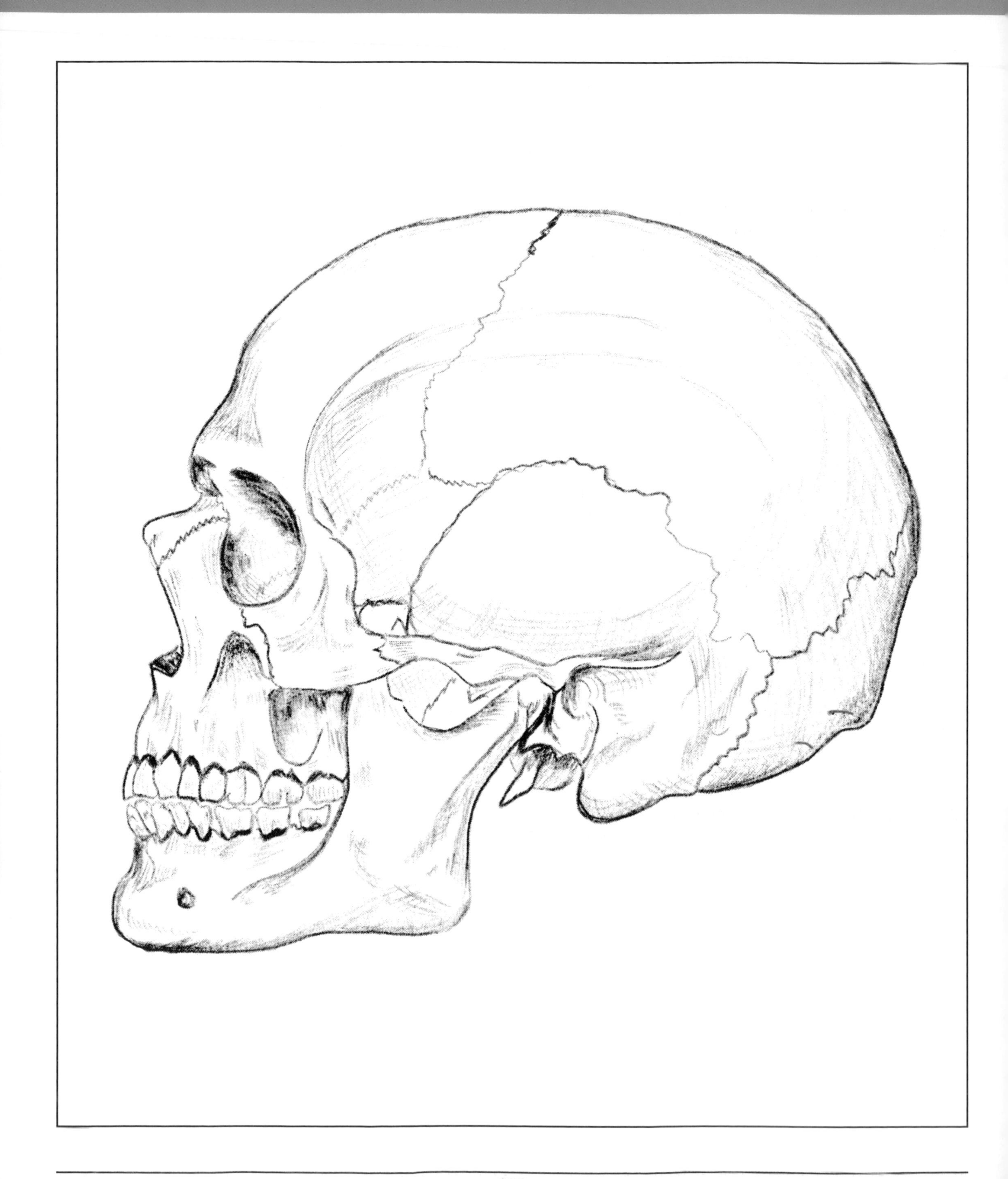

Appendix C • Companion CD-ROM Installation Instructions

The companion CD-ROM provides a Special Edition of Exercise Explorer made for this book. Exercise Explorer includes complete muscle anatomy that mirrors the content in the text, and it analyzes the book's exercises and activities in detail. In addition, the review questions at the end of each chapter in the book can be taken as an electronic quiz in Exercise Explorer, and skeletal drawings can be printed. The Exercise Explorer software must be installed on your computer before it can be used.

System Requirements

The software is designed to run in a Windows environment only. About 250 megabytes of hard disk space are needed, and speakers will allow you to hear the informative videos and muscle name pronunciations. Most Windows environments will work well with the software.

Installation

The installation of the companion software requires that several system components are installed as well as the Exercise Explorer software. The components needed include Microsoft Data Access Objects (DAO) and the QuickTime® video viewer application. The installation procedure is automated, and should start automatically when the CD is inserted in the drive. Please wait a few seconds after inserting the CD for the installer to initialize.

If the installer does not start automatically, then the installation can be started manually by running the Setup program on the install CD. To run Setup, click on the Windows Start button, select Run..., and enter "D:\setup.exe" (where "D" is the drive letter of the CD-ROM drive), then click OK. Or, you can open My Computer; *right* click on the CD drive that has the Exercise Explorer CD in it, and select Open. The contents of the CD will be displayed. Find the Setup.exe application and double click on it to install the software.

1. When the installer starts, the Welcome Screen is displayed. Click the View Professional Demo button to review a demo of the Professional Edition of Exercise Explorer, or click Next to continue and install the Academic Edition of the software.

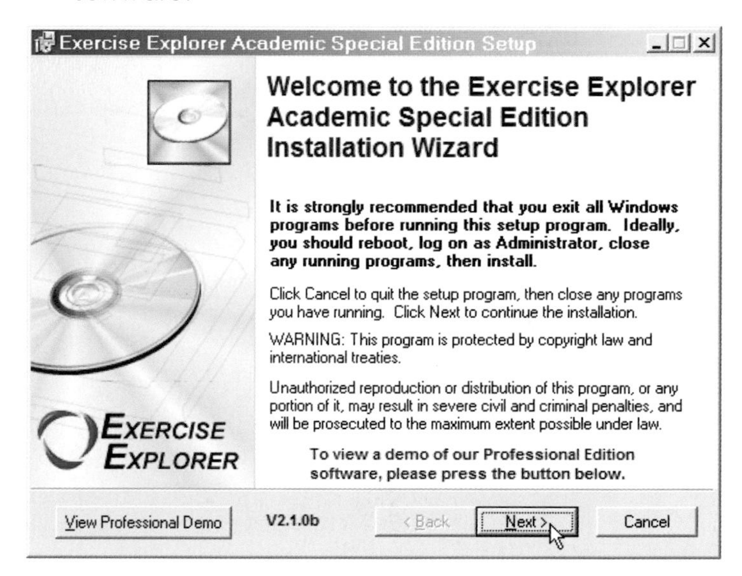

2. The End User License Agreement screen is displayed. Read the agreement and check the "I accept the license agreement" radio button, then click Next to continue.

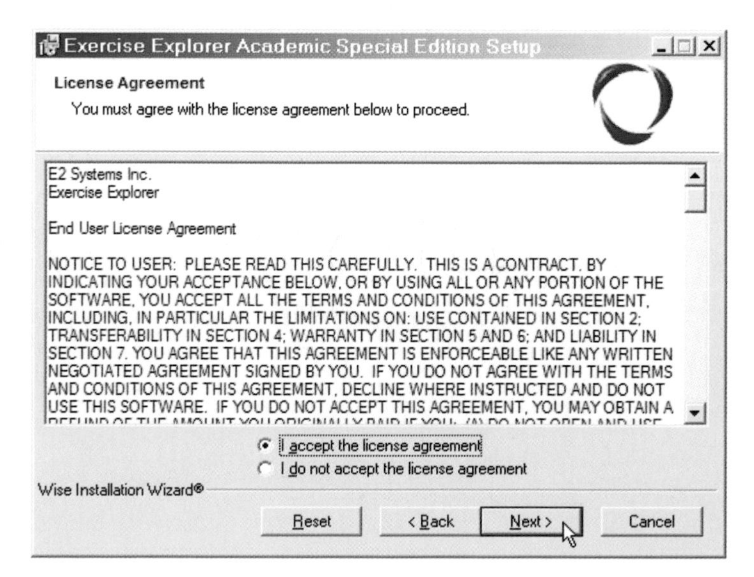

3. The User Information screen is displayed. Enter your name, company if appropriate, and installation serial number. This serial number is located inside the back cover of the book. The serial number must be valid. Make sure it is entered correctly. Click Next to continue.

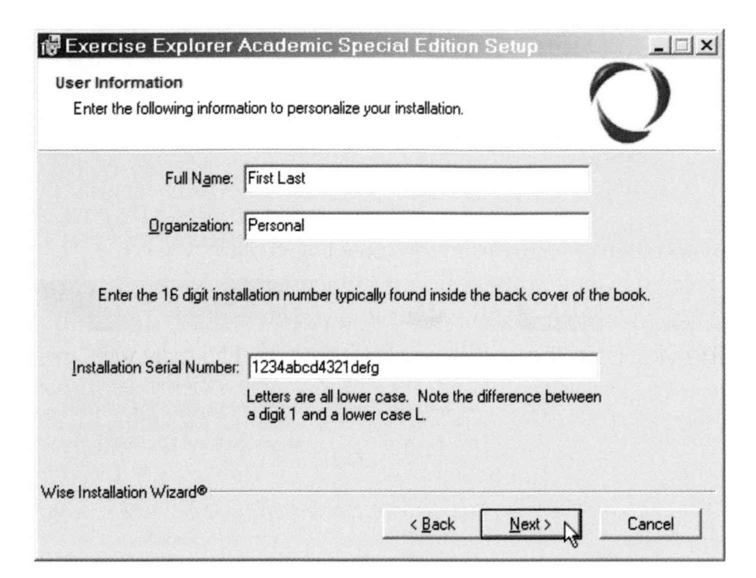

4. The Destination Folder screen is displayed. The default location to install the software is displayed. You may change the default location; however, it is recommended that you keep the default setting. Click the Browse button to change the folder. If security issues are a problem, choosing a public folder to install to may be a solution. Click Next to continue.

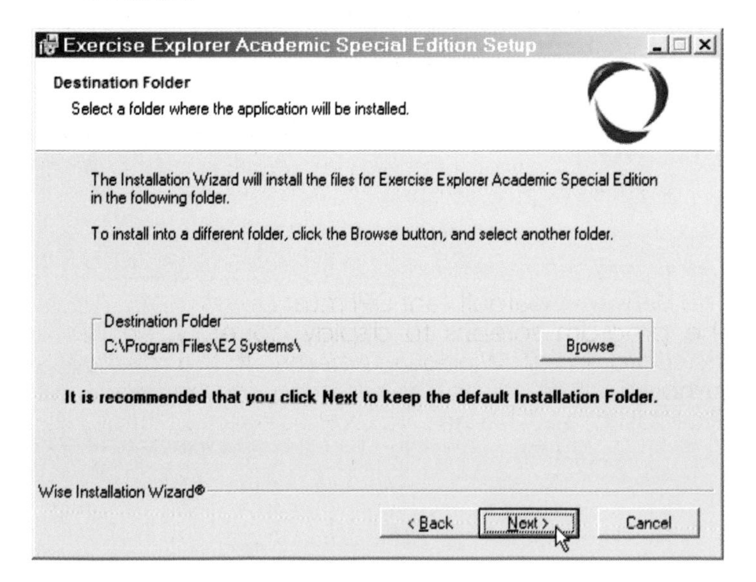

5. The QuickTime installation screen is displayed. Select the "QuickTime 4 or higher already installed" radio button to skip the QuickTime installation. If you do not already have QuickTime installed on your computer, it is recommended that you install the latest version from the Apple website after the Exercise Explorer installation is complete. If you prefer to install QuickTime from the CD-ROM, select the "Install QuickTime 4" radio button. The installer will start after the Exercise Explorer files have been installed. Click Next to continue. Refer to the Installing QuickTime section in this appendix for more details about the QuickTime installation.

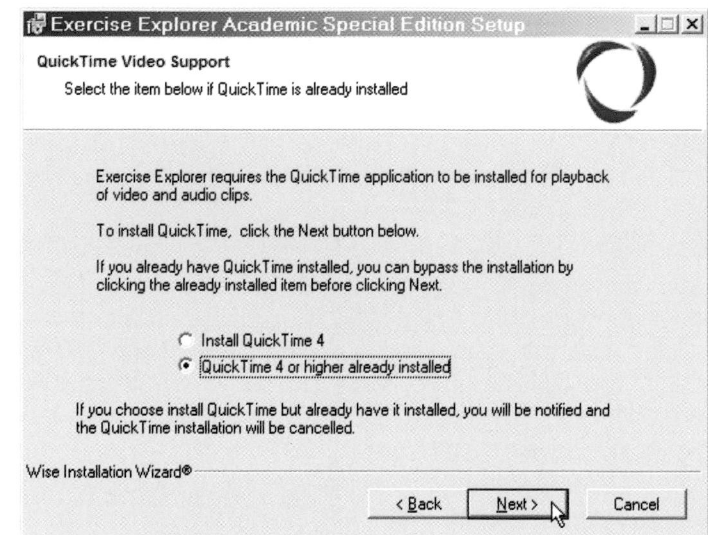

6. The Ready to Install screen is displayed. The software is ready to install. Click the Next button to begin installation and copying of files, or press the Back button to make any changes.

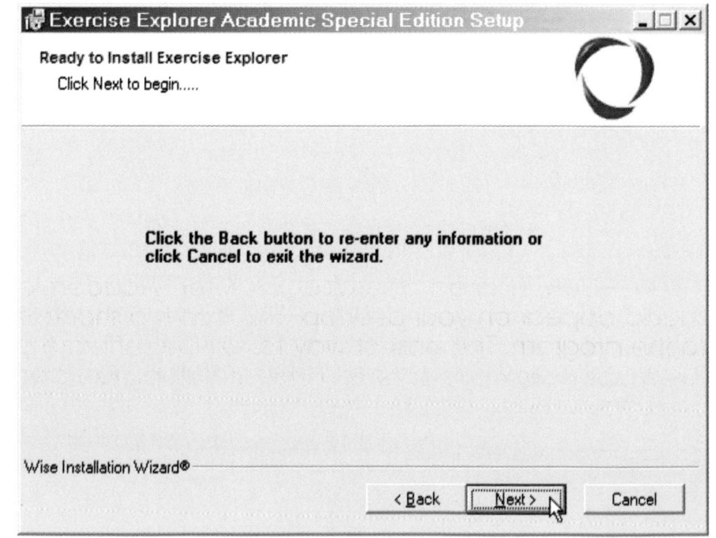

7. The Updating System screen displays the files being copied and the estimated time remaining until complete. This operation can take a few minutes.

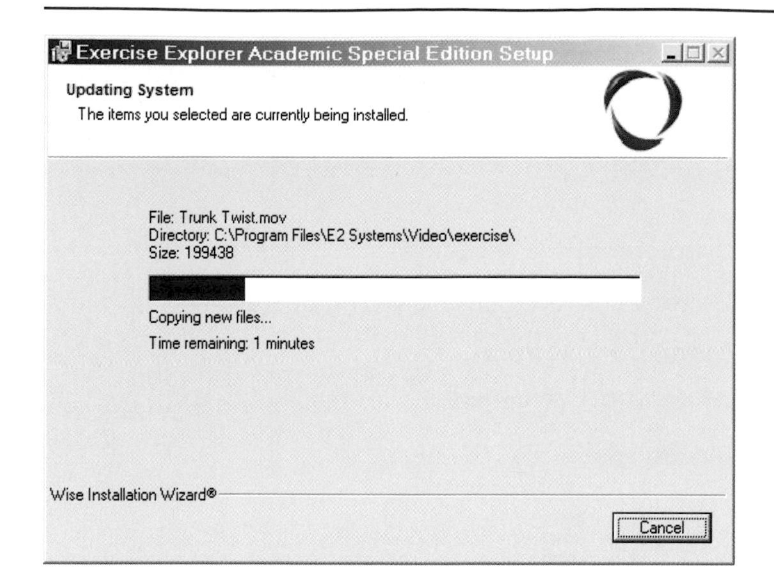

8. After the files are copied, the QuickTime installation will be started if selected, and when complete, the finished screen will be displayed. It may take an additional minute for the final screen to be displayed, so please wait a little while. Click Finish and the installer will close.

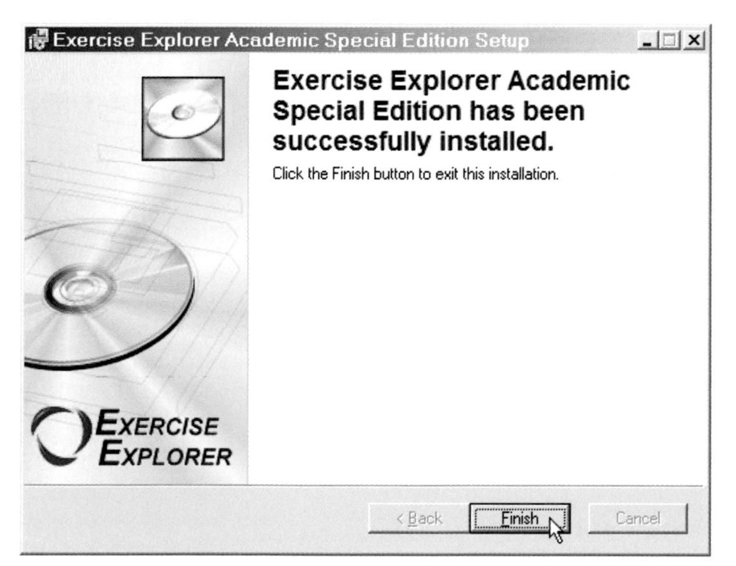

An icon named Exercise Explorer Academic should appear on your desktop. This icon is a shortcut to the program. The easiest way to run the software is to double click the icon on the desktop. Exercise Explorer can also be run by clicking on the Windows Start button, selecting Programs...E2 Systems, and selecting Exercise Explorer Academic. Use the Windows Add/Remove Programs application in the Windows Control Panel to uninstall Exercise Explorer.

Installing QuickTime from the CD-ROM

Note: It is recommended that you visit the Apple website to download and install the latest version of QuickTime, instead of installing it from the CD-ROM, especially if you are having any trouble playing videos.

The QuickTime installation is a product of Apple Computer and is distributed under license. You are not required to register or enter any information.

Eight screens will be displayed when installing QuickTime from the CD-ROM. On the first screen, click Next to continue. Select the default buttons as Next, Agree, then Next until the files install. No information needs to be entered to install QuickTime. It is recommended that you install QuickTime with all the default settings.

After clicking through the screens, QuickTime files will install. When the installation is complete, you will be asked to preview the readme file and play a sample movie. Uncheck both boxes and click Next.

The QuickTime installer will exit, and the folder for QuickTime will be displayed. This window can be closed to return to the Exercise Explorer Installation screen.

QuickTime Video Settings

The videos in the software are formatted to play with the QuickTime player. On some systems, the videos may not play properly or could cause the program to stop. Changing the following setting for the QuickTime player will usually fix these problems (*Note*: Depending on the version of QuickTime installed, this process may be slightly different than described):

- Exit the software, if running.
- Open the Windows Control Panel.
- Find the QuickTime item, and open it. This item is sometimes in the Other Control Panel Options area. Switching to Classic View makes it easier to find.
- Select the option "Safe mode (GDI only)." This option is usually found on the Advanced tab in the Video section, or related to the Video Display.
- Press the OK button, and close the Control Panel.

When the program starts again, problems with the videos should be resolved. If not, recheck the QuickTime setting in the Control Panel, and possibly make other changes in settings particular to your system.

Display Problems

The Windows default font DPI must be set to 96 DPI for the program screens to display correctly. 96 DPI is standard in all Windows systems. If the display is jumbled, then check this setting in the Windows Control Panel. Choose Display Properties, Settings tab, Advanced button. The General tab should display the DPI setting as: Normal size (96 DPI). Changing the setting will fix this display problem. The Display Properties Appearance tab also has font size settings that may affect the software screens.

Exercise Explorer Software Updates

To check for and download updates to the Exercise Explorer software, it needs to be installed, running, and displaying the Main screen, and your computer needs to be connected to the Internet. In Exercise Explorer, press the System Menu button (the leftmost of the four small buttons at the top right corner of the screen) and select Check For Updates to go to the website upgrade page.

Index

INDEX

INDEX

Muscle details and illustrations (continued)

About the Authors

Lawrence A. Golding, Ph.D., FACSM, has been a professional in the field of kinesiology and physical education for the past 50 years. He has been a continuous member of the American Alliance for Health, Physical Education, Recreation, and Dance (AAHPERD) since 1955, and of the American College of Sports Medicine since 1961. He is a Fellow of both organizations and holds the highest certifications from the ACSM — Program Director and Health and Fitness Director. Currently, Dr. Golding is a Distinguished Professor at the University of Nevada, Las Vegas — a rank held by only six other faculty. Dr. Golding is the chair of the planning committee for the ACSM Health and Fitness Summit and Exposition, and was the Southwest Regional Chapter ACSM Executive Director for 8 years. In 1996, he was appointed Editor-in-Chief for the new ACSM Health & Fitness Journal. Dr. Golding was, for 4 years, the chair of the Medical Committee of the IFBB (International Federation of Bodybuilders) and attended the Mr. Universe contest held throughout the world during those years. In 1993, Dr. Golding received the "Healthy American Award", an award the President's Council on Physical Fitness and Sports bestowed on 10 individuals who had done the most for the health of the nation. He has conducted adult fitness classes, workshops, and certifications for the last 40 years, has certified over 6,000 YMCA Physical Fitness Specialists, and currently consults at six fitness specialist workshops each year.

Scott M. Golding, MS, has been a professional in the software and computer technology industry for the past 25 years. He has been vice president of engineering and technology with several companies that develop specialized production equipment and software. In 2000, he founded E2 Systems, Inc., an organization that focuses on the development of exercise-related software.